# MA ... E
# CHEMISTRY

GCSE edition

John Gro...

*...ad of Scien... and Technology*
*...bury Comprehensive School*
*...fordshire*

a... ansfield

*...ad of ...ence and Technology*
*...e Har...y Grammar School*
*Folkeston...*

Longman

# Contents

# Preface

This book is about chemistry. We have tried to make it interesting and straightforward. The way chemistry is used in the real world is discussed throughout the book, often with photographs, and there are many diagrams to make things clearer than a lot of words. Technical words are printed in **bold** the first time they are used, and there is a short chemical dictionary at the end of the book to help you.

Each chapter in the book is divided into sections. At the end of each chapter is a summary of the main points, and some questions to test whether you have understood the work. Most chapters also contain extension material in the form of further reading and additional questions.

The methods the book shows you for doing calculations, especially those using scale factors, should be carefully studied. You will find a calculator very helpful.

This book will give you a good basis for examinations such as GCSE. When you have read it, you should have a good understanding of the importance of chemistry in our lives.

John Groves
David Mansfield

## Information for teachers

In the preparation of this textbook, great care has been taken to ensure that the reading level is accessible to the entire target age range for GCSE. The mathematical approach has also been carefully considered to ensure that it is consistent with that adopted in current mathematics courses. The student's confidence is built up by a staged introduction to technical terms.

The discussions of the application of chemistry to the solution of socio-economic problems should both stimulate and sustain the student's interest, and meet TVEI specifications.

The text fully covers the requirements of all current GCSE syllabuses, and teachers will be able to select material appropriate to their students' requirements. It is not essential that the chapters are studied in sequence.

# 1 Only one Earth

## 1.1 Using up the Earth's resources

Chemistry is about changing substances. It's about taking materials and transforming them into new materials that are more useful and more valuable. Much of the chemical industry is occupied in changing the natural resources of the Earth into substances to meet our basic human needs for food, water, warmth and shelter.

**Figure 1.1** *How many different resources are used to provide the materials we use in our homes?*

**Figure 1.2** *High-technology living requires more complex materials*

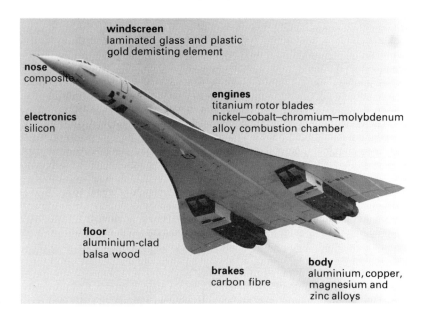

People in a modern society also require many more advanced materials to provide for the demands of high-technology living.

We all take for granted many of the materials that chemistry has provided us with. Nevertheless we should remember that there is only one planet Earth, and one day its resources will run out.

### How fast are we using up the Earth's resources?

The graphs in figure 1.3 show the rates at which we are using two very different substances: aluminium and oil. In both cases the overall trend is upwards. Political factors have affected the price of oil and hence the demand for it. Metal prices can also be affected by political decisions. For example, the price of tin has changed greatly in a way that has little to do with the amount available.

**Figure 1.3** *Trends in consumption worldwide of (a) aluminium, (b) oil.*

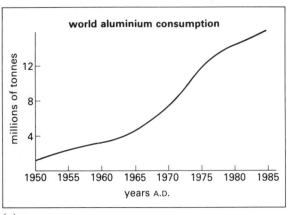

(a)

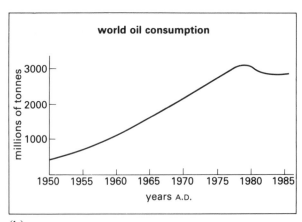

(b)

## Why are we using up the Earth's resources more quickly?

The graphs show that we are using up (or **consuming**) these resources more and more quickly. One reason for this is the general increase in world population. Figure 1.4 shows how rapidly this has risen in the last century or so. There are simply more people to consume the resources.

Another reason is the increasing wealth, and therefore consumption, of the industrialized nations (see figure 1.5).

**Figure 1.4** *How the world's population is rising*

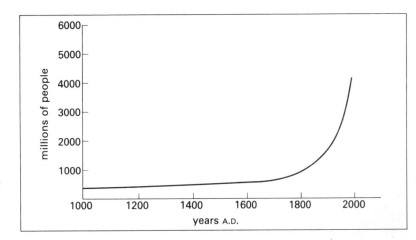

**Figure 1.5 (below)** *Many of the world's resources come from the poorer countries. This diagram shows that most are used in the richer ones*

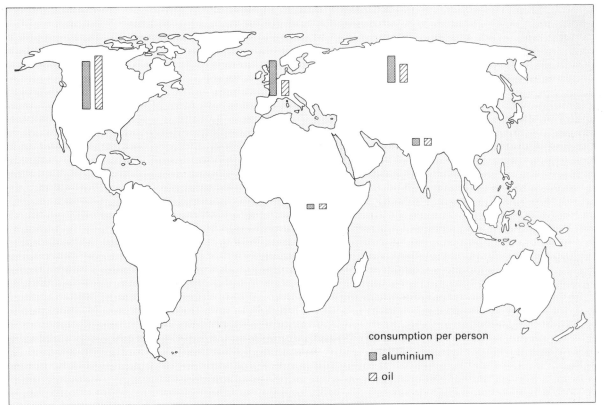

consumption per person

aluminium

oil

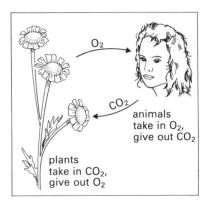

**Figure 1.6** *The oxygen that you breathe is quickly recycled*

**Figure 1.7 (above)** *Oil is a non-renewable resource*

**Figure 1.8** *Already there are only small amounts of metals left, and some fuels will run out within your lifetime*

## 1.2 Renewable and non-renewable resources

We all use oxygen from the air to breathe but there is no immediate danger of us running out of it. This is because it forms part of a **cycle** in which animals (including people) consume oxygen, turning it into carbon dioxide, and plants consume carbon dioxide and produce oxygen (see figure 1.6). This is a natural example of **recycling**. As this process does not take very long we say that oxygen has a short recycling time. Oxygen can be considered a **renewable resource**.

Oil takes millions of years to form. It can be burned in seconds. The gases we get by burning oil may eventually change back into oil. If this did happen it would take millions of years. Oil clearly has a very long recycling time. It is considered a **non-renewable resource**. Once we have used it there will be no more.

### How long will the non-renewable resources last?

We are likely to run out of some resources quite soon. For example, the world's oil is going to be largely used up within twenty years or so. Other substances, such as sand, are unlikely to run out for many lifetimes.

Fuels and metals are the resources that are in most immediate danger of running out. The bar chart in figure 1.8 shows estimates of how long some of these may last.

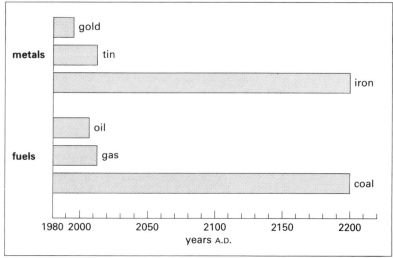

## 1.3 Predicting how long a resource will last

The chart in figure 1.8 shows rough estimates only. It is very difficult to say exactly when we will run out of a particular resource material.

**Figure 1.9** *The high price of oil made it worth facing up to the engineering problems of drilling in the hostile conditions of the North Sea*

**Figure 1.10** *This mine building was once part of a flourishing tin-mining industry in Cornwall but when the world price of tin fell it became uneconomic. When tin becomes scarcer, the price could rise again and the Cornish mines might re-open*

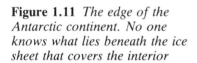

**Figure 1.11** *The edge of the Antarctic continent. No one knows what lies beneath the ice sheet that covers the interior*

One reason for this is that as the material runs short its price goes up. This encourages chemical companies to cope with lower-quality raw materials or to face hostile weather conditions (figure 1.9) in order to obtain supplies.

Another reason is that new resources are still being discovered. Antarctica is thought to have the largest coalfield in the world. It also has high-grade iron ore, and deposits of copper, nickel, chromium, silver and gold have already been found. Only the outer edge of this continent has so far been explored for minerals. At the moment it is thought too hostile a place to work in.

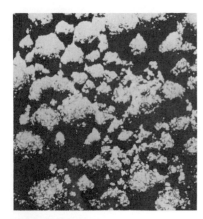

**Figure 1.12** *'Manganese nodules', seen here on the sea bed, contain many valuable minerals*

**Figure 1.13** *Only about 10 per cent of glass is recycled in the United Kingdom. In Holland the figure is nearer 50 per cent*

The sea bed is another place that we are only just beginning to find out about. Some deep-sea areas are known to be covered in 'manganese nodules' (see figure 1.12). These are potato-sized rocks rich in manganese, nickel, copper and cobalt.

## 1.4 Recycling

The oxygen that we breathe is recycled by the natural action of plants. Many other substances can be recycled. This avoids waste and pollution, as well as saving energy and ensuring that the Earth's resources last longer.

The most obvious example of recycling is the milk-bottle. This can be used and re-used many times. Even non-returnable glass bottles can be recycled via bottle banks (figure 1.13). Melting down and re-using the glass uses much less energy than making new glass. Aluminium soft drink and beer cans can also be recycled in this way.

Most of the iron and steel from household rubbish could be recycled. Items like food cans are easily separated from other materials using an electromagnet.

**Figure 1.14** *In Sweden a deposit is charged on all aluminium cans, to encourage recycling and prevent litter*

## 1.5 Where do our resources come from?

Only a tiny proportion of the Earth's resources is available to us. The Earth is about 12 800 km in diameter, but we can only obtain resources from the outermost layer. This layer is called the Earth's **crust** or the **lithosphere**, and it is about 10 km deep. Even our deepest mines and oil wells go only about 3–5 km deep. The sea and the air are also possible resources.

Our four main sources of chemicals are

1 the rocks in the lithosphere,
2 the sea,
3 the atmosphere,
4 living resources from plants, animals and microbes.

**Figure 1.15** *The elements present in rocks, sea and air*

The pie charts in figure 1.15 show you the main elements that are present in the rocks, the sea and the atmosphere.

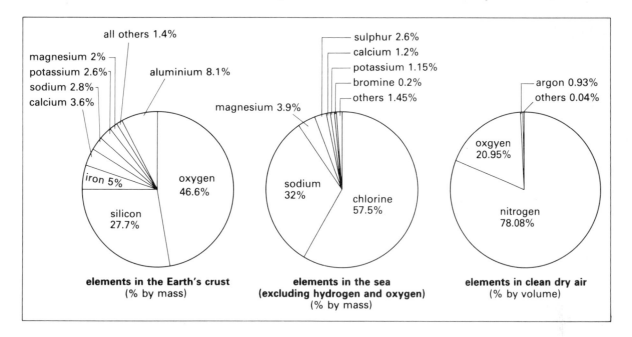

elements in the Earth's crust
(% by mass)

elements in the sea
(excluding hydrogen and oxygen)
(% by mass)

elements in clean dry air
(% by volume)

Fortunately the elements in the lithosphere are not evenly distributed. Some rocks contain very much more than the percentages shown in the pie chart (see figure 1.16).

**Figure 1.16** *There is only a small percentage of nickel in the Earth's crust but this rock in Canada contains a high percentage of nickel*

(a)

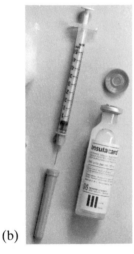

(b)

**Figure 1.17** *Bacteria like those in (a) can be changed by giving them new genes, so that they can make chemicals that we need. Insulin (b), which is needed by many diabetics, can now be made in this way*

Living resources include such materials as wood, and chemicals like sugar from sugar beet. They also include bacteria which, through genetic engineering, can be used to manufacture very complicated compounds (see figure 1.17).

## Summary: Only one Earth

★ The Earth contains only a finite amount of resource materials.
★ We are using the Earth's resources at an ever-increasing rate.
★ Renewable resources are those that can be recycled in a short time.
★ Non-renewable resources either are not recycled or have very long recycling times.
★ Fuels and some metals are resources that we may run short of in the near future.
★ The rocks, the sea, the air and plants provide most of our resources.

## Questions

1 Choose words from the list below to help you copy and complete the sentences.

    renewable   fuels    lithosphere   metals    rocks
    oxgyen    sodium  silicon      chlorine  nitrogen

(a) The outer layer of the Earth is known as the _____ .
(b) Two types of resource that we will soon be running short of are _____ and _____ .
(c) Resource materials that have a short recycling time are said to be _____ .
(d) The two most abundant (plentiful) elements in the Earth's crust are _____ and _____ .
(e) The two most abundant elements in the air are _____ and _____ .

Answer the following in full sentences.

2 Which country consumes most resources per person?
3 What is meant by a low-grade ore? Why might it be profitable to mine low-grade ores in the next century?
4 The main elements dissolved in sea water are chlorine and sodium. In fact neither is present as the element. Find an explanation for this (see chapter 5).
5 Present-day electronics depends heavily upon 'silicon chips'. Are we likely to run out of silicon in the near future? Explain your answer.
6 Is water a renewable or non-renewable resource? Explain your answer.

## A load of old rubbish?

You could say that we don't use up our mineral resources: we just spread them out. Consider the baked bean can! The steel could have come from Welsh iron ore, and the tin possibly from a Malaysian tin mine. Where will this steel and tin end up, once you have eaten the beans? Probably on a landfill waste tip – one of thousands across the country.

This may seem unimportant until you think that we use about 10 000 million food cans a year in Britain. Without some form of recycling, all this valuable steel and tin is lost for ever. The table below shows what effect recycling would have upon the time that our resources will last, supposing that we go on using them up at our present rate.

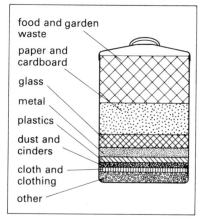

food and garden waste

paper and cardboard

glass

metal

plastics

dust and cinders

cloth and clothing

other

**(above)** *The average dustbin*

**(below)** *A waste separation unit for converting rubbish into solid fuel pellets*

| Percentage recycled | Lifetime of reserves will increase by this many times |
|:---:|:---:|
| 25 | 1.33 |
| 50 | 2 |
| 75 | 4 |
| 80 | 5 |
| 90 | 10 |
| 99 | 100 |

About 90 per cent of our household waste is dumped in landfill tips. Your dustbins are full of valuable materials, but we are just throwing them away.

There is, of course, an alternative to just dumping our waste. The diagram shows a waste separation unit.

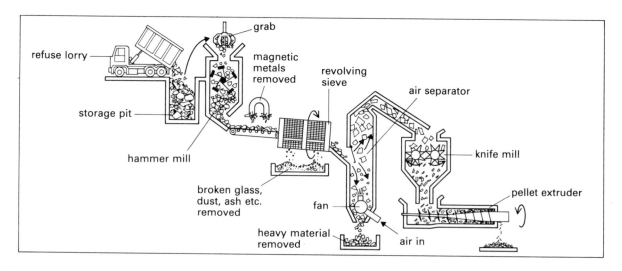

refuse lorry

grab

magnetic metals removed

revolving sieve

air separator

storage pit

hammer mill

knife mill

broken glass, dust, ash etc. removed

fan

pellet extruder

heavy material removed

air in

The rubbish is brought to the plant in lorries or by rail and crushed in a hammer mill. Food cans are magnetically separated early in the process. Usually they are sold to iron foundries which melt them down and use them to make cheap-quality iron for things like manhole covers. In some plants, however, the cans are broken up and soaked in alkali first. The tin coating, which is normally only about a fifth of a millimetre thick, dissolves off leaving the steel which can be added to newly made iron for steelmaking. Much of the steel that we use contains such 'clean' scrap steel – as much as 30 per cent. The tin is also recovered by a process called electrolysis. (You can read about electrolysis in chapter 15.) Although each can only contains about half a gram of tin this becomes a considerable amount in a plant that processes many thousands of cans.

Paper, card and other light waste is separated from the rest by blowing air through it. It is cut into tiny fragments and then made into useful fuel pellets. Even the other waste separation products need not be scrapped. Some manufacturers of cement and concrete have been investigating how broken glass, ash and the heavier materials can be incorporated into concrete.

The idea that your dustbin contains nothing but waste is a load of rubbish!

## Questions

7 Name four useful products that can be obtained from domestic waste.

8 Name one type of material that you think will make up an increasing proportion of household waste. Give reasons for your suggestion.

9 Draw a line graph of the time our resources will last (vertical scale) against the percentage recycling (horizontal scale).

10 What does the shape of the graph that you have drawn tell you?

11 Without recycling we are expected to run out of tin in the year 2010. How long should it last if we recycle 35 per cent of our tin?

12 Approximately how many tonnes of tin a year are used to coat British food cans? (1 tonne = 1000 kg.)

13 Coal-fired power stations, as well as making electricity, produce about 12 million tonnes of ash per year in the UK. Try to find out what they do with it.

# 2 Chemists and chemical reactions

## 2.1 Chemists at work

Almost everything you use comes from something that a chemist invented, improved or tested. Many people work as chemists. In a chemist's shop, chemists supply tablets and medicines, and sometimes make them up as well. Chemists who do this job are called **pharmacists**. A lot more chemists work in industry or for the Government. Here are some of the types of work they do.

**Figure 2.1** *This chemist is involved in testing new drugs*

**Figure 2.2** *Police chemists play an important part in the detection of crime*

**Figure 2.3** *These chemists are developing a new animal food protein*

### How to be good at chemistry

Chemistry is a branch of science. To be good at it you need to do more than just learn facts and ideas. You need to develop certain skills. Some of these are listed below.

observing – noticing things
using measuring instruments
tabulating – putting information in table form
using charts and graphs
pattern-seeking – noticing similarities and trends
predicting – being able to tell what will happen
hypothesizing – thinking up explanations for observations
designing experiments – planning experiments and choosing apparatus
applying knowledge – making some useful product

This book will help you to improve all these skills.

## 2.2 Chemical reactions

When you mix substances together you often get a **chemical reaction**. Chemistry is about chemical reactions. Table 2.1 shows some chemical reactions. Can you see how they are all alike?

In every reaction in the table a new substance is formed. This is true of all chemical reactions. The substances that you start off with are usually called the **reactants**. The substances you have at the end are called the **products**.

A lot of chemical reactions are quite difficult to reverse. You cannot turn a forest back into trees after a fire. Nor can you piece together a bomb after it has exploded.

The processes shown in figure 2.4 are not chemical reactions. Can you see why?

When you boil a kettle of water, you get steam. But this process is easily reversed. You just cool the steam to change it back to water. And an ice-lolly that has melted can easily be frozen again.

**Figure 2.4** *Some processes that are not chemical reactions*

an electric kettle boiling

an ice-lolly melting

a bulb giving out light

| Reaction | Starting substances (reactants) | Substances formed by the reaction (products) |
|---|---|---|
| forest fire | wood and oxygen in air | smoke and ashes |
| car rusting | steel and oxygen in air | rust |
| magnesium burning | magnesium and oxygen in air | white ash |
| plants rotting | plants | compost |
| firework | gunpowder | gases |
| chips cooking | potato and fat | chips |
| descaling a kettle | kettle fur and descaler | dissolved solids and a gas |
| bomb exploding | explosive | gases |

**Table 2.1** *Some chemical reactions*

When a bulb gives out light, a chemical reaction is not taking place because no new substances are formed. The bulb is glass and metal at the beginning and glass and metal at the end.

There are huge numbers of chemical reactions. If you look at a lot of reactions, you will see that many of them have two other things in common:

**1** heat, or some other form of energy, is often needed to start reactions off;
**2** once the reactions are going, they often give out heat. Lighting a bonfire is an example of this.

## 2.3 The units chemists use

When chemists study a chemical reaction, they don't just watch the changes taking place. They want to **measure** what is happening. They measure the masses of substances reacting and their temperature, the time the reaction takes to complete and the masses of the products. They then often want to be able to tell other chemists what they have found. To make sure that they will easily understand each other, chemists all try to use the same **units** for their measurements.

Apart from the units for pressure and temperature, all units used in this book are based upon a system called the 'Système International d'Unités', which is usually abbreviated to 'SI'. The six SI base units we use are listed in table 2.2.

| Quantity | SI unit | Symbol |
|---|---|---|
| length | metre | m |
| mass | kilogram | kg |
| time | second | s |
| electrical current | ampere | A |
| temperature | kelvin | K |
| amount of substance | mole | mol |

**Table 2.2** *The SI base units*

(The **mole** is a particularly important unit and we shall look at it more closely in chapter 7.)

Sometimes a quantity expressed in the basic SI unit would be numerically very large or very small. For instance, we might want to do an experiment using just a pinch of salt, which weighs about 0.0005 kg. Or we might need to measure a period of time of an hour or two – perhaps four or five thousand seconds. To avoid handling clumsy numbers like these, a suitable prefix may be used. Table 2.3 lists some useful examples.

| Prefix | Symbol | Multiplication factor | |
|--------|--------|----------------------|---|
| kilo- | k | 1000 | or $10^3$ |
| deci- | d | 1/10 | or $10^{-1}$ |
| centi- | c | 1/100 | or $10^{-2}$ |
| milli- | m | 1/1000 | or $10^{-3}$ |

**Table 2.3** *Prefixes for SI units*

Some examples of the use of these prefixes are:

$$25.7 \, cm \quad = \quad 25.7 \times 1/100 \, m \quad = \quad 0.257 \, m$$
$$13.6 \, km \quad = \quad 13.6 \times 1000 \, m \quad = \quad 13\,600 \, m$$
$$45 \, mg \quad = \quad 45 \quad \times 1/1000 \, g \quad = \quad 0.045 \, g$$

Further units may be derived by combining the basic SI units. Some of these are listed in table 2.4.

| Quantity | Name | Symbol |
|----------|------|--------|
| energy | joule | J |
| power | watt | W |
| electrical charge | coulomb | C |
| electrical potential | volt | V |
| volume | cubic metre | $m^3$ |

**Table 2.4** *Some derived units*

Some units are used a great deal in chemistry. Some of these are illustrated below.

The most common units of **mass** are:

milligrams (mg)　　grams (g)　　kilograms (kg)

Units of **volume** used in chemistry are:

cubic centimetres ($cm^3$)　　cubic decimetres ($dm^3$)

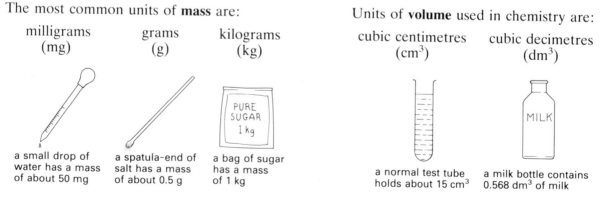

a small drop of water has a mass of about 50 mg

a spatula-end of salt has a mass of about 0.5 g

a bag of sugar has a mass of 1 kg

a normal test tube holds about 15 $cm^3$

a milk bottle contains 0.568 $dm^3$ of milk

Note that there are 1000 $cm^3$ in 1 $dm^3$.
　　Volumes of liquids are measured using a measuring cylinder, a burette or a pipette. Volumes of gases are measured in a gas syringe.

Two **temperature** scales are used by chemists:

> the Kelvin scale
> the Celsius scale

One degree on the Celsius scale (1 °C) is the same as one kelvin (1 K). The Kelvin scale starts 273 degrees below the Celsius scale, however. So

> Kelvin temperature   =   Celsius temperature   +   273
> Celsius temperature   =   Kelvin temperature   −   273

The units of **energy** used in chemistry are:

> joules           kilojoules
>  (J)               (kJ)

it takes about 60 kJ to raise the temperature
of a cup of water from 20 °C to boiling

The units used in studies of **electricity** are volts (V) and amperes (A), often abbreviated to amps. Volts and amps (or mV and mA) may be read directly from instruments.

   Another important unit is the coulomb, which is used for measuring electrical charge:

> charge in       current in       time in
> coloumbs  =       amps      ×   seconds
>    (C)             (A)             (s)

## 2.4 Presenting experimental results: charts and graphs

Often it is easier for chemists to explain their results to other people if they can show them as a chart or a graph instead of just a string of numbers.

   When you are doing chemistry you will need to be able to draw and interpret the following:

**1** bar charts,
**2** pie charts,
**3** line graphs.

We will look at each of these in turn.

### Bar charts

These are the simplest. In figure 2.5, a bar chart is used to compare the ways in which the country's energy supplies are

**Figure 2.5** *United Kingdom energy consumption*

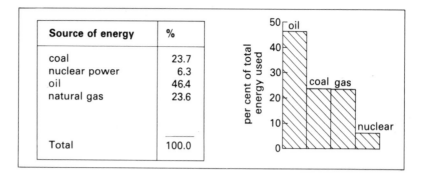

| Source of energy | % |
|---|---|
| coal | 23.7 |
| nuclear power | 6.3 |
| oil | 46.4 |
| natural gas | 23.6 |
| | |
| Total | 100.0 |

produced. It is obvious straight away that oil is by far the most important. Natural gas and coal are about equally used. A smaller (but growing) proportion of energy is produced in nuclear power stations.

When you are drawing a bar chart, there are a few things that you must remember to check.

1 Choose a scale that makes good use of your paper.
2 Include units (where appropriate).
3 Label each item clearly.

## Pie charts

These are easy to interpret but are a little more difficult to draw accurately. They usually give information in the form of percentages and so these must be calculated before the pie chart can be drawn.

An example of a pie chart is shown in figure 2.6. It represents the masses of three gases obtained from some liquid air.

The stages of drawing a pie chart (shown in the table in figure 2.6) are as follows:

1 Add up the total amounts.
2 Calculate the percentages (divide each item by the total, and multiply by 100).
3 Calculate the angles for each (multiply the percentage by 3.6).

**Figure 2.6** *The composition of air*

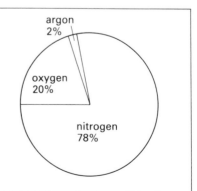

| Gas | Mass (tonnes) | Percentage | Angles |
|---|---|---|---|
| nitrogen | 39 | $(39 \div 50) \times 100 = 78\%$ | $78 \times 3.6 = 281°$ |
| oxygen | 10 | $(10 \div 50) \times 100 = 20\%$ | $20 \times 3.6 = 72°$ |
| argon | 1 | $(1 \div 50) \times 100 = 2\%$ | $2 \times 3.6 = 7°$ |
| Total | 50 | | 100% | 360° |

argon 2%

oxygen 20%

nitrogen 78%

### Line graphs

These are the most common and the most important graphs in chemistry. The graph in figure 2.7, for instance, shows the progress of a chemical reaction. A gas is given off during the reaction, and is collected. Every 10 seconds an automatic balance weighs the amount that has been collected. The mass of gas given off is then plotted against the time.

**Figure 2.7** *The progress of a reaction*

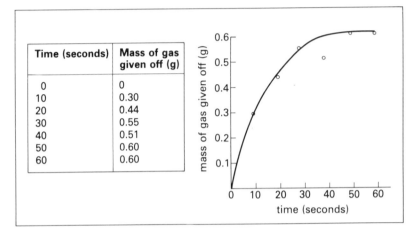

| Time (seconds) | Mass of gas given off (g) |
|---|---|
| 0 | 0 |
| 10 | 0.30 |
| 20 | 0.44 |
| 30 | 0.55 |
| 40 | 0.51 |
| 50 | 0.60 |
| 60 | 0.60 |

The steps to be taken in plotting a line graph are as follows:

1 Choose sensible scales so as to use most of the graph paper.
2 Write the numbers along each axis. Remember that on any one axis one square must *always* be the same amount.
3 Label each axis (mass of gas given off, and time, in figure 2.7).
4 Put the units on each axis (grams and seconds in figure 2.7).
5 Carefully plot each point.
6 Draw a straight line or a sensible curve through the points. (Remember that no one gets perfect results, so it is unlikely that your line will pass right through every point. Notice that in figure 2.7 the line is not drawn through the point for 40 seconds which, it is assumed, represents an experimental error.)

## 2.5 Using the scale factor method

As you work through your chemistry course you will often want to do calculations using the measurements you have made. Each time you meet a new calculation method in this book, it will be explained in full. Here is the first of these: the scale factor method. You will use it over and over again in chemical (and other) calculations. It is really just common sense. Try an example to begin with.

## *Example 1*

To make 10 scones requires 200 g of flour. How much flour is needed to make 5 scones?

You have probably already worked out the answer almost without thinking about it. But what are the stages that you went through? When the numbers get harder you will need to be clearer about these. Here are the actual steps you took in your calculation.

**1** 10 scones need 200 g.
**2** 5 scones will need less. I must *scale down* the amount of flour.
**3** I have scaled down the number of scones by 5/10, which equals 1/2.
**4** I must scale down the amount of flour by the same amount:

$$\frac{1}{2} \times 200 = 100$$

**5** Answer = 100 g

This can be set out like this:

| Number of scones | Mass of flour (in g) |
|:---:|:---:|
| 10 | 200 |
| scale-down factor = 5/10 or 1/2 | multiply by 5/10 or 1/2 |
| 5 | 100 |

So 5 scones will need 100 g of flour.

## *Example 2*

20 nails cost me 50 pence. What will 80 nails cost me?

| Number of nails | Cost (in p) |
|:---:|:---:|
| 20 | 50 |
| scale-up factor = 80/20 or 4 | multiply by 80/20 or 4 |
| 80 | 200 |

So 80 nails will cost 200p.

Chemical calculations using scale factors are no more difficult than these. The numbers themselves may be more awkward but even that doesn't matter if you can use a calculator. Follow through the working of the examples below. Use a calculator if you wish.

## *Example 3*

When 1 g of magnesium reacted with acid, 1000 cm³ of hydrogen gas was produced. How much gas will 0.2 g of magnesium give?

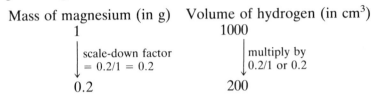

Mass of magnesium (in g)   Volume of hydrogen (in cm³)
1                                         1000

scale-down factor              multiply by
= 0.2/1 = 0.2                    0.2/1 or 0.2

0.2                                      200

So 0.2 g of magnesium gives 200 cm³ of hydrogen.

## *Example 4*

160 g of iron ore gave 112 g of iron metal. How much iron metal could we get from 480 g of iron ore?

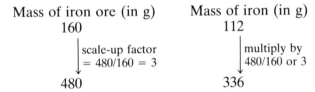

Mass of iron ore (in g)    Mass of iron (in g)
160                                    112

scale-up factor                 multiply by
= 480/160 = 3                  480/160 or 3

480                                     336

So 480 g of iron ore will give 336 g of iron.

## Summary: Chemists and chemical reactions

* Chemists do many different jobs.
* Most things you use are made from material that was invented, improved or tested by a chemist.
* To be a good chemist you need many different skills.
* In a chemical reaction, at least one new substance is always formed.
* The substances at the start of a chemical reaction are the reactants.
* The substances at the end of a chemical reaction are the products.
* Most chemical reactions are difficult to reverse.
* Many chemical reactions need heat or some other form of energy to start them off.
* Many chemical reactions give out heat once they have started.
* Chemists use a system of units called SI units.
* The most commonly used units in chemistry are the gram (g), the cubic centimetre (cm³), the cubic decimetre (dm³) and the joule (J).
* To convert temperatures from the Celsius scale to the Kelvin scale add 273.
* Bar charts, pie charts and line graphs are all important ways of communicating facts in chemistry.
* Using scale factors is a simple way of scaling up or down the amount of chemicals needed for a reaction.

## Questions

1 Choose from the words in the list the one that best fits each of the following descriptions:

> reversible
> chemical reaction
> energy
> reactant
> products

(a) a process in which a new substance is formed
(b) the chemical used at the start of a chemical reaction
(c) the chemicals formed during a chemical reaction
(d) this is often needed to get a reaction started
(e) this word does not apply to most chemical reactions

2 Are the following processes chemical reactions? Answer yes or no.

(a) Lemon juice was added to some washing soda. The mixture fizzed and frothed.
(b) Some salt was added to water and stirred. The salt slowly disappeared.
(c) Some pieces of candle wax were warmed in a test tube. They melted. When the tube cooled down it contained a piece of wax that was the same shape as the tube.
(d) Two colourless solutions were mixed together. A bright yellow solid was formed.
(e) When I mixed some garden lime and fertilizer together, there was a horrid smell. Neither the lime nor the fertilizer had this smell before I mixed them.

3 (a) What is 2.36 kg in grams?
(b) What is 45 mg in grams?
(c) What is $2\,dm^3$ in cubic centimetres?
(d) What is 298 K in degrees Celsius?

4 Estimate the volume of liquid that each of the following can hold:

(a) a teaspoon
(b) a cup
(c) a normal-size test tube
(d) a boiling tube

5 The table on the left gives the densities (in $g/dm^3$) of a group of elements (known as the halogens) when they are in the gaseous state. Plot a bar chart of this data.

| | |
|---|---|
| fluorine | 1.70 |
| chlorine | 3.21 |
| bromine | 7.59 |
| iodine | 11.27 |

6 The table below describes the main users of water in the United Kingdom. Draw a pie chart to illustrate this data.

| | |
|---|---|
| Domestic users | 41% |
| Central Electricity Generating Board | 38% |
| Other industries | 20% |
| Agriculture | 1% |

7 An experiment was set up in which air was passed back and forth over a heated metal. The volume of the air decreased as it reacted with the metal and a pupil took the readings in the table below. Plot a line graph of these readings.

| Time (minutes) | 0 | 1 | 2 | 3 | 4 | 5 | 6 | 7 |
|---|---|---|---|---|---|---|---|---|
| Volume (cm$^3$) | 50 | 47.2 | 45.6 | 44.9 | 43.4 | 42.5 | 41.8 | 41.4 |

8 Joanne asked her school to arrange a period of work experience with a firm where she could use her chemistry. She was surprised to learn how many different kinds of work in chemistry there are: pharmaceutical, analytical, organic, inorganic, physical, and many more. She got a placement in an analytical laboratory where she had to estimate the mass of calcium dissolved in various water samples.

Here are some of her results.

| Sample number | Volume of water (cm$^3$) | Mass of calcium (g) |
|---|---|---|
| 1 | 10 | 0.10 |
| 2 | 20 | 0.16 |
| 3 | 5 | 0.04 |
| 4 | 25 | 0.18 |

(a) How many cm$^3$ are there in 1 dm$^3$?
(b) Joanne had different volumes of water to analyse but she then had to calculate the amount of calcium that would have been in 1 dm$^3$ of each sample of water.

Use the scale factor method to do this for each of the samples 1 to 4. Show your working clearly.
(c) Draw a bar chart showing the amount of calcium (in g/dm$^3$) in the water samples.
(d) Five types of chemistry are mentioned above. Find out what each one is about.

# Chemistry in the kitchen

There are often very different views about the role of chemicals in the kitchen. Most people are happy to use detergents, bleaches and disinfectants. The same cannot be said about the chemicals used in food.

**INGREDIENTS:**
WHOLEWHEAT, VINE FRUITS, SUGAR,
DRIED COCONUT, BANANA, APPLE,
HAZELNUTS, SALT, MALT FLAVOURING,
VEGETABLE OIL, NIACIN, IRON,
VITAMIN B$_6$, RIBOFLAVIN (B$_2$),
THIAMIN (B$_1$), FOLIC ACID,
VITAMIN D, VITAMIN B$_{12}$.

CONTAINS NO ARTIFICIAL
FLAVOURING OR COLOURS

**TYPICAL NUTRITIONAL COMPOSITION PER 100 GRAMMES**

| | |
|---|---|
| Energy | 338 kcal |
| | 1440 kJ |
| Protein (N x 5.7) | 8.1 g |
| Fat | 3.5 g |
| Dietary Fibre | 6.4 g |
| Available Carbohydrate | 73.2 g |
| Vitamins: | |
| Niacin | 18.0 mg |
| Riboflavin (B$_2$) | 1.5 mg |
| Thiamin (B$_1$) | 1.0 mg |
| Folic Acid | 250 µg |
| Vitamin D | 2.8 µg |
| Vitamin B$_{12}$ | 1.7 µg |
| Iron | 6.7 mg |
| Vitamin B$_6$ | 1.8 mg |

Should this product in any way fall below the high

Ingredients:
Sugar, Modified starch, Hydrogenated vegetable oil,
Emulsifiers: E477, E322; Gelling agents: E339,
E450(a); Fat-reduced cocoa, Lactose, Caseinates,
Flavourings, Whey powder, Colours: Caramel (E150),
E102, E122, E142, E160(a); Antioxidant: E320.

Ingredients: Cured pork and pork liver (Pork, Pork liver, Salt, Dextrose,
Sugar, Preservative: E250), Water, Caseinates, Wheatflour, Salt,
Spices, Flavour enhancer: 621; Di and triphosphates, Acidity
regulator: E339, Onion powder, Marjoram. (Minimum 73% Pork).

Keep refrigerated. Once pack is opened, eat within 3 days

*Detailed nutritional labelling has made consumers aware of the range of additives in everyday foods*

Many kitchen processes, such as making meringues or sauces, are quite subtle chemical experiments in which natural foods are blended and reacted together. At the other extreme we have *Surimi*. This is a kind of Japanese minced fish from which all strong flavours and natural colours have been removed. Add some chemicals called polyphosphates, together with some salt and sugar, and you have a bland unidentifiable seafood flavour. The product now only needs re-flavouring, re-colouring and re-texturing to give us 'crab sticks'.

People have become increasingly conscious that food is being 'chemically tampered with'. Perhaps their biggest concern has been over colourings. Their worries can be summarized by the comment, 'if they have to add colouring to blackcurrant juice then what on earth have they done to it?' It is well known that a few people are allergic to some colourings, such as the yellow dye tartrazine, which is used in many foods, including some brands of orange squash. It is called by the code name 'E102' on food labels. Does this mean tartrazine should be banned? If so, should we also ban strawberries, shellfish and milk – all of which cause allergies in quite a lot of people?

Another common group of additives are the preservatives. These check the growth of bacteria and moulds. The use of nitrates and nitrites in cured meats gives some cause for concern but apart from these many people accept the need for preservatives. The bacterium *Salmonella* can cause serious food poisoning – it is definitely worse than a trace of sulphur dioxide!

The oxygen in the air is very reactive. Without antioxidants fats and oils react with oxygen and soon go rancid, producing chemicals that can cause illness and even cancer.

But shouldn't we all be eating healthy, natural food? Although ... don't almonds contain cyanide, which interferes with respiration? And doesn't the oxalic acid in rhubarb remove calcium from the blood? And what about the substances called psoralens in celery, which can cause genetic mutations? *And* bananas – they contain 5-hydroxytryptamine, which is known to interfere with brain biochemistry.

Perhaps it all depends upon how much of them we eat!

### Questions

**9** Are food additives necessary? Give reasons for your answer.

**10** Why is listing additives on food labels a good idea?

**11** Make a table of the additives shown on the labels of five different foods. Find out everything you can about these.

# 3 Experimental techniques and measuring

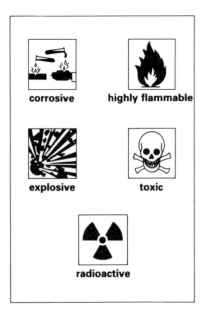

**Figure 3.1** *Some standard hazard symbols*

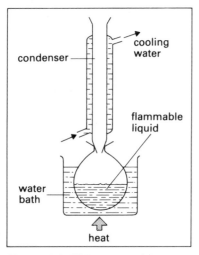

**Figure 3.2** *Using a condenser to stop flammable vapour escaping*

## 3.1 Safety

Whatever the experiment, there are always five safety factors to consider:

1 Are these substances **corrosive**?
2 Are they **flammable** or **explosive**?
3 Is **pressure build-up** likely?
4 Are any reactants or products likely to be **poisonous**?
5 Are they **radioactive**? (It is unlikely that you will be expected to handle radioactive substances at school, however.)

You should be sure you can recognize the standard hazard symbols shown in figure 3.1. They are often used as warnings of danger.

### Care with corrosives

There are two common groups of **corrosive** chemicals (that is, substances that will eat away, or burn, things that they touch). These are **acids** and **alkalis**. All experiments using these should be carried out wearing safety spectacles. If concentrated solutions are being used then it is a good idea to wear plastic gloves as well.

### Keeping cool over flammables

Most organic chemicals are **flammable**, that is, they catch fire easily. They must obviously be kept well away from flames!

Liquids like these must always be heated on a water bath, not over a bunsen burner. If temperatures above 100 °C are needed then an oil bath can be used. Any apparatus in which more than a few cm$^3$ of flammable liquid are being heated must have some method of condensing the flammable vapours (see figure 3.2).

### Handling explosive gas–air mixtures

Some flammable gases, such as hydrogen and natural gas, form highly explosive mixtures with air.

**Figure 3.3** *Precaution in using a flammable gas*

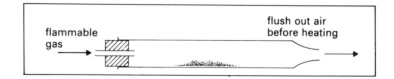

Such mixtures often form in experiments using combustion tubes. In such experiments it is **essential** to let the gas pass through the tube for a minute or two to flush out any air. Only then is it safe to start any heating.

## Preventing pressure build-up

The glassware used in laboratories is unable to withstand much pressure build-up. To avoid this it is essential that apparatus is open to the atmosphere at some point to allow for expansion and production of gases.

**Figure 3.4** *Apparatus for preparing gases: (a) a safe design, (b) a dangerous design – there is nowhere for gases or vapours to escape to*

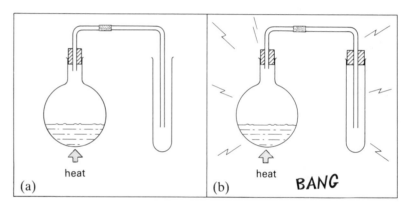

Figure 3.4 illustrates a safe and a dangerous apparatus design. The only experiments likely to need totally closed systems are those which use gas syringes. Here any pressure build-up is not dangerous, because it is relieved by the movement of the syringe plunger (see figure 3.5).

**Figure 3.5** *A gas syringe provides a way of preventing pressure build-up*

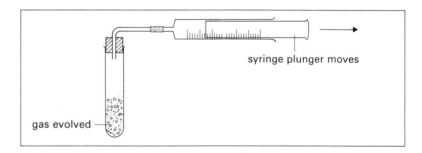

## Preventing suck-back

In many experiments a gas is produced by heating a substance or mixture, and the gas is then collected over water. Great care is needed in these experiments to avoid

**Figure 3.6** *(a) Suck-back is hard to stop; (b) adding a small length of rubber tube can help prevent suck-back, and adding a Bunsen valve to the delivery tube makes it even safer*

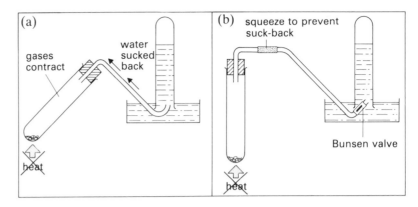

the process called 'suck-back'. This usually happens if the heating is stopped without taking the delivery tube out of water, or if the apparatus is cooled by draughts. The cooling makes the gases inside the apparatus contract, and cold water is sucked into the apparatus. The results can be dramatic – and dangerous, since the hot glass will probably shatter when touched by the cold water.

### Avoiding getting poisoned

Always assume that chemicals are harmful. They often are! You don't have to eat a chemical to be poisoned by it. Some of the most dangerous can be breathed in. Others can be absorbed through the skin. Here are some obvious precautions to take.

1 Never handle chemicals with your fingers. Use a spatula.
2 Never taste chemicals.
3 Be very cautious about smelling chemicals.
4 Never pipette poisonous chemicals by mouth.
5 Do all work with poisonous gases in a fume cupboard.

## 3.2 Handling chemicals

Substances of different kinds are handled in different ways.

### Handling solids

In many ways these are the easiest substances to handle. They are transferred using a spatula.

Small amounts of solids can often be heated safely in an open ignition tube. Larger quantities require a porcelain or nickel crucible for heating.

### Handling liquids

Small amounts of liquids are conveniently transferred using a small dropping pipette (dropper).

Quite often it is necessary to add a liquid to some other piece of apparatus. Two ways of doing this are shown in

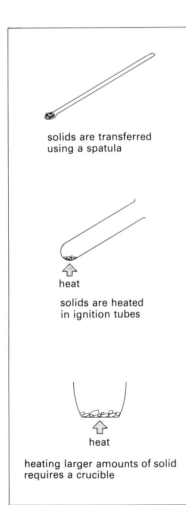

solids are transferred using a spatula

solids are heated in ignition tubes

heating larger amounts of solid requires a crucible

**Figure 3.7** *Handling solids*

**Figure 3.8** *Handling liquids*

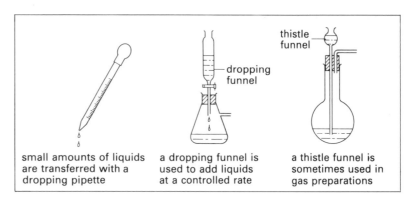

small amounts of liquids are transferred with a dropping pipette

a dropping funnel is used to add liquids at a controlled rate

a thistle funnel is sometimes used in gas preparations

**Figure 3.9** *Heating liquids safely*

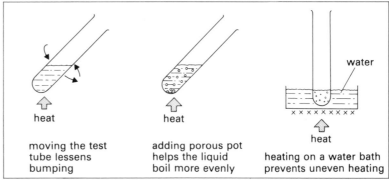

moving the test tube lessens bumping

adding porous pot helps the liquid boil more evenly

heating on a water bath prevents uneven heating

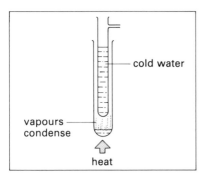

**Figure 3.10** *A simple condenser*

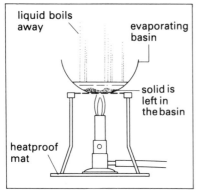

**Figure 3.11** *Using an evaporating basin*

figure 3.8. If a thistle funnel is being used then its stem must extend below the liquid in the flask, to prevent gases from escaping up the funnel.

When heating liquids it is important to avoid local overheating as this can cause 'bumping' – that is, liquid comes shooting out of the apparatus. There are three ways of cutting down the risk of bumping (see figure 3.9).

**1** Don't heat just one part of the apparatus for long.
**2** Add porous pot or 'anti-bumping granules' to the liquid.
**3** Use a water bath for heating.

During heating, liquids will, of course, boil away (evaporate). This can be prevented using a condenser as shown in figure 3.2. Evaporation of a small quantity of liquid can be prevented by heating it in a boiling tube fitted with a side-arm test tube filled with cold water as a simple condenser (see figure 3.10).

You may, of course, want to evaporate the liquid away. If so, the best piece of apparatus to use is an evaporating basin. Evaporating basins are shallow and wide, allowing the vapours to escape easily (see figure 3.11).

## Handling gases

In school laboratories most gases are made in arrangements like those in figure 3.6(b) and the two right-hand diagrams in figure 3.8.

**Figure 3.12 (right)** *Three ways of drying gases*

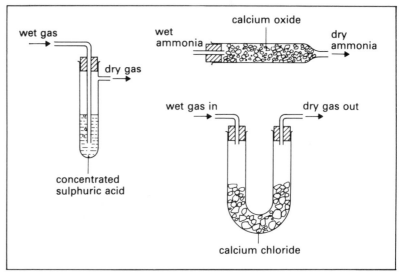

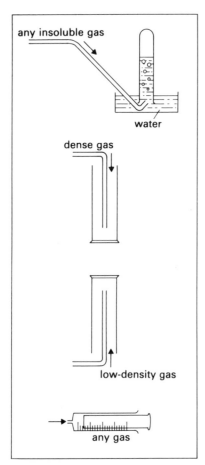

**Figure 3.13** *Collecting gases*

Apart from their preparation, the two most common operations with gases are drying them and collecting them.

To dry gases, they are passed over – or through – a substance that absorbs water strongly. For instance, many gases are dried by bubbling them through concentrated sulphuric acid. Ammonia cannot be dried in this way, because it reacts with acids. Instead it is passed through a tube containing large lumps of calcium oxide. Calcium chloride is another solid that is useful for drying gases.

Four ways of collecting gases are shown in figure 3.13.

## 3.3 Measuring out chemicals

Solids are nearly always measured by using a balance to find their mass (in grams). Liquids and gases are usually measured by their volume (in cubic centimetres). Figure 3.14 shows suitable apparatus for doing this.

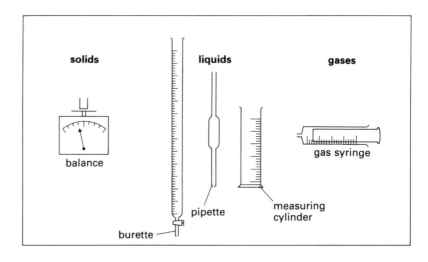

**Figure 3.14** *Measuring out chemicals*

## 3.4 Reading scales

Many pieces of apparatus and many measuring instruments have scales upon them. Mis-reading such scales is quite easy unless you are careful. There are four stages in taking a scale reading correctly.

1 Check what units are being used.
2 Check whether there are any decimal points in the numbers.
3 (a) Choose two numbers that are next to each other.
  (b) Count the number of lines from one number to the next.
  (c) Calculate how much each line represents.
4 Now take your reading.

Some examples will make this clear.

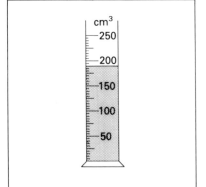

**Figure 3.15** *Example 1*

### Example 1

Figure 3.15 shows a measuring cylinder. How much liquid does it hold?

| | |
|---|---|
| 1 Units | $cm^3$ |
| 2 Decimal points? | none |
| 3 (a) Numbers next to each other = | 150 and 200 |
| (b) Number of lines between = | 10 |
| (c) Each line equals $(200 - 150) \div 10 =$ | 5 |
| 4 Reading $= 150 + (8 \times 5) =$ | $190 \, cm^3$ |

### Example 2

Figure 3.16 shows an ammeter. What current is passing through it?

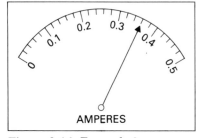

**Figure 3.16** *Example 2*

| | |
|---|---|
| 1 Units | amps |
| 2 Decimal points? | yes |
| 3 (a) Numbers next to each other = | 0.3 and 0.4 |
| (b) Number of lines between = | 5 |
| (c) Each line equals $(0.4 - 0.3) \div 5$ | |
| $= 0.10 \div 5 =$ | 0.02 |
| 4 Reading $= 0.3 + (3 \times 0.02) =$ | 0.36 amps |

### Example 3

Figure 3.17 shows a thermometer. What temperature does it register?

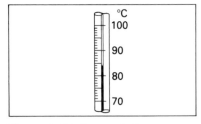

**Figure 3.17** *Example 3*

| | |
|---|---|
| 1 Units | °C |
| 2 Decimal points? | none |
| 3 (a) Numbers next to each other = | 80 and 90 |
| (b) Number of lines between = | 10 |
| (c) Each line equals $(90 - 80) \div 10 =$ | 1 |
| 4 Reading $= 80 + (4 \times 1) =$ | 84 °C |

## 3.5 Using a calculator – answering to four significant figures

Some calculators look very complicated. For your chemistry at this level, however, you really only need one that adds, subtracts, multiplies and divides. The one difficulty arises because calculators insist on trying to give you very accurate answers.

Suppose you want to divide 10 g of a chemical equally between seven test tubes. You need to work out how much to put in each tube. To do this you key in

$$10 \div 7 =$$

Out comes the answer: 1.428 5714.

This may be very accurate – indeed, it is accurate to the nearest ten-millionth of a gram – but it is also ridiculous! You can't possibly divide up your sample as accurately as that.

To overcome this problem, chemists don't give all these figures when they record their results. Chemists in schools usually work to what is called four significant figures. Examples of how to do this are given below. It is a three-stage process.

### *Example 4*

What is 1.428 5714 correct to four significant figures?

| | | |
|---|---|---|
| **1** Write down the first four non-zero numbers | | 1428 |
| **2** Check the next two figures after these: 57 | | |
| (a) if below 50, then ignore it | | – |
| (b) if above 50, then add 1 to the last number in **1** | | 1428 becomes 1429 |
| **3** Add zeros (if necessary) and correctly show the position of the decimal point | | 1.429 |

**Answer:** 1.429

### *Example 5*

What is 0.001 0121 correct to four significant figures?

| | | |
|---|---|---|
| **1** Write down the first four non-zero numbers | | 1012 |
| **2** Check the next two figures after this (here we add a zero to get the second figure) | | 10 |
| (a) if below 50 then ignore it | | (below 50) |
| (b) if above 50 then add 1 to the last number in **1** | | – |

**3** Add zeros (if necessary) and correctly
show the position of the decimal
point                                          0.001 012

**Answer:** 0.001 012

## Example 6

What is 96 482 correct to four significant figures?

**1** Write down the first four non-zero
numbers                                        9648
**2** Check the next two figures after this
(here we add a zero to get the second
figure):                                       20
(a) if below 50 then ignore it                 (below 50)
(b) if above 50 then add 1 to the last
number in **1**                                 –
**3** Add zeros (if necessary) and correctly
show the position of the decimal point 96 480

**Answer:** 96 480

## Summary: Experimental techniques and measuring

- ★ Chemicals may be corrosive, flammable, poisonous and/or radioactive.
- ★ Safety spectacles, and possibly plastic gloves, must be worn when corrosive substances are being used.
- ★ Flammable substances must always be heated on a water bath or an oil bath.
- ★ A condenser must be used when flammable substances are heated.
- ★ The air must be flushed out of combustion tubes before they are heated.
- ★ Apparatus must normally have an outlet for gases, to prevent pressure build-up and explosions.
- ★ Apparatus must be designed so that suck-back is impossible.
- ★ Chemicals must not be touched, tasted or breathed in, in case they are poisonous.
- ★ Solids are transferred with a spatula.
- ★ Liquids are transferred using a dropper. They are added to other apparatus using a dropping funnel or a thistle funnel.
- ★ Precautions must be taken to avoid 'bumping' when liquids are heated.
- ★ Evaporation of a liquid during heating can be prevented by using a condenser. Evaporation can be encouraged by heating in an evaporating basin.
- ★ Gases are dried by passing them over a substance that absorbs water strongly, such as concentrated sulphuric

acid or solid calcium chloride. Ammonia is dried over calcium oxide.
* Gases are collected over water, by displacing air, or in a gas syringe.
* Solids are measured by mass. Liquids and gases are measured by volume.
* Most measurements are taken by reading a scale.
* The answers to calculations in chemistry are usually rounded to three or four significant figures.

## Questions

1 Draw a labelled sketch of the apparatus you would choose for each of the following tasks.
(a) Transferring a solid from a jar to a test tube.
(b) Heating a reaction mixture containing alcohol. (Alcohol is flammable.)
(c) Collecting a water-soluble gas.
(d) Reducing the volume of a solution by boiling off some of the water.
(e) Adding 20 cm$^3$ of liquid to a flask at a steady rate.
(f) Accurately measuring out 5 cm$^3$ of a liquid.
(g) Drying ammonia gas.
(h) Keeping 5 cm$^3$ of solution at 80 °C for three hours, making sure that the water is not allowed to evaporate.
(i) Collecting a gas in an experiment where suck-back is known to happen frequently.

2 Name three pieces of safety equipment that you would expect to find in any lab.

3 Look at the diagrams below. They show instruments for measuring length, the volume of a gas sample and a period of time. Write down the reading on each scale.

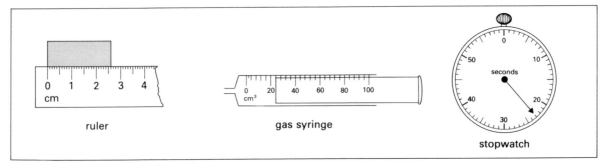

ruler          gas syringe          stopwatch

4 In some chemical calculations, the following answers were obtained. Round off each number to three significant figures.
(a) 2.6576
(b) 22 412.0
(c) 0.006 7558
(d) 158.099
(e) 0.04 998

**5** The apparatus in the diagram was designed by a pupil trying to find out what volumes of dry hydrogen gas are obtained when different lengths of magnesium ribbon react with hydrochloric acid.

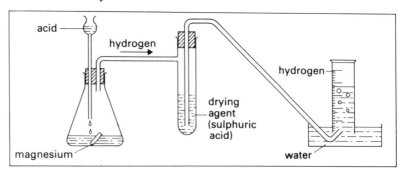

(a) There are several mistakes in his apparatus design. Make a note of all those that you can see.

(b) Draw a labelled diagram of an apparatus suitable for carrying out the experiment.

**6** Some pupils were given the following design brief:

'Phosphoric acid is added to sodium bromide to produce misty fumes of hydrogen bromide. This is a dense gas that is highly soluble in water.

Design an experiment to make and collect a dry sample of hydrogen bromide.'

Five of the designs are shown below. Each one has a fault. Explain what the fault is in each case.

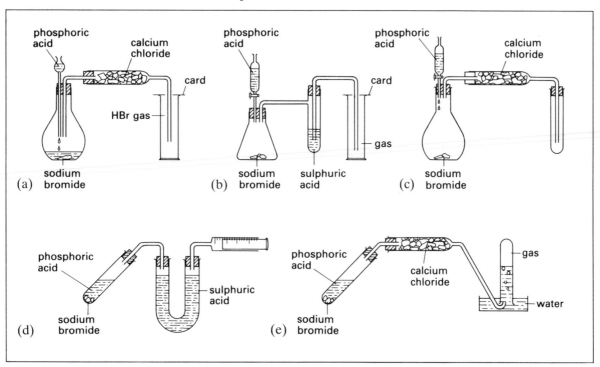

## Towards the robot chemist?

Over the last few decades consumers have come to demand more reliable products. This has meant that manufacturers need to carry out quality checks more frequently, and so faster methods of analysing chemical products have had to be developed.

In order to achieve this, automatic analysis machines have been invented. These are generally more accurate than humans can be, and also much faster – one machine may deal with more samples than 100 technicians could cope with in the same time. Automatic analysers are now used in every field of chemistry.

Most of the machines follow a similar sequence. First there is a device which automatically measures out some definite volume of the solution to be tested. This is usually done by using a peristaltic pump, of the kind shown in the diagram.

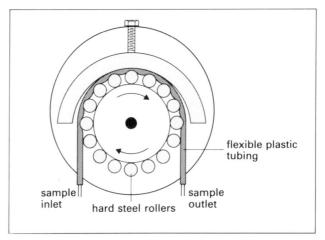

*A peristaltic pump*

What happens in the next stage will depend upon what substance the machine is trying to measure. For example, if it were checking the concentration of salt in estuary water then seeing how well the water conducts might give a reasonable measure of the salt content.

If it were checking the amount of nickel in a solution then a chemical would be added which reacts with nickel to turn the solution red. The more nickel there is present, the redder the solution would become. The amount of colour can be measured by shining a light through the solution on to a photocell. The current that this produces will depend upon the amount of nickel present.

A machine to check the acid in effluent water from a chemical factory might measure the pH value of the water.

*Properties of blood such as pH, haemoglobin content and oxygen levels can now be measured automatically*

The final stage in the operation of such machines involves an electrical signal being fed into a computer that is programmed to calculate the answer and print out the result.

Some automatic analysers can do over 3000 analyses per hour. The days of the routine analytical chemist may seem numbered. There will, however, always be a need for chemists to design or modify automatic analysers that will make new, faster or more accurate measurements.

## Questions

7 What is meant by 'analysis'?

8 Explain how a peristaltic pump could be used to measure out a definite volume of solution.

9 Draw a flow chart for an automatic analyser for measuring nickel concentrations.

10 Solutions containing sodium give a yellow colour when added to a non-luminous flame. Suggest a design for an automatic analyser for measuring sodium.

# 4 Separating mixtures

**Figure 4.1** *Oil gushing out of the sinking 200 000 tonne Amoco Cadiz oil tanker. The oil and water do not mix*

Chemists often have to separate chemicals that are mixed up together. How they separate them depends on what is in the mixture. It depends on whether the chemicals are solids, liquids or gases. It is also important to know if the chemicals mix together easily or not.

## 4.1 Separating liquids from each other

### Separating immiscible liquids

Petrol and water do not mix easily. They are **immiscible**. Even petrol and water that have been stirred together soon form separate layers again. These immiscible liquids are separated using a **separating funnel** (see figure 4.2). The

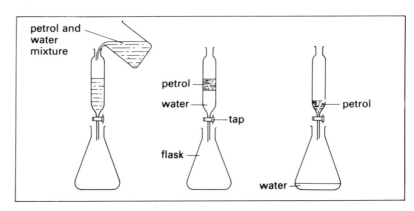

**Figure 4.2** *Using a separating funnel*

mixture is poured into the funnel. When the layers have separated, the tap is opened. Liquid runs out until only the top layer is in the funnel. Then the tap is closed.

## Separating miscible liquids

Water and car antifreeze are **miscible**. Once they have been stirred together they do not form separate layers. **Distillation** (see figure 4.3) must be used to separate such mixtures.

Distillation works because the liquids have different boiling points. For example, you can use distillation to separate water and antifreeze because water boils at 100 °C and antifreeze very much higher. The mixture is poured into a round-bottomed flask. Then the flask is heated. When the temperature reaches 100 °C, the water vapour (steam) moves up past the thermometer and enters the condenser. Here it cools down and changes back into liquid water. This drips from the condenser and is collected in a flask. Any liquid which drips from the condenser is known as **distillate**. All the water eventually becomes distillate. The antifreeze remains in the round flask. It will not distil unless it is heated far above 100 °C.

The fractionating column shown in figure 4.4 gives better separation than simple distillation. For mixtures of liquids with very different boiling points it is not really necessary.

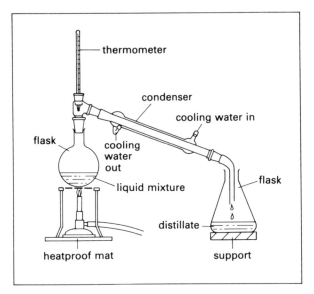

**Figure 4.3 (above)** *Separating miscible liquids by simple distillation*

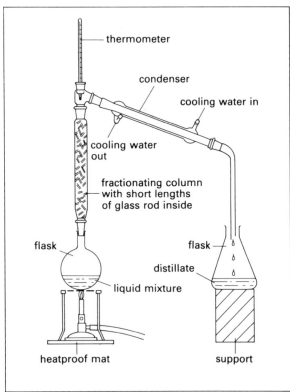

**Figure 4.4 (right)** *A fractionating column gives improved separation*

But it is important if the liquids have similar boiling points. Alcohol and water form such a mixture. Alcohol boils at 80 °C. Water boils at 100 °C. When this mixture is heated, the vapour produced is mostly alcohol with some steam. Water has the higher boiling point and so it condenses more easily than alcohol. This is what happens in the fractionating column. The water condenses in the column and drips back into the round-bottomed flask. When all the alcohol has distilled, the reading on the thermometer rises from 80 °C to 100 °C. The water then distils over, into the collecting flask.

**Figure 4.5** *Fractional distillation is used to produce spirits such as whisky and vodka. It is also used in the petroleum industry*

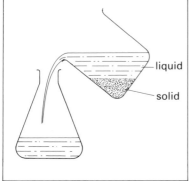

**Figure 4.6** *Decanting*

filter paper

funnel

filtrate (the liquid that drips through the paper)

solid trapped in the paper

**Figure 4.7** *Filtering*

## 4.2 Separating solids from liquids

### Separating an insoluble solid from a liquid

Sand is **insoluble** in water. This means that sand will not dissolve in water. You can stir sand and water together, but the sand will sink once you stop stirring. The water can be separated by carefully pouring it off. This method is called **decanting** (see figure 4.6).

Soil, like sand, is insoluble in water. But it is less dense than sand. So soil does not sink easily when it is stirred into water. Instead it makes a cloudy mixture called a **suspension**. To separate this mixture you have to strain off the bits of soil. You can do this by pouring the mixture into filter paper. The liquid (water) drips through the tiny holes in the paper. The particles of solid (soil) remain trapped. This is known as **filtering** (see figure 4.7).

**Figure 4.8** *A simple centrifuge*

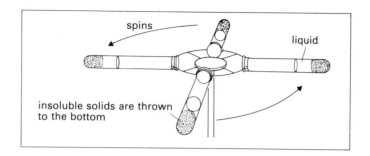

Another method of separating an insoluble solid from a liquid is to use a **centrifuge**. This is a machine which speeds up the rate at which solids settle out. Tubes containing the mixture are spun at very high speeds. The solids are thrown to the bottom of the tubes and stay there when the centrifuge stops. The liquid can then be decanted.

### Separating a soluble solid from a liquid

Salt is **soluble** in water. When salt is stirred with water it dissolves to give a clear solution. Any solid that is dissolved in a liquid is called a **solute**. The liquid that dissolves it is called the **solvent**. The mixture is called a **solution**.

The simplest way to obtain salt from salt solution is to pour the mixture into an evaporating basin and evaporate the water, as shown in figure 3.11 on page 27.

Water is not the only liquid that can be used as a solvent. Alcohol and many other liquids are also used. A mixture containing alcohol could not be boiled in an evaporating basin. It would burst into flames. Even if you heated it on a water bath, it would fill the laboratory with fumes. There could be an explosion. So instead you would use distillation (see figure 4.3). The solute would stay in the round flask. The alcohol would collect in a separate flask after cooling down in the condenser.

## 4.3 Other ways of separating mixtures

### Paper chromatography

Paper chromatography is used to separate mixtures of substances in a solution. It is very useful for coloured substances. An ink may contain a red dye X and a blue dye Y dissolved in water. Small amounts of ink can be separated into these dyes using paper chromatography. A small spot of ink is put on to a sheet of special paper. The paper is placed standing up in a beaker containing a little solvent (see figure 4.9, overleaf). In this case the solvent is water. As the water moves up the paper, the dyes get carried up with it. Usually the dyes do not move as fast as the solvent. They also do not

**Figure 4.9** *Paper chroma-*
*tography. The black ink has*
*separated into two dyes, X*
*and Y*

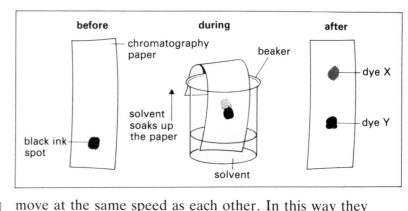

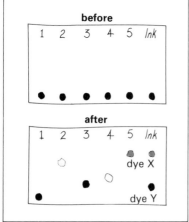

**Figure 4.10** *Using paper*
*chromatography to identify dyes*

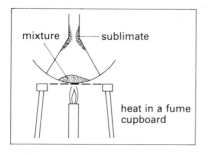

**Figure 4.11** *Sublimation*

move at the same speed as each other. In this way they
become separated. The **chromatogram** in figure 4.9 shows
how this ink contains two different dyes, X and Y.

Paper chromatography can also be used to identify the
dyes in the ink. To find out what substances you had in the
ink, you would have to repeat the experiment. This time you
also put spots of known dyes (numbered 1 to 5 in figure 4.10)
near the bottom of the paper. The solvent is allowed to rise
up the paper, as before. Then the chromatogram is
examined.

Dye Y has moved the same distance up the paper as the
known dye 3. So it is probably the same substance. Which of
the known dyes (1 to 5) is dye X in the ink?

Although chromatography is easier with coloured
substances, it can be used for colourless ones. At the end of
the experiment the paper is sprayed with a chemical that
colours the spots. This chemical is called a **locating agent**.

## Sublimation

When a solid is heated, it usually turns into a liquid. If you
then heat the liquid, it becomes a gas. Some solids do not do
this. Iodine and ammonium chloride are examples. When
either of these is heated, the solid turns straight into a gas.
Cooling makes the gas turn back into a solid again. This
process is called **sublimation**. Substances that sublime can be
separated from other solids using the apparatus shown in
figure 4.11. The gas turns back into a solid, on the cold
upper part of the funnel. The solid is called the **sublimate**.

## Summary: Separating mixtures

* ★ Liquids that do not mix are immiscible.
* ★ Liquids that do mix are miscible.
* ★ A separating funnel separates immiscible liquids.
* ★ Distillation separates miscible liquids.
* ★ Miscible liquids with similar boiling points can be
  separated better with a fractionating column than by
  simple distillation.

* A solid that does not dissolve in a liquid is insoluble.
* A solid that will dissolve in a liquid is soluble.
* A liquid used to dissolve a solid is called a solvent.
* The solid that dissolves in a solvent is called a solute.
* The mixture of solute and solvent is called a solution.
* An insoluble solid is separated from a liquid by decanting, filtering or centrifuging.
* A soluble solid is separated from a liquid by evaporation or distillation.
* Chromatography is used to separate and identify substances.
* Solids that change directly into gases can be separated from other solids by sublimation.

## Questions

1 Choose a word from those below to help you complete the following sentences.
>    solvent
>    miscible
>    filtrate
>    suspension
>    distillate

(a) A liquid or solution that drips through a filter paper is a ＿＿＿＿＿＿＿ .

(b) A liquid that drips from the end of a condenser during a distillation is called the ＿＿＿＿＿＿＿ .

(c) A cloudy mixture of a solid and liquid is a ＿＿＿＿＿＿＿ .

(d) A description of two liquids that mix together is ＿＿＿＿＿＿＿ .

(e) A liquid that is used to dissolve another chemical is called a ＿＿＿＿＿＿＿ .

2 Here is some information about the mixture that is obtained in the manufacture of aspirin. Explain, using both words and diagrams, how you would separate the mixture to obtain all three chemicals in as pure a state as possible.

| Name of chemical | Melting point (°C) | Boiling point (°C) | Solubility in water |
|---|---|---|---|
| water | 0 | 100 | – |
| aspirin | 135 | – | insoluble |
| ethanoic acid | 17 | 118 | miscible |

## Extracting caffeine

Many people know that coffee contains a mild stimulant called caffeine. You may have seen 'decaffeinated' coffee in the shops. This is coffee with the caffeine taken out of it.

Caffeine is also present in one of the most popular sparkling drinks on the market. It is possible to separate the caffeine from this drink. The method works because caffeine dissolves well in an organic liquid called 1,1-dichloromethane. This liquid is immiscible with water.

One difficulty is that the drink also contains a preservative called benzoic acid and this too dissolves in 1,1-dichloromethane. This problem is overcome by adding sodium carbonate to the drink. This neutralizes the acid, changing it into a salt which does not dissolve. It is usual to start with about $200\,cm^3$ of the drink and to add about two spatulas of solid sodium carbonate, shaking the liquid until the sodium carbonate has dissolved.

The next stage is to add about $60\,cm^3$ of 1,1-dichloromethane (which should not be used outside of a fume cupboard) and swirl the mixture for at least five minutes. During this process the caffeine is extracted from the water into the 1,1-dichloromethane. The immiscible liquids are now separated from each other, using a separating funnel (the 1,1-dichloromethane solution becomes the lower layer).

The caffeine solution usually still contains a small amount of water at this stage. This is removed by adding a solid called magnesium sulphate, which soaks up water. The solid is now separated from the solution.

In the final stage, the volatile 1,1-dichloromethane is separated from the very much less volatile caffeine. Care must be taken to avoid overheating the caffeine. It is also most important that the very unpleasant 1,1-dichloromethane vapours are prevented from escaping into the lab.

### Questions

3 Look up and explain the meanings of the following words:
   (a) immiscible     (b) insoluble     (c) extracted
   (d) organic     (e) volatile     (f) involatile
4 Draw a flow chart making clear the main stages followed in separating the caffeine from the drink.
5 After the drink has been swirled with 1,1-dichloromethane the immiscible liquids are separated. Draw a diagram of an apparatus for doing this.
6 Draw a suitable apparatus for separating the magnesium sulphate from the caffeine solution.
7 The dichloromethane is finally separated from the caffeine. Draw a fully labelled diagram of how this could be done. Explain any precautions you would take.

# 5 Elements, compounds and mixtures

## 5.1 Pure substances

What does the word *pure* mean? The label on the carton shown in figure 5.1 says that it contains pure orange juice. This means that it contains only the juice of oranges. No colouring or flavouring has been added.

A chemist would not consider orange juice to be pure. To a chemist something that is pure contains only one substance. Orange juice contains water, sugar, citric acid and many other substances. To a chemist, orange juice is a **mixture**.

**Figure 5.1** *To a chemist, pure orange juice is not pure – it is a mixture*

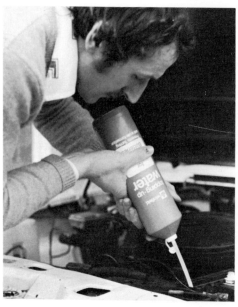

**Figure 5.2** *Distilled water is pure – it contains no other substances. It is used in car batteries*

### Is it pure?

How can you tell if a substance is pure or not? If it is pure then all of it should behave in the same way. When a pure liquid is distilled it all boils at the same temperature. If a pure solid is slowly heated it all melts at the same temperature. A pure dye will give a single spot in a chromatography experiment. All the dye moves the same

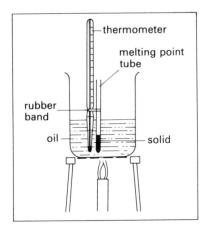

**Figure 5.3** *Finding a melting point*

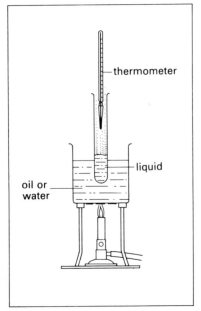

**Figure 5.4** *Finding a boiling point*

distance along the paper. Finding a melting point, finding a boiling point and chromatography are all methods of checking purity.

### Finding melting points

Figure 5.3 shows the apparatus used to measure the melting point of a substance. The way a solid melts tells you a lot about its purity. Pure solids melt sharply. That means that they melt at a particular temperature. Impure solids usually melt over a range of several degrees. They often melt well below the melting point of the pure substance.

### Finding boiling points

You can measure the boiling point of a liquid, with a thermometer. You hold the thermometer above the liquid (see figure 5.4). As you heat the liquid, the reading on the thermometer goes up slowly. When the liquid starts to boil, the reading increases sharply. Then the reading stays steady at some particular temperature. This temperature is the boiling point of the liquid.

This method can be dangerous with flammable liquids. Another method of finding the boiling point of a liquid is distillation. Distillation is safer for flammable liquids. You could use the apparatus shown in figure 4.3. If there is more than one substance present, the reading on the thermometer will change during the distillation.

## 5.2 Elements, compounds and mixtures

### Atoms

Everything is made up of billions of tiny particles called **atoms**. Atoms are much too small to be seen. Twenty million of them in a row would only stretch about 1 cm.

### Elements

Atoms are not all the same. There are more than one hundred different kinds. Some substances contain only one type of atom. These substances are called **elements**. The element carbon contains only carbon atoms. Oxygen contains only oxygen atoms. Hydrogen contains only hydrogen atoms.

Because an element is made of only one type of atom, you cannot make the element into any simpler substance. Elements are already the most elementary (simplest) substances of all.

The names of some common elements are given in table 5.1. All the elements have special symbols that are used as a sort of shorthand. The symbols are also in the table. A larger list can be found in appendix 1.

| Element | Symbol |
|---------|--------|
| hydrogen | H |
| carbon | C |
| nitrogen | N |
| oxygen | O |
| sodium | Na |
| magnesium | Mg |
| aluminium | Al |
| chlorine | Cl |
| potassium | K |
| calcium | Ca |
| iron | Fe |
| copper | Cu |
| zinc | Zn |

**Table 5.1** *Some elements and their symbols.*

## Molecules

Sometimes groups of atoms join together to form particles called **molecules**. For example, the atoms in hydrogen gas are always joined in pairs. Because hydrogen gas always contains two atoms in each molecule, it is given the formula $H_2$. Oxygen gas contains two oxygen atoms in each molecule and so it is given the formula $O_2$. Phosphorus atoms join together in groups of four. The phosphorus molecule has the formula $P_4$. Figure 5.5 shows how the atoms of these molecules are arranged.

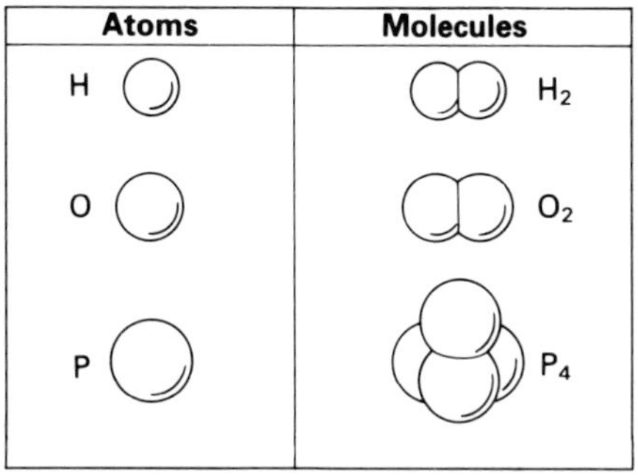

**Figure 5.5** *Some atoms and the molecules they form*

## Compounds

Sometimes different types of atoms join together. These substances cannot be elements because they are made from different atoms. They are **compounds**.

**Figure 5.6** *Sugar is a pure compound. It is made of atoms of carbon, hydrogen and oxygen*

Water is a very simple compound. Its molecules each contain two hydrogen atoms and one oxygen atom. Its formula is $H_2O$.

Sugar is a much more complicated compound. Its molecules each contain 12 carbon atoms, 22 hydrogen atoms and 11 oxygen atoms. Its formula is $C_{12}H_{22}O_{11}$.

## Compound formation

When a chemical reaction occurs between two or more elements, a compound is made. Usually the compound is very different from the elements that make it. For example, water is very different from its elements, hydrogen and oxygen. Hydrogen is a gas which burns easily. Oxygen is a gas in which many other substances burn fiercely. If hydrogen and oxygen are mixed together, no change appears to take place. But if a flame is put into this mixture, there is an explosion. The two gases react together and water is formed (see figure 5.7). Water is a liquid which does not burn or help other substances to burn.

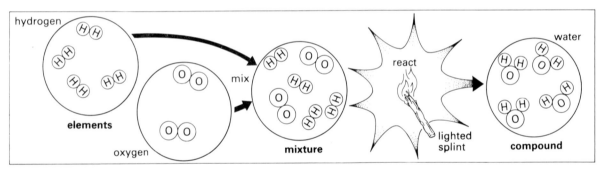

**Figure 5.7** *What happens when hydrogen and oxygen react together*

A second example is the compound sodium chloride. Sodium is a very **reactive** metal. For example, it fizzes and sometimes catches fire when put in water. Chlorine is a poisonous green gas. When chlorine is passed over heated sodium a chemical reaction takes place. A lot of heat is given out as the sodium and chlorine join together. The white solid formed is common salt. Just think how different salt is from the elements which go to make it!

Fluorides are a third example. Fluorine is the most reactive of all the non-metallic elements. But tooth enamel, which is the hardest and most unreactive material in the body, is made up of fluorides. It becomes even harder when it is brushed regularly with fluoride toothpaste (see figure 5.8).

**Figure 5.8** *Your teeth might well catch fire in the **element** fluorine! Small amounts of fluorine compounds, however, will reduce tooth decay*

## Mixtures

Earlier in this chapter you saw that orange juice is a mixture. This is because it is made from different ingredients. Water, sugar and citric acid are some of them. These

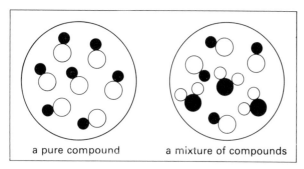

**Figure 5.9** *A pure compound and a mixture of compounds*

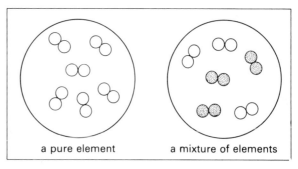

**Figure 5.10** *A pure element and a mixture of elements*

substances are all compounds. So you could describe orange juice as a mixture of compounds (see figure 5.9).

Sea water is another mixture of compounds. Its main ingredients are water and sodium chloride (salt).

As well as having mixtures of compounds, you can have mixtures of elements (see figure 5.10). Air is a mixture of the elements nitrogen and oxygen along with very small amounts of other gases.

## Differences between mixtures and compounds

All compounds contain more than one element. But this does not mean that compounds are mixtures. Mixtures contain *different* types of molecule. Compounds contain only *one* type of molecule. Some more differences between mixtures and compounds are shown in table 5.2.

| Mixtures | Compounds |
| --- | --- |
| can usually be separated easily into their ingredients | are not easily separated into their elements |
| do not give out heat when they are formed | often give out heat when they are formed |
| can contain various amounts of each ingredient | always contain a definite amount of each element |
| look and behave like an 'average' of their ingredients | usually look and behave quite differently from the elements of which they are made |

**Table 5.2** *Differences between mixtures and compounds*

## Summary: Elements, compounds and mixtures

* In chemistry, 'pure' means 'containing only one substance'.
* A pure substance always has a definite melting point and boiling point.
* All substances are made from particles called atoms.
* There are just over a hundred different types of atom.
* Elements contain only one kind of atom.
* Molecules are groups of atoms joined together.
* Compounds contain different types of atoms joined together.
* Compounds are formed when two or more elements react together.
* A compound is a pure substance.
* A mixture contains more than one substance.

## Questions

1 Copy out the following sentences and fill in the missing words.
   (a) Elements are substances that contain only one type of _____ .
   (b) The substances that form when elements react together are called _____ .
   (c) A small particle consisting of two or more atoms joined together is called a _____ .
   (d) If a substance melts sharply at a particular temperature then it is probably a _____ substance.
   (e) Compounds are often very _____ from the elements from which they were made.

2 A pupil was given an unlabelled white solid to investigate. She took a little of the solid to find its melting point. It changed from a solid to a liquid sharply at 185–186°C. Next she heated some more of the solid strongly in a test tube. Droplets of clear liquid condensed on the upper part of the tube. She thought this was probably water. The solid at the bottom of the tube turned into a black crumbly residue. She thought this might be carbon. She removed some of the black solid and heated it very strongly in a bottle top. It glowed red and slowly disappeared.
   (a) Was the original white solid an element or a compound? Give reasons for your answer.
   (b) Draw a diagram of an apparatus suitable for finding the melting point of the solid.
   (c) Find out how you could show that the colourless liquid was water.
   (d) The black solid, which was indeed carbon,

disappeared when strongly heated in air. Can you suggest what happened to it?
(e) What elements do you think the white solid contained?

## Magnesium from sea water

*Magnesium is very light and its alloys are used in aircraft manufacture*

Quite a lot of compounds are dissolved in sea water. One is common salt, of course. There is also quite a lot of a substance called magnesium chloride. There are smaller amounts of many other compounds, including even gold chloride – although not much of this!

Magnesium is a very light metal that is frequently used in alloys for aircraft construction. This magnesium is extracted from sea water.

The first step in the extraction is to add calcium hydroxide. This is usually obtained from limestone. Sometimes it is made by roasting oyster shells, if there is a convenient source nearby. The calcium hydroxide does not affect the common salt but reacts with magnesium chloride, precipitating solid magnesium hydroxide. This solid is separated and then heated with carbon and chlorine. Water, carbon monoxide and magnesium chloride are formed. The magnesium chloride is separated and melted. Melted magnesium chloride conducts electricity. At the same time it decomposes into magnesium metal and chlorine gas.

### Questions

**3** Write down the three elements mentioned in the description. Also write down their symbols.
**4** Make a list of all the compounds mentioned.
**5** Make a simple flow chart showing the main steps in getting magnesium from sea water.
**6** Write down the meanings of the following words: *alloy, element, compound, precipitate, extraction, decompose, electrolysis*.
**7** Suggest how the precipitated magnesium hydroxide could be separated from the sea water.
**8** Chlorine is produced at the same time as magnesium. What do you think is done with this poisonous gas?
**9** The famous German chemist Fritz Haber tried extracting gold from sea water. Unfortunately it is only possible to get 0.000 008 grams of gold from 1 tonne of sea water! See if you can find out how genetic engineering could one day make the extraction of gold from sea water an economic process.

# 6 Heating substances

## 6.1 Types of change brought about by heat

When you heat a substance, there are several changes which might take place. They are discussed here with simplified diagrams of the structures of the substances. (More details about structures are given in chapter 14.)

### No change occurs

Many substances do not change even when heated to quite high temperatures. Sodium chloride (common salt) is an example. Sand is another. Figure 6.1 shows a third.

### Changes in physical state

Ice changes from a solid to a liquid when it is heated. Both solid ice and liquid water contain $H_2O$ molecules. What is different is the arrangement of these molecules (see figure 6.2).

When water is heated to its boiling point, it changes from a liquid into a gas. Water molecules leave the liquid and spread out into the air (see figure 6.3).

Melting and boiling are changes of **physical state**. The physical state of a substance tells you whether it is a solid, liquid or gas. Neither melting nor boiling is a chemical reaction. Melted ice can easily be refrozen and steam readily changes back to water.

**Figure 6.1** *Some substances, like the suits of protective clothing worn by these firemen, are specially treated so that they are not affected by heat*

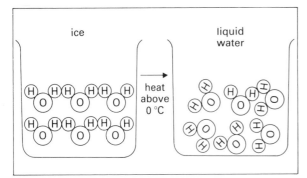

**Figure 6.2** *What happens when ice melts – there is no change in mass*

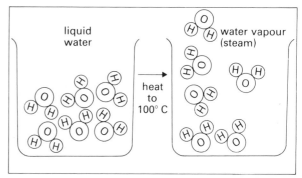

**Figure 6.3** *What happens when water boils – the mass decreases as water vapour leaves the beaker*

## Decomposition

When mercury oxide is heated it **decomposes**. This means that it breaks up into simpler materials. In this case, mercury metal and oxygen gas are formed (see figure 6.4).

## Combining with gases in the air

When copper is heated in air it becomes coated with a black solid. This is because copper atoms at the surface join with oxygen gas from the air. Together they make copper oxide which is black (see figure 6.5).

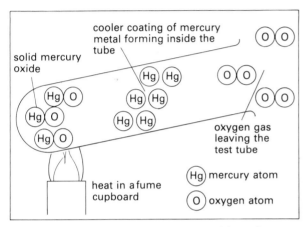

**Figure 6.4** *Decomposing mercury oxide – the mass decreases as oxygen gas leaves the test tube*

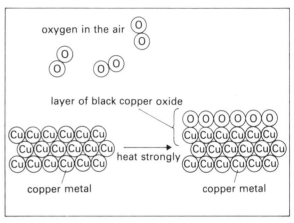

**Figure 6.5** *Forming copper oxide – the mass of the solid increases as oxygen from the air combines with the copper atoms*

When you heat carbon it reacts with oxygen in the air to form carbon dioxide gas (see figure 6.6). If you carry on heating it long enough, all of the carbon will react in this way. Then there is no solid left at the end of the reaction.

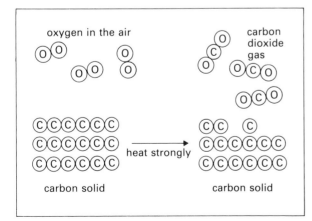

**Figure 6.6** *Forming carbon dioxide – the mass decreases as the carbon atoms form carbon dioxide, which goes into the air*

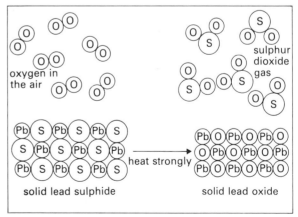

**Figure 6.7** *Heating lead sulphide in air – the mass decreases slightly as sulphur atoms in the solid are replaced by lighter oxygen atoms*

### 'Swapping' atoms with the air

When you heat lead sulphide, it changes from a black solid to a yellow solid and gives off a gas with a choking smell. The yellow solid is lead oxide. The gas is sulphur dioxide. The lead has 'swapped' a sulphur atom for an oxygen atom (see figure 6.7, on the previous page). The mass of the solid is a bit less than before because each oxygen atom has only half the mass of a sulphur atom.

## 6.2 The effect of heat on common substances

In section 6.1 we looked at the different types of change that can take place when you heat a substance. In this section we shall look at how some substances are affected by heat.

### Substances that give off gases

Many substances give off gases when they are heated. Sometimes it is useful to know what type of substance gives off a particular gas. It can help you identify the substance being heated.

Table 6.1 lists some common gases. It also tells you:

**1** how to recognize these gases,
**2** which compounds these gases may have come from.

| Gas | How to recognize the gas | Some compounds that give off this gas when heated |
|---|---|---|
| oxygen | colourless gas; relights a glowing splint | some metal oxides; manganates(VII) (permanganates); nitrates (may also give off brown fumes) |
| nitrogen dioxide | brown fumes | nitrates (given off along with oxygen) |
| carbon dioxide | colourless gas; turns limewater chalky | carbonates and hydrogencarbonates; any compound containing carbon, when it is burnt |
| steam | turns cobalt chloride paper pink | hydrated compounds (see section 10.4); any compound containing hydrogen, when it is burnt |

**Table 6.1** *Identifying some compounds by the gases they give off*

## How heat affects particular substances

Table 6.2 describes what happens to some particular substances when they are heated.

| Substance | What happens when it is heated |
|---|---|
| ammonium chloride | the white solid sublimes; it disappears from the bottom of the test tube and reappears on the cooler part of the tube |
| cobalt chloride crystals | the purple-pink crystals give off steam; as they do so they turn blue |
| copper sulphate crystals | the blue crystals give off steam; as they do so they turn white |
| copper metal | the reddish metal glows red and then becomes covered in a layer of black solid (copper oxide) |
| iodine | the dark crystals change into a purple vapour; the crystals reappear on the cooler part of the tube |
| lead nitrate | the colourless crystals give off a brown/yellow gas (nitrogen dioxide + oxygen) |
| red lead | the red solid darkens and gives off oxygen gas; a yellow/orange powder (lead(II) oxide) remains |
| magnesium ribbon | the silvery metal burns with a brilliant flame; a white ash (magnesium oxide) remains |
| potassium manganate(VII) | the purple crystals darken to a black powder; oxygen is given off |
| zinc oxide | the white solid turns yellow when heated but returns to white when cooled |
| zinc carbonate | the white solid gives off carbon dioxide gas; zinc oxide is left in the tube |

**Table 6.2** *How heat affects some substances*

## Summary: Heating substances

★ When a substance is heated there are seven possibilities:
  1 nothing happens,
  2 a solid melts,
  3 a liquid boils,
  4 the substance decomposes,
  5 the substance reacts with a gas from the air, usually oxygen,
  6 the substance 'swaps' atoms with a gas from the air, usually oxygen,
  7 the substance sublimes.
★ Some metal oxides give off oxygen when they are heated.
★ Most carbonates give off carbon dioxide when heated.
★ Most nitrates give off oxygen and often nitrogen dioxide when heated.
★ Many substances can be recognized by the way they behave when heated.

## Questions

1 This question is about the substances described in the table below. (The letters A to E are not chemical symbols for the substances.)
  (a) Which substance is a compound that decomposes to form a solid and a gas?
  (b) Which substance is not affected by heat?
  (c) Which substance changed entirely into a gas when heated?
  (d) Which substance melted when heated?
  (e) Which substance combined with a gas from the air to form a new solid?
  (f) Which of the substances A to E in the table below could not possibly be an element? Give reasons for your answer.

| Substance | Appearance before heating | Appearance during heating | Appearance after heating | Change in mass of the solid |
|---|---|---|---|---|
| A | white powder | white powder | white powder | none |
| B | white solid | colourless liquid | white solid | none |
| C | grey lumps | burns with a red flame | white powder | increases |
| D | black powder | glows red and slowly disappears | nothing left | decreases |
| E | green powder | powder darkens | black powder | decreases |

2

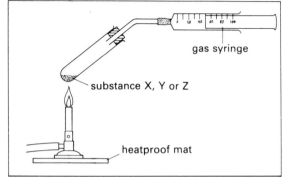

Samples of three substances were labelled X, Y and Z. These substances were heated in the apparatus shown in the diagram. At the start of each experiment the syringe held $50\,cm^3$ of air. After heating each substance the boiling tube was allowed to return to room temperature before the syringe was read. The results obtained are shown in the table below.

| Substance | Appearance before heating | Appearance after heating | Syringe reading before heating ($cm^3$) | Syringe reading after heating ($cm^3$) |
|---|---|---|---|---|
| X | pinkish-red metal | black powder | 50 | 40 |
| Y | red powder | yellow powder | 50 | 82 |
| Z | white solid | yellow powder but white when cool | 50 | 76 |

Substances X, Y and Z are all described in section 6.2.
(a) What is substance X?
(b) What is substance Y?
(c) What is substance Z?
(d) Why was the apparatus allowed to cool down after the experiment before the syringe reading was taken?

3 A pupil carried out an experiment in which she heated a green solid. The apparatus she used is shown below.

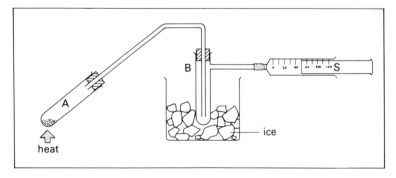

The green substance turned black and at the same time some clear liquid collected in tube B and a gas collected in syringe S. The following readings were taken:

| | |
|---|---|
| mass of tube A at start | 24.30 g |
| mass of tube A at end | 23.80 g |
| mass of tube B at start | 25.90 g |
| mass of tube B at end | 26.36 g |
| syringe reading at start | 0 cm$^3$ |
| syringe reading at end | 24 cm$^3$ |

(a) How can you tell that a chemical reaction has occurred in this experiment?

(b) What type of chemical reaction has occurred?

(c) Could the green substance be an element? (Give reasons.)

(d) What mass of gas was collected? (Show your working.)

(e) The liquid in tube B changed the colour of blue cobalt chloride to pink. What is the liquid?

(f) The gas from the syringe turned limewater milky. What was the gas?

(g) What should the pupil know about the green solid?

## Polythene and diamonds – how chemists use heat

All solids melt when they are heated, provided that they do not catch fire or decompose first. This idea is particularly important in the plastics industry, where chemists classify plastics according to how they behave when heated.

Some plastics decompose, burn or char before they start to melt and are called **thermosetting** plastics (**thermosets** for short). The resin used in glassfibre boats is a thermoset. Others melt – or at least soften – before they decompose or burn. They are called **thermoplastics**. Polythene is a thermoplastic.

Thermosets are manufactured by mixing the reactants in a mould which produces the article in its final shape. They tend to be more rigid than thermoplastics, usually resisting softening until the moderately high temperatures at which they start to char.

Thermoplastics can be manufactured as granules in one factory and then transported elsewhere to be softened or melted and shaped. Thermoplastics are also easily coloured by adding pigments to the liquid plastic. The pigment quickly spreads evenly throughout the liquid.

*Thermoplastics can be re-softened and shaped in this vacuum forming machine*

Another process involving heat and colouring is the manufacture of synthetic gemstones. Synthetic rubies, for example, are made by melting pure aluminium oxide (melting point over 2000 °C) in an oxygen–hydrogen flame and adding a few per cent of chromium oxide to the melt. This provides the red colour. The molten mixture is then cooled very slowly to give the atoms time to arrange themselves into the regular pattern of a ruby crystal. So rubies, like thermoplastics, are made by adding a colouring substance to a colourless melt.

Gemstones are known for their beauty but they are also extremely hard and it is their hardness which makes them useful as bearings in precision instruments – nowadays more are used for this purpose than in jewellery. High-quality synthetic gemstones are also needed for making lasers.

Glass manufacture is, in some ways, similar to the manufacture of synthetic gemstones. To make glass, silicon oxide is heated with sodium carbonate and calcium carbonate. The carbonates decompose during this process, leaving a molten mixture of the oxides. This cools to give glass.

Several early attempts to make synthetic gems led to the discovery of other important substances. In 1891, for example, a chemist called Acheson was trying to make synthetic diamonds, but instead discovered silicon carbide (**carborundum**). This is the third hardest substance known and is an important abrasive. Synthetic diamonds were eventually produced in 1955 (although many people claimed to have made them before that date), by heating carbon to a very high temperature at enormous pressure in the absence of air. More recently researchers have found how to make diamonds at low temperatures and pressures, starting with hydrogen and methane gas.

One way of making glass is by the float process, which produces very large sheets

### Questions

4 Explain the meaning of each of the following words: decompose, combustion, thermoset, thermoplastic, synthetic.
5 State one advantage of thermosets over thermoplastics.
6 State one advantage of thermoplastics over thermosets.
7 What gas is given off from the calcium carbonate during glass-making?
8 What gas would you get if you burned diamonds?
9 Is melting a thermoplastic a chemical reaction?
10 Is glass-making a chemical reaction?
11 Glass is really a supercooled liquid rather than a solid. Find out what this means.
12 Rubies are used in some lasers. Lasers are already used extensively in hospitals for such things as surgery and eye treatment. Can you think of any other possible uses?

# 7 Counting particles – the mole

## 7.1 How heavy are particles?

### How heavy are atoms?

Atoms are not all alike. Some atoms have a greater mass than others. It is useful to be able to compare their masses, and for this we use the **relative atomic mass** (r.a.m.).

| Element | Relative atomic mass |
|---------|---------------------|
| hydrogen | 1 |
| helium | 4 |
| carbon | 12 |
| oxygen | 16 |
| magnesium | 24 |

**Table 7.1** *The approximate relative atomic masses of some elements*

The values in table 7.1 are obtained by comparing the atoms with a particular type of carbon atom known as carbon-12. This is given the relative atomic mass of 12.

A hydrogen atom has one-twelfth the mass of an atom of carbon-12: the r.a.m. of hydrogen is $\frac{1}{12}$ of 12 = 1.

A helium atom has one-third the mass of an atom of carbon-12: the r.a.m. of helium is $\frac{1}{3}$ of 12 = 4.

A magnesium atom has twice the mass of an atom of carbon-12: the r.a.m. of magnesium is $2 \times 12 = 24$.

### How heavy are molecules?

The **relative molecular mass** (r.m.m.) of a compound allows us to compare the masses of molecules in the same way as those of atoms.

To work out the r.m.m. we write down the formula of the compound and add together the r.a.m.s of all the elements that go to make up the molecule. (In these examples, we have used the list of r.a.m.s of the elements in appendix 1, at the end of this book.)

### Example 1

Calculate the r.m.m. of $O_2$.

$$O_2$$
$$2 \times 16 = 32$$

### Example 2

Calculate the r.m.m. of $CO_2$.

$$CO_2$$
$$12 + (2 \times 16) = 44$$

### Example 3

Calculate the r.m.m. of $C_2H_5OH$.

$$C_2H_5OH$$
$$(2 \times 12) + (5 \times 1) + 16 + 1 = 46$$

### Example 4

Calculate the r.m.m. of $MgSO_4$.

$$MgSO_4$$
$$24 + 32 + (4 \times 16) = 120$$

### Example 5

Calculate the r.m.m. of $Na_2CO_3.10H_2O$.

$$Na_2CO_3.10H_2O$$
$$(2 \times 23) + 12 + (3 \times 16) + [10 \times (2 + 16)] = 286$$

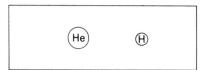

**Figure 7.1** *A helium atom has 4 times the mass of a hydrogen atom*

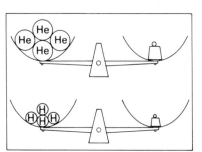

**Figure 7.2** *To get the same number of atoms we will need 4 times the mass of helium*

## 7.2 Getting equal numbers of particles – the mole

We saw in section 7.1 that a helium atom has 4 times the mass of a hydrogen atom. It follows that 1 g of hydrogen will contain more atoms than 1 g of helium does (see figure 7.1).

If we wanted equal numbers of hydrogen atoms and helium atoms, we would need 4 times as much helium as hydrogen. For instance, 1 g of hydrogen and 4 g of helium have the same numbers of atoms (see figure 7.2).

It is known from experiments that 1 g of hydrogen contains $6.02 \times 10^{23}$ atoms (approximately 6 with 23 noughts after it!). So 4 g of helium will also contain $6.02 \times 10^{23}$ atoms. Similarly, since carbon atoms are 12 times as heavy as hydrogen atoms, we would need 12 g of carbon in order to get $6.02 \times 10^{23}$ carbon atoms.

You have probably realized by now that the relative atomic mass of any element in grams is going to contain $6.02 \times 10^{23}$ atoms. This amount of an element is called a **mole of atoms** of the element. The number of atoms in it ($6.02 \times 10^{23}$) is called the **Avogadro constant**.

| Element | Relative atomic mass | Mass of 1 mole of atoms (g) | Number of atoms in 1 mole |
|---|---|---|---|
| hydrogen | 1 | 1 | $6.02 \times 10^{23}$ |
| helium | 4 | 4 | $6.02 \times 10^{23}$ |
| carbon | 12 | 12 | $6.02 \times 10^{23}$ |
| oxygen | 16 | 16 | $6.02 \times 10^{23}$ |

**Table 7.2** *A mole of atoms of some elements*

Remember:

> **a mole of atoms = relative atomic mass in grams**, and
> **equal numbers of moles means equal numbers of particles.**

## From moles to masses

Quite often you may have to calculate the mass of an amount given in moles. You simply use the scale factor method described in section 2.5.

### *Example 6*

What is the mass of 0.6 mole of titanium? (The relative atomic mass of titanium is 48.)

Amount of titanium (in mol)          Mass of titanium (in g)
1                                    48
| scale-down factor                  | multiply
| = 0.6                              | by 0.6
0.6                                  28.8

So 0.6 mol of titanium has a mass of 28.8 g.

### *Example 7*

What is the mass of 2.5 moles of neon? (The relative atomic mass of neon is 20.)

Amount of neon (in mol)              Mass of neon (in g)
1                                    20
| scale-up factor                    | multiply
| = 2.5                              | by 2.5
2.5                                  50

So 2.5 mol of neon have a mass of 50 g.

### From masses to moles

Again, we use the scale factor method, in a similar way as above.

### *Example 8*

How many moles is 2 g of calcium atoms? (The relative atomic mass of calcium is 40.)

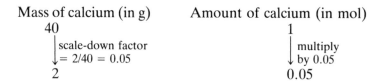

| Mass of calcium (in g) | Amount of calcium (in mol) |
|---|---|
| 40 | 1 |
| scale-down factor<br>= 2/40 = 0.05 | multiply<br>by 0.05 |
| 2 | 0.05 |

So 2 g of calcium is 0.05 mol.

### A mole of oxygen?

We saw in section 5.2 that the atoms in oxygen gas are joined together as $O_2$ molecules. We have to be careful when we talk about 'a mole of oxygen'.

A mole of oxygen *atoms* contains $6.02 \times 10^{23}$ oxygen *atoms* and is the relative *atomic* mass of oxygen in grams – 16 g.

A mole of oxygen *molecules* contains $6.02 \times 10^{23}$ oxygen *molecules* and is the relative *molecular* mass in grams – 32 g.

Clearly when talking about moles of elements that exist as molecules we must be sure to say whether we are talking about atoms or molecules of the element.

| 1 mole of oxygen atoms | 1 mole of oxygen molecules |
|---|---|
| $6.02 \times 10^{23}$ atoms | $6.02 \times 10^{23}$ molecules |
| 16 g | 32 g |

### A mole of a compound

We have seen that equal numbers of moles means equal numbers of particles. So a mole of ammonia, say, contains about $6 \times 10^{23}$ molecules of ammonia, and its mass will equal the relative molecular mass of ammonia in grams:

1 mole = $6 \times 10^{23}$ $NH_3$ molecules

1 mole = $14 \times 3 = 17$ g

A mole of ammonia molecules will, of course, contain one mole of nitrogen atoms and three moles of hydrogen atoms.

We can calculate the mass of a mole of any compound by working out its relative molecular mass in grams, as indicated by its formula. Calculations for compounds are done just like those for elements. The only difference is that we use the relative molecular mass instead of the relative atomic mass.

## 7.3 The concentration of a solution

Chemists usually express the concentration of a solution in one of two ways:

grams of solute per $dm^3$ of solution (remember $1 dm^3 = 1000 cm^3$), or
moles of solute per $dm^3$ of solution.

The second way is often more useful, as it tells us more about the number of particles present.

### From solution concentration to moles

Again, we use the scale factor method.

### Example 9

How many moles of copper sulphate are present in $200 cm^3$ of a solution containing 0.5 moles per $dm^3$?

| Volume of copper sulphate solution (in $cm^3$) | Amount of copper sulphate (in mol) |
|:---:|:---:|
| 1000 | 0.5 |
| scale-down factor = 200/1000 = 0.2 | multiply by 0.2 |
| 200 | 0.1 |

So $200 cm^3$ of copper sulphate solution containing 0.5 moles per $dm^3$ contains 0.1 mole of copper sulphate.

This procedure can be summarized by the equation:

$$\text{number of moles} = \frac{\text{volume of solution } (cm^3)}{1000} \times \frac{\text{concentration of solution}}{\text{(moles per } dm^3)}$$

### From solution concentrations to grams

These calculations simply require you to change the moles into grams at the end.

### Example 10

What mass of sodium chloride (NaCl) will there be in $100 cm^3$ of a solution containing 0.2 moles per $dm^3$?
(Relative atomic masses: Na = 23; Cl = 35.5.)

*Stage 1*:

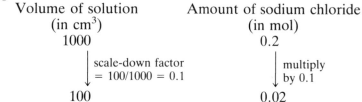

Volume of solution (in cm³): 1000 → scale-down factor = 100/1000 = 0.1 → 100

Amount of sodium chloride (in mol): 0.2 → multiply by 0.1 → 0.02

*Stage 2*:

1 mole = 23 g + 35.5 g = 58.5 g.

0.02 mole = 0.02 × 58.5 = 1.17 g.

So 100 cm³ of a solution containing 0.2 moles per dm³ contain 1.17 g of sodium chloride.

## Molar solutions

There is a useful shorthand way of writing the concentration of solutions.

> A solution containing 1 mole of solute in 1 dm³ of solution is called a 1 molar (1M) solution.

Similarly,

0.5 mole of solute in 1 dm³ of solution gives a 0.5M solution,

2 moles of solute in 1 dm³ of solution gives a 2M solution,

1 mole of solute in 500 cm³ of solution gives a 0.5M solution,

and so on.

The concentration of a solution expressed in this way is sometimes called its **molarity**.

## Summary: Counting particles – the mole

★ The relative atomic mass (r.a.m.) of an element is a measure of how heavy one atom of that element is, compared with an atom of carbon-12.

★ The r.a.m. of hydrogen is 1.

★ The relative molecular mass of a compound is equal to the sum of the r.a.m.s of all the atoms that make up a molecule of the compound.

★ Moles provide an easy way of counting out equal numbers of particles.

★ A mole of atoms of an element contains $6.02 \times 10^{23}$ atoms.

★ A mole of any compound contains $6.02 \times 10^{23}$ molecules or units of that compound.

★ A molar (1M) solution contains 1 mole of solute per dm³ of solution.

## Questions

In some of these questions you will need to use values of relative atomic masses. You will find these in appendix 1.

**1** Calculate the relative molecular mass of each of the following compounds:
(a) $H_2S$     (b) $NH_3$     (c) $C_8H_{18}$     (d) $C_{12}H_{11}O_{22}$
(e) $CuSO_4.5H_2O$     (f) $C_6H_{12}O_6$     (g) $KAl(SO_4)_2.12H_2O$
(h) $Al_2(SO_4)_3$     (i) $Na_2SO_4$     (j) $CuSO_4.4NH_3$

**2** What is the mass of 1 mole of each of the following?
(a) copper atoms                    (b) aluminium atoms
(c) fluorine atoms                   (d) sodium atoms
(e) calcium atoms

**3** What is the mass of each of the following?
(a) 2 moles of magnesium atoms
(b) 0.5 mole of boron atoms
(c) 0.1 mole of iodine atoms
(d) 10 moles of lithium atoms
(e) 0.002 mole of neon atoms

**4** How many moles is each of the following?
(a) 28 g of iron atoms              (b) 1.4 g of nitrogen atoms
(c) 3.0 g of carbon atoms          (d) 9.6 g of copper atoms
(e) 5.4 g of silver atoms

**5** What is the mass of each of the following?
(a) 1 mole of HCl                     (b) 4 moles of $BF_3$
(c) 0.2 mole of $NO_2$               (d) 2 moles of $C_4H_9OH$
(e) 0.01 mole of $H_2SO_4$

**6** How many moles is each of the following?
(a) 50 g of $CaCO_3$                 (b) 64 g of $CH_4$
(c) 1.1 g of $CO_2$                   (d) 44 g of $MgCO_3$
(e) 2.3 g of $C_2H_5OH$

**7** Remember that equal numbers of moles means equal numbers of particles. Which of the following pairs contain equal numbers of particles? (Show your working.)
(a) 34 g of $H_2S$ and 98 g of $H_2SO_4$
(b) 71 g of $Cl_2$ and 88 g of $CO_2$
(c) 2.8 g of CO and 4.4 g of $CO_2$
(d) 0.34 g of $NH_3$ and 0.34 g of $PH_3$
(e) 8.1 g of HBr and 12.8 g of HI

**8** Calculate the number of moles present in:
(a) $500 \, cm^3$ of 1M solution
(b) $250 \, cm^3$ of 0.1M solution
(c) $25 \, cm^3$ of 0.2M solution
(d) $2 \, dm^3$ of 2M solution
(e) $50 \, cm^3$ of 4M solution

**9** Calculate the mass of solute present in:
(a) $1000 \, cm^3$ of 0.05M KBr solution
(b) $10 \, cm^3$ of 0.1M $MgSO_4$ solution
(c) $400 \, cm^3$ of 0.25M $CuSO_4.5H_2O$ solution
(d) $25 \, cm^3$ of 0.025M $CoCl_2$ solution

# 8 Using formulae and equations

## 8.1 Word, diagram and formula equations

In any chemical reaction, a new substance is always formed. The chemicals that you start off with are the reactants. The new substances formed are the products. A **chemical equation** is a short way of describing exactly what happens during a reaction.

### Word equations

The simplest chemical equation is the word equation. Some examples of equations are given below.

The names on the left-hand side tell you what you started with. The names on the right tell you what has been formed.

---

The reaction of hydrogen with oxygen:

hydrogen + oxygen $\longrightarrow$ water

The reaction of carbon with oxygen:

carbon + oxygen $\longrightarrow$ carbon dioxide

The decomposition of calcium carbonate:

calcium carbonate $\longrightarrow$ calcium oxide + carbon dioxide

The reaction of lead sulphide with oxygen:

lead sulphide + oxygen $\longrightarrow$ lead oxide + sulphur dioxide

---

**Figure 8.1** *In some rocket engines, hydrogen reacts with oxygen to form water*

### The formulae of elements

The elements in table 8.1 (over the page) all exist as molecules with two atoms. This is shown in the formula by writing a small 2 after, and slightly below, the symbol for the element. These are called **diatomic** molecules.

The atoms of most other elements join up to make larger, more complicated structures. For these elements, we do not try to show how many atoms are joined up together. Instead we just write the symbol for the element without any numbers after it. A few examples are given in table 8.2.

| Element | Symbol | Formula of molecule |
|---------|--------|--------------------|
| hydrogen | H | $H_2$ |
| nitrogen | N | $N_2$ |
| oxygen | O | $O_2$ |
| fluorine | F | $F_2$ |
| chlorine | Cl | $Cl_2$ |
| bromine | Br | $Br_2$ |
| iodine | I | $I_2$ |

**Table 8.1** *Some diatomic molecules*

| Element | Symbol |
|---------|--------|
| carbon | C |
| iron | Fe |
| sulphur | S |
| zinc | Zn |
| calcium | Ca |

**Table 8.2** *The symbols of some common elements*

## The formulae of compounds

The formula of water is $H_2O$. This tells us that water contains the elements hydrogen and oxygen only – two hydrogen atoms for each oxygen atom.

All compounds have a definite formula and some examples of these are given in table 8.3.

| Compound | Formula |
|----------|---------|
| calcium carbide | $CaC_2$ |
| calcium oxide | $CaO$ |
| carbon dioxide | $CO_2$ |
| carbon monoxide | $CO$ |
| hydrochloric acid | $HCl$ |
| iron sulphide | $FeS$ |
| water | $H_2O$ |
| zinc chloride | $ZnCl_2$ |

**Table 8.3** *The formulae of some compounds*

Later on in this book you will learn how to work out the formulae of compounds. For now, you just need to know what they mean.

## Diagram equations

Now that you know about formulae, you can draw a diagram of what is happening in a chemical reaction. First we shall look at the reaction of hydrogen ($H_2$) with oxygen ($O_2$) to form water ($H_2O$):

hydrogen + oxygen $\longrightarrow$ water

To start, draw in atoms joined up as shown in the formulae.

This tells you what the reactants and products are like. But it does not explain the reaction completely. For one thing, the reactants contained two atoms of oxygen. The products only seem to contain one.

To use up both oxygen atoms you need to form two water molecules for each oxygen molecule.

To form two water molecules you must have two hydrogen molecules. So now the equation is balanced.

The next diagram correctly shows what happens in the reaction. Each oxygen molecule reacts with two hydrogen molecules to form two water molecules.

## Chemical equations using formulae

Chemists do not usually draw diagrams for equations. Instead they use the formulae of the substances. For example, in the equation

$$2H_2 + O_2 \longrightarrow 2H_2O$$

the 2s in front of the 'H$_2$' and 'H$_2$O' show that when one oxygen molecule reacts with 2 hydrogen molecules, 2 molecules of water are formed.

## Writing chemical equations

Writing out a chemical equation is done in four stages. Here are some examples showing how to do this.

## *Example 1*

Write an equation for the reaction of iron with sulphur to give iron sulphide, FeS.

**1** Write out a word equation

      iron  + sulphur $\longrightarrow$   iron sulphide

**2** Write in the formulae in place of the words

      Fe  +   S  $\longrightarrow$     FeS

**3** How many atoms on each side of this equation?

      one Fe +   one S $\longrightarrow$ one Fe and one S

**4** Balance the equation so that there are equal numbers of each type of atom on either side

      This equation already has equal numbers of atoms on either side.

**Final equation:**

      Fe  +   S  $\longrightarrow$     FeS

## Example 2

Write an equation for the reaction of zinc with hydrochloric acid to form hydrogen gas and zinc chloride, $ZnCl_2$.

**1** Write out a word equation

zinc + hydrochloric acid $\longrightarrow$ hydrogen + zinc chloride

**2** Write in the formulae in place of the words

$Zn$ + $HCl$ $\longrightarrow$ $H_2$ + $ZnCl_2$

**3** How many atoms on each side of the equation?

one zinc + one hydrogen and one chlorine $\longrightarrow$ two hydrogen + one zinc and two chlorine

**4** Balance the equation

The zinc atoms are 'balanced'. There is one on each side of the equation. Neither the hydrogen atoms nor the chlorine atoms are balanced. We need one more of each on the left-hand side. This would be be corrected if we took not one but two HCl molecules.

**Final equation:**

$Zn$ + $2HCl$ $\longrightarrow$ $H_2$ + $ZnCl_2$

## Example 3

Write an equation for the reaction of calcium oxide with carbon to form calcium carbide, $CaC_2$, and carbon monoxide, CO.

**1** Write a word equation

calcium oxide + carbon $\longrightarrow$ calcium carbide + carbon monoxide

**2** Replace words by formulae

$CaO$ + $C$ $\longrightarrow$ $CaC_2$ + $CO$

**3** How many atoms on each side?

one calcium and one oxygen + one carbon $\longrightarrow$ one calcium and two carbon + one carbon and one oxygen

**4** Balance the equation

The calcium atoms and oxygen atoms are balanced. The carbon atoms are not balanced. There are three on the right-hand side and only one on the left. So we need not one but three carbons, on the left.

**Final equation:**

$CaO$ + $3C$ $\longrightarrow$ $CaC_2$ + $CO$

Some chemical equations are difficult to balance. Sometimes it helps to draw out a diagram equation. This can show just what you need more of.

### State symbols

Some chemical equations include extra symbols after the formulae. These usually tell you the physical state of the substances. Table 8.4 lists the symbols used.

| Physical state | Symbol used |
|----------------|-------------|
| solid | (s) |
| liquid | (l) |
| gas | (g) |
| aqueous solution (solution of substance in water) | (aq) |

**Table 8.4** *State symbols used in equations*

$H_2O(l)$ means that the water is a liquid,
$H_2(g)$ means the hydrogen is a gas,
$Fe(s)$ means the iron is a solid,
$ZnCl_2(aq)$ means the zinc chloride is dissolved in water.

## 8.2 Formulae and percentage compositions

Quite often it is necessary to find the formula of a new compound by finding out how much of each element is in it. This section shows you how to do these calculations.

### Calculating formulae

There are always three stages in the calculation.

**1** Find the mass of each element.
**2** Scale the mass of one of the elements up to 1 mole.
**3** Scale the mass of the other element by the same factor and deduce the formula.

### *Example 4*

2.4 g of magnesium reacted with oxygen to give 4.0 g of magnesium oxide. What is the formula of magnesium oxide? (Relative atomic masses: Mg = 24; O = 16.)

**1** Mass of magnesium = 2.4 g      Mass of oxygen = 4.0 − 2.4
                                                              = 1.6 g

**2** Mass of magnesium (in g)       Mass of oxygen (in g)
            2.4                                       1.6

| scale-up factor = 24/2.4 = 10        | multiply by 10

1 mole = 24                                  16 = 1 mole

**3** For every mole of magnesium atoms in magnesium oxide there is one mole of oxygen atoms.

**Answer:** The formula of magnesium oxide is MgO.

### Example 5

10.8 g of aluminium reacted with excess dilute nitric acid.
The solution was evaporated to dryness and the remaining
solid was heated strongly. It left 20.4 g of aluminium oxide.
What is the formula of this oxide? (Relative atomic masses:
(Al = 27; O = 16.)

**1** Mass of aluminium = 10.8 g   Mass of oxygen = 20.4 − 10.8
                                                      = 9.6 g

**2** Mass of aluminium (in g)       Mass of oxygen (in g)
         10.8                             9.6

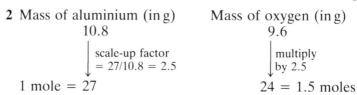

        | scale-up factor         | multiply
        | = 27/10.8 = 2.5         | by 2.5

  1 mole = 27                         24 = 1.5 moles

**3** For every one mole of aluminium atoms in aluminium
oxide, there are $1\frac{1}{2}$ moles of oxygen atoms, that is, for
every two moles of aluminium there are 3 moles of oxygen.

**Answer:** The formula of aluminium oxide is $Al_2O_3$.

### Example 6

0.28 g of silicon reacted with excess chlorine to form 1.70 g
of silicon chloride. What is the formula of the chloride?
(Relative atomic masses: Si = 28; Cl = 35.5.)

**1** Mass of silicon = 0.28 g   Mass of chlorine = 1.70 − 0.28
                                                      = 1.42 g

**2** Mass of silicon (in g)       Mass of chlorine (in g)
         0.28                           1.42

        | scale-up factor         | multiply
        | = 28/0.28 = 100         | by 100

  1 mole = 28                         142 = 4 moles

**3** For every one mole of silicon atoms in silicon chloride
there are four moles of chlorine atoms.

**Answer:** The formula of silicon chloride is $SiCl_4$.

## Calculating percentage composition

Sometimes it is useful to know how much of a particular
element there is in a compound. For example, you might
want to know how much iron could be obtained from some
iron oxide, or how much nitrogen there is in a particular
fertilizer.
  There are three stages in this calculation.

**1** Calculate the relative molecular mass of the compound.
**2** Calculate the fraction of this figure that comes from the
element.
**3** Multiply the fraction by 100 to get a percentage.

*Example 7*

What is the percentage of sulphur in sulphur dioxide, $SO_2$?

1 Relative molecular mass = $32 + (2 \times 16) = 64$.
2 Fraction that is sulphur = $32/64 = \frac{1}{2}$.
3 Percentage sulphur = $\frac{1}{2} \times 100 = 50$ per cent.

*Example 8*

What is the percentage of iron in iron oxide, $Fe_2O_3$?

1 Relative molecular mass = $(2 \times 56) + (3 \times 16) = 160$.
2 Fraction that is iron = $112/160 = 0.70$.
3 Percentage iron = $0.70 \times 100 = 70$ per cent.

*Example 9*

One form of iron sulphate is hydrated, that is, it contains water: its formula is $FeSO_4.7H_2O$. What percentage of this compound is water? (This calculation is slightly different from the last two, in that we are calculating the percentage of a compound instead of an element. The approach is just the same, however.)

1 Relative molecular mass = $56 + 32 + (4 \times 16) + 7 \times (2 + 16) = 278$.
2 Fraction that is water = $7 \times (2 + 16)/278 = 126/278 = 0.4532$.
3 Percentage water = $0.4532 \times 100 = 45.32$ per cent.

## 8.3 From equations to masses

This section is about calculating the amounts of reactants and products in reactions. First, let's look at the reaction of hydrogen with oxygen to form water:

| hydrogen | + | oxygen | $\longrightarrow$ | water |
|---|---|---|---|---|
| $2H_2$ | + | $O_2$ | $\longrightarrow$ | $2H_2O$ |
| 2 moles of $H_2$ | + 1 mole of $O_2$ | give | 2 moles of $H_2O$ |
| 4 g of $H_2$ | + 16 g of $O_2$ | give | 36 g of $H_2O$ |

The equation tells us how many moles of each reactant are needed. This is easily converted into grams. You just have to multiply by the relative molecular mass (see section 7.1): so we need 16 g of oxygen to react with 4 g of hydrogen.

How much would have been needed to react with 1 g of hydrogen? You have probably already worked out that you would need 4 g of oxygen. You have taken the answer found from the equation and used a scale-down factor (see section 2.5).

Almost all calculations about the mass of reactants and products have four stages.

**1** Write a balanced chemical equation.
**2** Write down the number of moles of each chemical.
**3** Change moles into masses in grams by multiplying by the relative molecular mass.
**4** If necessary, use a scale factor to answer the question asked.

Read through the following examples carefully. Make sure you understand the calculations. (Relative atomic masses are listed in appendix 1.)

## *Example 10*

Calcium carbonate, $CaCO_3$, decomposes when it is heated, forming calcium oxide, $CaO$, and carbon dioxide, $CO_2$. What mass of calcium oxide is formed when 20 g of calcium carbonate is decomposed?

**1** $CaCO_3(s) \longrightarrow CaO(s) + CO_2(g)$
**2**    1 mole            1 mole        1 mole
**3**     100 g             56 g          44 g

**4** Mass of $CaCO_3$ (in g)              Mass of CaO (in g)
            1 mole = 100                      1 mole = 56

                    scale-down factor                     multiply
                    = 20/100                              by 20/100

                    20                                    11.2

**Answer:** 20 g of calcium carbonate gives 11.2 g of calcium oxide.

## *Example 11*

What mass of magnesium chloride, $MgCl_2$, is formed when 48 g of magnesium reacts with an excess of chlorine gas?

**1** $Mg(s) + Cl_2(g) \longrightarrow MgCl_2(s)$
**2** 1 mole     1 mole              1 mole
**3**  24 g       71 g                95 g

**4** Mass of Mg (in g)                  Mass of $MgCl_2$ (in g)
        1 mole = 24                          1 mole = 95

                    scale-up factor                       multiply
                    = 48/24                               by 48/24

                    48                                    190

**Answer:** 48 g of magnesium gives 190 g of magnesium chloride.

## *Example 12*

On heating hydrated copper sulphate, $CuSO_4.5H_2O$, it loses its water. What mass of water is driven off when 25 g of hydrated copper sulphate is heated?

1 $CuSO_4.5H_2O(s) \longrightarrow CuSO_4(s) + 5H_2O(g)$
2         1 mole             1 mole     5 moles
3          250 g             160 g       90 g

4     Mass of hydrated           Mass of water
   copper sulphate (in g)       lost (in g)
       1 mole = 250          5 moles = 90

               scale-down factor          multiply
               = 25/250            by 25/250

             25                    9

**Answer:** 25 g of copper sulphate loses 9 g of water.

## Evidence for equations

So far we have assumed that the equations for the reactions are known. Next we describe two methods that can be used to find or check the equations for a reaction.

*From the masses of chemicals reacting* When iron filings are added to copper sulphate solution the solution loses its blue colour. Pink copper powder appears at the bottom of the test tube. The following reaction has occurred:

    iron + copper sulphate $\longrightarrow$ iron sulphate + copper

But this does not give us enough information to write an equation for this reaction. There are two types of iron sulphate: one has the formula $FeSO_4$ and the other is $Fe_2(SO_4)_3$. To know which one has been formed in this reaction we need information about the masses of the chemicals that reacted.

Suppose that in such an experiment 5.6 g of iron filings reacted with excess copper sulphate to form 6.4 g of copper. From this we can work out which of the two iron sulphates was formed. The two possible equations are shown below.

iron   + copper sulphate $\longrightarrow$ iron(II) sulphate +    copper
$Fe(s)$  +    $CuSO_4(aq)$   $\longrightarrow$    $FeSO_4(aq)$   +    $Cu(s)$
1 mole       1 mole              1 mole         1 mole
56 g iron                 gives                 64 g copper

or

iron   + copper sulphate $\longrightarrow$ iron(III) sulphate +   copper
$2Fe$   +    $3CuSO_4$    $\longrightarrow$    $Fe_2(SO_4)_3$   +   $3Cu$
2 moles     3 moles            1 mole        3 moles
112 g iron              gives               192 g copper

As you see, the reaction producing iron(II) sulphate fits best with the experimental results. 1 mole of iron gives 1 mole of copper.

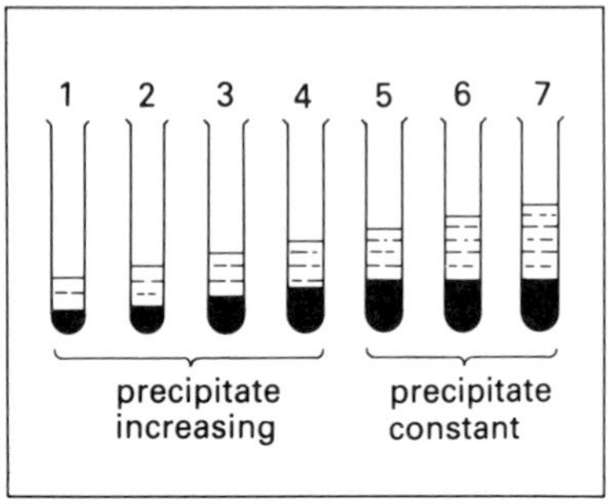

*From heights of precipitates* Many reactions take place in solution in water. In some of these, the product does not dissolve in water and it **precipitates**, that is, it appears as solid particles. These usually settle out at the bottom of the tube or flask. The amount of precipitate will obviously depend upon the amounts of reactants used.

When finding the equation for a reaction from precipitate heights it is usual to add different amounts of one reactant to a fixed amount of the other reactant. A typical set of results for this is shown in figure 8.2.

Suppose that in an experiment $5 \text{ cm}^3$ of a solution of potassium iodide, KI, was added to each of seven tubes. The volumes of a solution of lead nitrate, $Pb(NO_3)_2$, shown in the table were then added to these tubes and the precipitate was allowed time to settle. (All solutions contained 1 mole of solute per $dm^3$.) The heights of precipitate were found to be as shown in table 8.5.

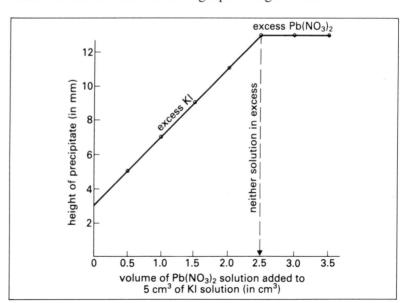

**Figure 8.2** *The height of precipitate can help us find the equation for the reaction*

| Test tube number | 1 | 2 | 3 | 4 | 5 | 6 | 7 |
|---|---|---|---|---|---|---|---|
| **Volume of KI (cm$^3$)** | 5 | 5 | 5 | 5 | 5 | 5 | 5 |
| **Volume of Pb(NO$_3$)$_2$ (cm$^3$)** | 0.5 | 1.0 | 1.5 | 2.0 | 2.5 | 3.0 | 3.5 |
| **Height of precipitate (mm)** | 6 | 7 | 9 | 11 | 13 | 13 | 13 |

**Table 8.5**

These results are shown on a graph in figure 8.3.

**Figure 8.3** *Plotting the results from figure 8.2*

To start with, adding more lead nitrate gives us more precipitate. This is because there is not enough lead nitrate in the first few test tubes to react with all the potassium iodide. The maximum height of precipitate is reached in tube 5. Beyond this adding extra lead nitrate makes no difference, so all the potassium iodide must have been used up.

It is in tube 5 that we have just the right amount of both solutions. This happens when we have:

$5.0\,cm^3$ of 1M potassium iodide solution $= 5/1000 \times 1$
$\qquad\qquad\qquad\qquad\qquad\qquad\qquad\qquad\quad = 0.005$ mole
$2.5\,cm^3$ of 1M lead nitrate solution $\qquad = 2.5/1000 \times 1$
$\qquad\qquad\qquad\qquad\qquad\qquad\qquad\qquad\quad = 0.0025$ mole

This is consistent with a 2:1 reaction, that is,

$$2KI(aq) + Pb(NO_3)_2(aq) \longrightarrow PbI_2(s) + 2KNO_3(aq)$$

## Summary: Formulae and equations

★ A chemical equation is a short way of describing what happens during a chemical reaction.
★ Some elements, such as hydrogen, exist as molecules containing two atoms. This is indicated in the element's chemical formula, for example, $H_2$.
★ All compounds have a definite chemical formula. This indicates the relative numbers of each type of atom that are present.
★ Chemical equations must always be balanced. This means that each side of the equation must contain the same number of each particular type of atom.
★ The formula of a compound can be calculated if the amounts of each element in it are known.
★ The equation for a reaction can be used to calculate the masses of reactants and products.
★ The equation for a reaction can be found, or checked, from the masses of the chemicals reacting or, where one of the products does not dissolve in water, from the heights of precipitates formed.

## Questions

1 Write word equations for each of the following reactions:
  (a) The magnesium burnt brightly in the oxygen to form magnesium oxide powder.
  (b) The zinc carbonate decomposed to give zinc oxide and carbon dioxide.
  (c) The calcium oxide reacted with the water to give calcium hydroxide.
  (d) The methane gas burnt in the oxygen to form carbon dioxide and water.

**2** The formulae of the substances mentioned in question 1 are given below. Use these to write a properly balanced equation for each reaction.

(a) Mg     $O_2$     MgO

(b) $ZnCO_3$     ZnO     $CO_2$

(c) CaO     $H_2$     $H_2O$     $Ca(OH)_2$

(d) $CH_4$     $O_2$     $CO_2$     $H_2O$

**3** Use the information below to calculate the formulae of the compounds formed in the following reactions:

(a) 6.4g of copper reacts with 1.6g of oxygen.

(b) 0.78g of potassium reacts with 1.6g of bromine.

(c) 0.24g of carbon reacts with 0.32g of oxygen.

(d) 1.55g of phosphorus reacts with 8.88g of chlorine.

**4** Calculate the mass of each element in the compounds formed in the following reactions. From this calculate the formulae of the compounds.

(a) 12 g of carbon reacts to form 44 g of an oxide.

(b) 3.2 g of sulphur reacts to form 8.0 g of an oxide.

(c) 1.12 g of iron reacts to form 1.60 g of iron oxide.

(d) 0.14 g of nitrogen reacts to form 0.17 g of a compound of nitrogen and hydrogen.

**5** (a) 5.12 g of iodine reacted with 5.35 g of antimony. After all the iodine had been used up 3.63 g of antimony remained. What is the formula of antimony iodide?

(b) 8.65 g of mercury reacted with bromine. When all the mercury had reacted the excess bromine was removed by evaporating it. 15.57 g of mercury bromide remained. What is its formula?

**6** What is the percentage of each of the elements specified in the table below?

| Element | Compound | Formula |
|---------|----------|---------|
| carbon | ethanol | $C_2H_5OH$ |
| silicon | silicon dioxide | $SiO_2$ |
| sodium | sodium chloride | NaCl |
| aluminium | potash alum | $KAl(SO_4)_2.12H_2O$ |

**7** The most common ore of aluminium is bauxite, $Al_2O_3$.

(a) Calculate the percentage of aluminium in pure bauxite.

(b) Calculate how many kilograms of aluminium could be extracted from 1 tonne (1000 kg) of pure bauxite.

(c) A particular aluminium ore only contains 56 per cent of bauxite. What mass of aluminium could be extracted from 1 tonne of this ore?

**8** The following compounds are all used as fertilizers for adding nitrogen to the soil. Calculate the percentage of nitrogen in each compound.

| Compound | Formula |
|----------|---------|
| ammonium chloride | $NH_4Cl$ |
| ammonium nitrate | $NH_4NO_3$ |
| ammonium sulphate | $(NH_4)_2SO_4$ |
| calcium cyanamide | $CaCN_2$ |

**9** Calculate the percentage of water in each of the following hydrated compounds:
(a) $CaSO_4.2H_2O$
(b) $CuSO_4.5H_2O$
(c) $Cu(NO_3)_2.3H_2O$
(d) $Al_2(SO_4)_3.18H_2O$

**10** (a) Find the mass of calcium oxide formed when 10 g of calcium burns in air according to the equation

$$2Ca(s) + O_2(g) \longrightarrow 2CaO(s)$$

(b) Find the mass of carbon dioxide produced when 48 g of carbon burns in air according to the equation

$$C(s) + O_2(g) \longrightarrow CO_2(g)$$

(c) Find the mass of oxygen given off when 6.4 g of hydrogen peroxide decomposes according to the equation

$$2H_2O_2(l) \longrightarrow 2H_2O(l) + O_2(g)$$

(d) Find the mass of hydrogen chloride needed to convert 2.4 g of magnesium metal into the chloride according to the equation

$$Mg(s) + 2HCl(g) \longrightarrow MgCl_2(s) + H_2(g)$$

**11** When varying amounts of iron(III) chloride solution (concentration 1 mole per $dm^3$) were added to 5 $cm^3$ of sodium hydroxide solution (concentration 3 moles per $dm^3$), and the heights of the resulting precipitates measured, the following results were obtained:

| Volume of iron (III) chloride solution ($cm^3$) | 1.0 | 2.0 | 3.0 | 4.0 | 5.0 | 6.0 |
|---|---|---|---|---|---|---|
| Height of precipitate (mm) | 6 | 7 | 8 | 9 | 10 | 10 |

The formula for iron(III) chloride is $FeCl_3$ and that for sodium hydroxide is $NaOH$.

(a) Plot these results on a graph.

(b) Deduce and write the equation for the formation of the precipitate.

12 (a) What mass of copper is obtained when 19.9 g of copper oxide, $CuO$, is changed to copper metal by reaction with hydrogen?

(b) What mass of calcium hydroxide, $Ca(OH)_2$, is formed when 112 g of calcium oxide, $CaO$, reacts with water?

(c) What mass of oxygen is needed to react with 2 g of calcium to form calcium oxide?

(d) What mass of ammonia, $NH_3$, can be formed when 14 tonnes of nitrogen react with excess hydrogen?

## Hydrogen: superfuel of tomorrow?

As the world's oil reserves dwindle we shall have to adopt new fuels, especially for transport systems. One particularly promising suggestion here is hydrogen.

The oceans provide a virtually unlimited source of the element, provided an economic method of extracting it can be found.

One method would be to use electricity to decompose acidified water. Unfortunately this would be relatively expensive, given present methods of electricity generation.

*This car runs on hydrogen, which is stored in cylinders in its boot*

*Power from a nuclear reactor like the one shown above could provide the energy needed to decompose water into hydrogen and oxygen. The hydrogen thus produced could be a useful fuel for cars and aircraft*

Another method would be to decompose the water by heat (**thermal decomposition**). This would require temperatures of around 3000 °C, and such temperatures are not likely to be economically achieved until nuclear fusion reactors are operational.

Yet another method would be to use the heat available from the present nuclear fission reactors to bring about a series of reactions in which the overall products are hydrogen and oxygen. Temperatures of about 1500 °C are attainable for this. Some possible reaction schemes are described below.

**Scheme 1** In this scheme there are three stages.

*Stage 1*: Carbon reacts with water to form carbon monoxide and hydrogen (temperature needed = 700 °C).
*Stage 2*: The carbon monoxide reacts with an oxide of iron, $Fe_3O_4$. Carbon is re-formed and a new oxide of iron, $Fe_2O_3$, is produced (temperature needed = 250 °C).
*Stage 3*: When heated to 1400 °C the new iron oxide decomposes to the original iron oxide and oxygen.

You will see that overall the only permanent products are hydrogen and oxygen. The same is true of the second reaction scheme.

**Scheme 2** There are four steps in this scheme. All occur below 800 °C.

*Stage 1*: Calcium bromide reacts with water, forming calcium hydroxide and hydrogen bromide.
*Stage 2*: Hydrogen bromide reacts with mercury, forming mercury bromide and hydrogen.
*Stage 3*: Mercury bromide reacts with calcium hydroxide, forming calcium bromide, water and mercury oxide.
*Stage 4*: Mercury oxide decomposes upon heating to form mercury and oxygen.

## Questions

**13** Write a balanced chemical equation for the thermal decomposition of water.
**14** Write balanced chemical equations for each of the three stages described in reaction scheme 1.
**15** Why is it reasonable to describe the reactions in scheme 1 as corresponding to the decomposition of water?
**16** Here are the formulae of the compounds mentioned in scheme 2. With the aid of simple diagrams explain what each formula tells us about the relative numbers of each atom present.

$CaBr_2$    $H_2O$    $Ca(OH)_2$    $HBr$    $HgBr_2$    $HgO$

**17** Write balanced equations for each of the four stages described in reaction scheme 2.

# 9 Oxygen and the air

The ancient Greeks thought that air was an element. Today we know that it is a mixture of gases. This chapter is about air and the gases in it.

## 9.1 Gases from air

### The major gases in air

A Frenchman named Lavoisier was the first person to show that air contains more than one gas. A modern version of his experiment is shown in figure 9.1.

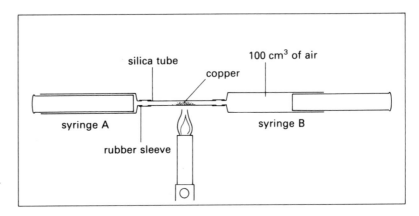

**Figure 9.1** *A modern version of Lavoisier's experiment*

At the start, syringe A contains air. Syringe B is empty. The air is passed backwards and forwards over the heated copper. Two things happen. The copper goes black on the outside. At the same time the volume of air shrinks. Some typical results might be as follows:

$$\text{Volume of air at start} \quad = 100\,\text{cm}^3$$
$$\text{Volume remaining at end} \quad = 79\,\text{cm}^3$$
$$\text{So volume of air used up} \quad = 21\,\text{cm}^3$$

$$\text{Percentage of air used up} = \frac{21 \times 100}{100} = 21\%$$

**Figure 9.2** *Firemen need their own supply of oxygen when working in smoke*

So the air must contain 21 per cent of an active gas. This gas is oxygen. The oxygen reacts with the copper to form black copper oxide.

$$copper \ + \ oxygen \longrightarrow copper \ oxide$$

or

$$2Cu \ + \ O_2 \longrightarrow 2CuO$$

The remaining 79 per cent of the air is much less reactive than oxygen. It is mostly nitrogen. There are also small amounts of other gases (see table 9.1).

| Gas | Approximate % by volume |
|---|---|
| nitrogen | 78 |
| oxygen | 21 |
| argon | 1 |
| water vapour | 0.5 (variable) |
| carbon dioxide | 0.03 |
| neon, helium, krypton and xenon (in total) | 0.002 |

**Table 9.1** *The gases in the air*

### Extracting gases from the air

Air is a mixture of gases. This mixture can be separated. The air is cooled down to a liquid and then distilled. This is explained below.

1 Water vapour and carbon dioxide are first frozen out of the air in a refrigeration unit. If this were not done, these substances would freeze and block up pipes later in the process.
2 The 'air' is then pumped into a container until it is under very high pressure. During this process the gas gets quite hot and has to be cooled down again.
3 This compressed air is allowed to escape into another larger container. You may have noticed that air escaping from a bicycle tyre always feels cold. The same idea is used here. As the compressed air escapes it gets cold: so cold that it liquefies.
4 The liquefied air is now allowed to warm up. The gases boil at different temperatures. The gas with the lowest boiling point distils off first, then the gas with the next lowest, and so on (see the description of fractional distillation in section 4.1). The boiling points of these gases are given in table 9.2.

| Gases | Boiling point °C | K |
|---|---|---|
| xenon | −108 | 165 |
| krypton | −152 | 121 |
| oxygen | −183 | 90 |
| argon | −186 | 87 |
| nitrogen | −196 | 77 |
| neon | −246 | 27 |
| helium | −269 | 4 |

**Table 9.2** *The boiling points of the gases in the air*

## 9.2 Some properties and uses of nitrogen and oxygen

In some ways oxygen and nitrogen are very alike. This can be seen from table 9.3.

| Property | Oxygen | Nitrogen |
|---|---|---|
| appearance | colourless transparent gas | colourless transparent gas |
| odour | has no smell | has no smell |
| density | $1.33 \, g/dm^3$ | $1.16 \, g/dm^3$ |
| acidity | neutral | neutral |
| reactivity | reacts well with most other elements and with many compounds | only reacts with a few elements and compounds |

**Table 9.3** *Some properties of oxygen and nitrogen*

The one very important difference between the gases is their chemical reactivity. Nitrogen is very **inert** (unreactive). Oxygen is very reactive.

### Some uses of nitrogen

Many uses of nitrogen depend upon its low reactivity. The space at the top of aircraft fuel tanks is kept filled with nitrogen. If this space became filled with air then the fuel/air mixture would become explosive.

Because it is so inert, nitrogen is useful in welding. A welder will often direct a stream of nitrogen gas across the weld. This stops the hot metal from reacting with oxygen from the air.

Oil tankers also make use of the inertness of nitrogen. Mixtures of oil vapours with air can be dangerously explosive. Empty or part-empty tanks are therefore flushed with nitrogen to prevent the formation of such mixtures.

Liquid nitrogen is used in the food industry. Because it is very cold and very unreactive it is ideal for freezing foods (see figure 9.3, opposite).

### Some uses of oxygen

Any substance that burns in air will burn much more fiercely in pure oxygen. This is how an oxy-acetylene burner works. Burning acetylene in oxygen produces a flame hot enough to weld or cut through steel (see figure 9.4).

The **respiration** of living organisms requires oxygen. In respiration the oxygen taken in by a plant or animal reacts

**Figure 9.3** *Liquid nitrogen is very cold. At this factory, 6 000 000 hamburgers a day are frozen in tunnels cooled by liquid nitrogen*

**Figure 9.4** *Acetylene burns in oxygen to give a very hot flame that can be used for welding metals together*

with glucose. Water and carbon dioxide are produced, together with the energy the organism needs in order to live and grow. Living things therefore cannot survive without oxygen. Oxygen is used by doctors to help sick people breathe and maintain their respiration. It is also used by climbers of very high mountains, by firemen (see figure 9.2) and by underwater swimmers, so that they can survive in places where there is not enough air for them to breathe.

**Figure 9.5** *You have to have oxygen to breathe. This girl has been in a water accident. She needs to have extra oxygen to help her to breathe properly*

## 9.3 Fires and fire-fighting

### Combustion

Combustion is another name for burning. When substances burn in air they are reacting with oxygen. The more oxygen

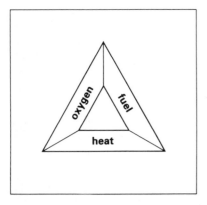

**Figure 9.6** *The fire triangle reminds us of the three things necessary for fire*

there is, the better they react. This gives you a test for oxygen. A wooden splint that is just glowing red will burst into flames when placed in oxygen.

### Fire!

Combustion can be controlled. Then it is very useful. It is used to power cars, to cook meals and to keep you warm. But accidental combustion can cause death and damage.

For a fire to start three things are needed (figure 9.6):

1 heat,
2 fuel,
3 oxygen.

So to prevent a fire you must stop heat, fuel and oxygen from coming together.

### Fire prevention

In a fire, the fuel is the substance that burns. Newspapers and wood shavings are fuels. One way to prevent fires is not to leave fuels lying around.

Some fuels burn very well. Petrol, methylated spirits and paint thinners are examples. So it is important that these are kept away from heat.

Keeping fire doors closed in buildings cuts down the flow of air. This can reduce the amount of oxygen getting to a fire.

### Fire-fighting

If you want to put out a fire you must remove the heat, the fuel or the oxygen – or all these!

If possible, the fuel supply should be cut off first. This might mean turning off a gas tap or stopcock.

To remove the supply of oxygen you must stop air from reaching the flames. Sometimes you can smother the fire with a fire blanket or mat. Earth or sand can also be used to cover the flames. Sand is used to put out burning metals.

**Figure 9.7** *If a lorry with this warning label is in an accident, the code 3YE tells the firemen what to do. They need to wear breathing apparatus to protect them from fumes. They should use foam and not water to put out any fire. People in the area should be moved away from the accident*

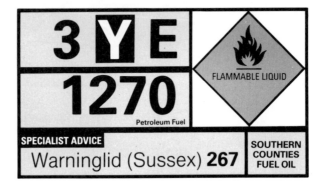

**Figure 9.8** *Using foam to fight a fire. The foam can be built up into a layer that will keep out the oxygen*

Fire extinguishers can be used. There are several types. They all work by smothering the fire with either a gas, a liquid, a powder or a foam (see figure 9.8). This stops oxygen from reaching the flames. Most types also cool down the fire (that is, they take away the heat).

There are four main types of fire:

1 burning solids like wood and paper,
2 burning liquids like cooking fat and petrol,
3 burning gases,
4 burning metals like magnesium or sodium.

Not all fire extinguishers are suitable for all fires. Table 9.4 shows when an extinguisher should be used.

| Type of extinguisher | What it contains | What it may be used on | What it should *not* be used on |
|---|---|---|---|
| soda-acid | sulphuric acid<br>water<br>sodium hydrogencarbonate | solids | liquids<br>gases<br>metals<br>electrical fires |
| carbon dioxide | carbon dioxide gas | solids<br>liquids | metals |
| foam | aluminium sulphate<br>sodium hydrogencarbonate<br>detergent and water | solids<br>liquids | metals<br>electrical fires |
| powder | sodium hydrogencarbonate | solids, liquids,<br>gases or metals | |
| volatile liquid | tetrachloromethane or<br>similar liquid | solids<br>liquids | metals<br>in enclosed spaces |

**Table 9.4** *Uses of fire extinguishers*

## 9.4 Oxygen in the laboratory

Most substances react with oxygen. The compounds formed are called **oxides**. This section tells you how oxygen can be made in the laboratory. It also tells you what happens when some substances react with oxygen.

### Making oxygen in the laboratory

Figures 9.9 and 9.10 show two methods of making oxygen. In both the gas is collected over water.

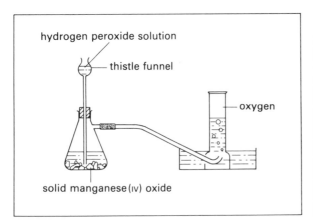

**Figure 9.9** *Making oxygen in the laboratory*

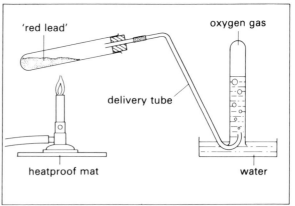

**Figure 9.10** *Heating 'red lead' (an oxide of lead)*

| Element | What happens | Equation | Oxide + water |
|---|---|---|---|
| calcium | the grey metal burns with a bright red flame to form a white powder | $2Ca + O_2 \longrightarrow 2CaO$ | alkaline |
| magnesium | the silvery ribbon burns with a brilliant white flame to form a white powder | $2Mg + O_2 \longrightarrow 2MgO$ | alkaline |
| sodium | the silvery metal burns with a bright yellow flame to form a white powder | $2Na + O_2 \longrightarrow 2Na_2O_2$ | alkaline |
| carbon | the black powder burns slowly; it glows red and forms a colourless gas | $C + O_2 \longrightarrow CO_2$ | slightly acid |
| phosphorus | the yellow solid burns with a very bright white flame; the white smoke settles out as a powder | $P_4 + 5O_2 \longrightarrow P_4O_{10}$ | acid |
| sulphur | the yellow solid burns with a bright blue flame, producing a choking colourless gas | $S + O_2 \longrightarrow SO_2$ | acid |

**Table 9.5** *Burning elements in oxygen*

### Making oxides in the laboratory

Table 9.5 (opposite) explains what happens when elements are heated and lowered into a gas jar of oxygen. The final column tells you what sort of solution is formed when the oxide is added to water.

Metals form oxides that give alkaline solutions if they dissolve. Most non-metals form oxides that give acid solutions. (Acids and alkalis are discussed in section 19.1.)

## 9.5 Corrosion and air pollution

Corrosion is the slow 'eating away' of a substance; it is caused by a chemical reaction. Two types of corrosion are easily seen: the corrosion of iron and steel and the corrosion of stonework (see figure 9.11).

### The rusting of iron

When iron corrodes, rust is formed. A bicycle left out in the rain soon rusts. You might think that rusting is a reaction between iron and water. The experiment shown in figure 9.12 proves that both water and air are involved in the reaction.

**1** The nail in tube 1 is in contact with both air and water. It is well rusted within a few days.
**2** The nail in tube 2 is in contact with air, but not water. It does not rust.
**3** The nail in tube 3 is in contact with water, but not air. It does not rust.

**Figure 9.11** *The stonework of a bridge across the river Thames in London being treated for corrosion*

**Figure 9.12** *An experiment to find out what is necessary to make iron rust*

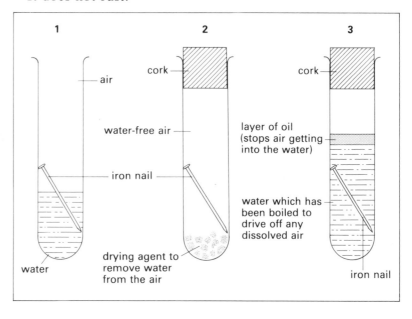

**Figure 9.13** *Experiment to find out which part of the air is used up when iron rusts*

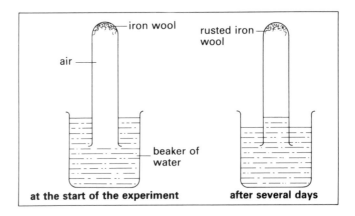

Another experiment is shown in figure 9.13. The water slowly rises up the tube as the iron rusts. After a few days the water stops rising. This happens when about 20 per cent of the air has been used up. From this you can tell that oxygen from the air is reacting with the iron. Rust is in fact iron oxide:

$$\text{iron} + \text{oxygen} \longrightarrow \text{iron oxide}$$

This reaction will only happen when water is present, however.

To prevent iron from rusting you have to keep water and air away from the metal. Coating the iron with oil or paint is the most common way of doing this.

## Air pollution

Polluted air is air which contains harmful substances. Each day in Britain over a thousand tonnes of smoke and poisonous gases go into the air. Most of this comes from burning fuels.

Smoke often contains tiny particles of carbon. These make buildings dirty. They can also make you ill by making catarrh, bronchitis and coughs worse.

Badly tuned car engines give out carbon monoxide gas. This is very poisonous. It reacts with the blood and stops the blood carrying oxygen around the body.

There is also nitrogen dioxide in car exhaust fumes. This, too, is a poisonous gas. Some car manufacturers now put an extra part inside the exhaust pipe which makes the nitrogen dioxide decompose into nitrogen and oxygen, which are harmless.

Burning coal produces a gas called sulphur dioxide. This gas reacts with water and oxygen in the air to form sulphuric acid. So the rain in cities contains dilute sulphuric acid! This speeds up the corrosion of stonework and metals, and damages growing plants (see figure 9.15).

Many cities have now introduced clean air regulations. In those cities, smokeless fuels must be used. Cars and lorries

**Figure 9.14** *Years ago 'smog' was common in England because there was little control over the waste gases and smoke allowed into the air*

**Figure 9.15** *Even now, we are not free from the effects of air pollution. Acid fumes carried on the south-westerly winds have damaged trees, lakes and crops in many parts of Europe. This photograph was taken in northern Germany*

are not allowed to pour out thick black smoke. Factories have to remove most of the fine grit from smoke. They also have to build high chimneys to spread out what smoke they do make. So now those cities are much cleaner places in which to live.

## Summary: Oxygen and the air

* Air contains 78 per cent nitrogen, 21 per cent oxygen and about 1 per cent other gases.
* Oxygen is a very reactive gas.
* Nitrogen is a rather inert gas.
* Gases can be removed from the air by liquefying them first and then distilling them.
* Nitrogen is used to provide an inert atmosphere.
* Nitrogen is also used for low-temperature freezing.
* Oxygen is needed for respiration.
* Oxygen is also used in oxy-acetylene burners.
* Combustion means burning in oxygen.
* Fires need heat, fuel and oxygen.
* Fire extinguishers prevent oxygen from reaching the flames.
* Oxygen can be made by decomposing hydrogen peroxide using manganese(IV) oxide, or by heating 'red lead'.
* Oxides are formed when substances react with oxygen.
* Metals form oxides that are neutral or alkaline in water.
* Non-metals form oxides that are acids in water.
* For iron to rust, it needs both oxygen and water.
* Polluted air may contain smoke and poisonous gases.
* Carbon particles, carbon monoxide, nitrogen dioxide and sulphur dioxide are some of the substances found in city air.

## Questions

1 Each of the sentences below contains a mistake. Decide what the mistake is and then write out a corrected sentence.
(a) Oxygen can be obtained by distilling and then liquefying air.
(b) Liquid hydrogen is used for fast freezing of foods.
(c) Sulphur dioxide comes mainly from car exhaust pipes.
(d) Most metals form oxides that give acid solutions.
(e) Keep fire doors open. A good flow of air will help to extinguish fires.

2 The diagram shows some experiments which investigate the rusting of iron.

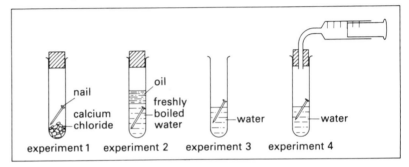

experiment 1    experiment 2    experiment 3    experiment 4

(a) What is the purpose of the calcium chloride in experiment 1?
(b) What is the purpose of the layer of oil in experiment 2?
(c) Will the nail rust most quickly in experiment 1, 2 or 3?
(d) Why was freshly boiled water used in experiment 2?
(e) Explain what you would expect to see happen in experiment 4.

3 When lead nitrate is heated strongly it decomposes giving lead oxide, nitrogen dioxide and oxygen. Some information about these substances is given in the table below.

| Substance | Formula | Melting point | Boiling point |
|---|---|---|---|
| lead nitrate | $Pb(NO_3)_2$ | decomposes | – |
| lead oxide | $PbO$ | $+888\,°C$ | unknown |
| nitrogen dioxide | $NO_2$ | $-11\,°C$ | $+21\,°C$ |
| oxygen | $O_2$ | $-218\,°C$ | $-183\,°C$ |

(a) What is the physical state of each substance at $25\,°C$?
(b) Design an apparatus to heat lead nitrate and collect nitrogen dioxide and oxygen separately. Draw a neat fully labelled diagram and add brief notes on what is happening in each part of the apparatus.

(c) Write a balanced equation for the decomposition of lead nitrate.

(d) A pupil burned samples of calcium, sulphur and carbon in separate tubes of the oxygen gas. She then added Universal indicator solution to each of the tubes. Describe the colour changes in each case and give reasons for your answers. (You may use table 19.1 on page 228 to help you answer this question.)

(e) The pupil tried burning carbon and sulphur in separate tubes of nitrogen dioxide gas. To her surprise they did burn; the carbon formed carbon dioxide and nitrogen, and the sulphur formed sulphur dioxide and nitrogen. Write equations for these two reactions.

## Yesterday's atmosphere

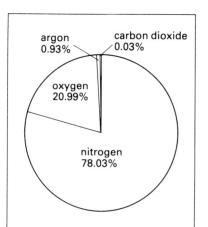

The present composition of the Earth's atmosphere

To trace the history of the Earth's atmosphere is particularly difficult. We have to rely upon the marks that it has left upon rocks and fossils.

Certain gases are unexpectedly absent from the Earth's atmosphere. This is taken to mean that the gases originally surrounding the Earth were swept away during the violent changes which followed the planet's formation 4500 million years ago.

A new atmosphere is thought to have arisen by the release of gases from the Earth's rocks. Even today, volcanoes provide spectacular evidence for such processes. At that time the atmosphere was mainly made up of methane, $CH_4$, and ammonia, $NH_3$.

Gradually the Earth's surface changed, and the atmosphere changed with it. By about 2000 million years ago the methane had been replaced by carbon dioxide, and nitrogen had replaced the ammonia.

At about this time free oxygen began to appear. From rock samples it is clear that there was enough oxygen to convert iron into iron oxide, $Fe_2O_3$. There must, however, have been very much less oxygen than there is now. Evidence for this comes from sediments containing an oxide of uranium called uraninite. Uraninite reacts quickly in our present atmosphere, gaining extra oxygen and forming a different oxide of uranium.

Scientists have disagreed about the cause of the appearance and build-up of free oxygen. It certainly seems strange that such a reactive gas should exist as the element. One theory to explain this oxygen suggests that ultra-violet rays acted on water in the upper atmosphere, decomposing it. Oxygen and hydrogen are supposed to have been formed. Most of this would have recombined but some of the hydrogen would escape from the atmosphere before recombination, thus leaving free oxygen.

Another theory suggests that the oxygen arose from photosynthesis. It is known that organisms capable of making oxygen existed 2000 million years ago. It is thought that living creatures invaded the land around 400 million years ago, when the oxygen level had reached about 10 per cent of its present-day value.

## Questions

4 Why is it surprising that the atmosphere should contain oxygen gas?

5 Why should hydrogen be more likely to escape from the atmosphere than oxygen?

6 Write a chemical equation to describe the formation of glucose, $C_6H_{12}O_6$, and oxygen from carbon dioxide and water. What name is given to this process?

7 How do we know that there was very little oxygen in the atmosphere 2000 million years ago?

8 Imagine you had to separate the gases present in the Earth's early atmosphere. Draw a well-labelled flow diagram of a suitable industrial process. Use the data in table below.

| Gas | Major or minor component | Melting point (°C) | Boiling point (°C) |
|---|---|---|---|
| methane | major | −182 | −164 |
| ammonia | major | − 78 | − 33 |
| hydrogen | minor | −259 | −252 |
| nitrogen | minor | −210 | −196 |

# 10 Water and hydrogen

**Figure 10.1** *Few things grow or live in the desert because there is so little water there*

Water is the commonest chemical compound. It is also one of the most important. Where there is plenty of water, living things usually flourish. In hot dry deserts, life is much more difficult (see figure 10.1).

Most living things contain a large amount of water. For example, about 60 per cent of your body is water.

Apart from the water that we drink and get from food, we each use about $150\,dm^3$ of water each day (about 30 gallons). Most of this is used for washing and for flushing lavatories.

Industry also uses large amounts of water. Most of this is used for cooling (see figure 10.2). It is also used as a solvent in many processes.

**Figure 10.2** *Vast amounts of river water are used for cooling at this power station. The water that is returned to the river is slightly warmer and contains less dissolved oxygen*

## 10.1 What is water?

### The elements in water

Have you ever noticed the steam that comes out of a car exhaust on a cold day? This steam is the water vapour produced when the petrol burns. Since we know that burning is a reaction with oxygen, it seems likely that water is an oxide.

If calcium metal is added to water, hydrogen gas is given off. This cannot have come from the calcium, because calcium is an element. It must have come from the water.

From these two facts we can tell that water is a compound containing hydrogen and oxygen. It is, in fact, hydrogen oxide. Chemists often write it as $H_2O$. This means that it contains twice as many hydrogen atoms as oxygen atoms (see figure 10.3).

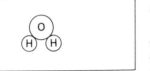

**Figure 10.3** *A water molecule*

### Testing for water

Sometimes, we need to test a liquid to see if it is water. Table 10.1 lists some tests that can be used.

| Property or test | Result |
|---|---|
| appearance | colourless liquid |
| boiling point | 100 °C |
| melting point | 0 °C |
| effect upon cobalt chloride paper | changes from blue to pink |
| effect upon anhydrous copper sulphate | changes from white to blue |
| acidity/alkalinity | neutral |
| density | 1 g/cm$^3$ |

**Table 10.1** *Testing for water*

You should be careful when you draw conclusions from the results of these tests. For example, a substance that turns cobalt chloride paper pink must contain water. It need not, however, be pure water. It might be vodka or sea water. Pure water will give the positive result for *all* the tests in table 10.1.

## 10.2 Impure forms of water

### The water cycle

It is very unusual to find really pure water. This is because water is such a good solvent.

Rain water is one of the purest forms of water, but even this contains dissolved gases.

As rain water trickles through the ground, it dissolves substances from the soil and rocks. Water from limestone areas contains a lot of dissolved solids.

Rain water always finds its way into rivers. These flow to the sea, taking the dissolved solids with them. Over millions of years, these solids have built up and this is what gives sea water its salty taste.

The water cycle, shown in figure 10.4, gives you an idea of how water circulates around the Earth. The Sun evaporates water from the sea. Winds blow these tiny

**Figure 10.4** *The water cycle*

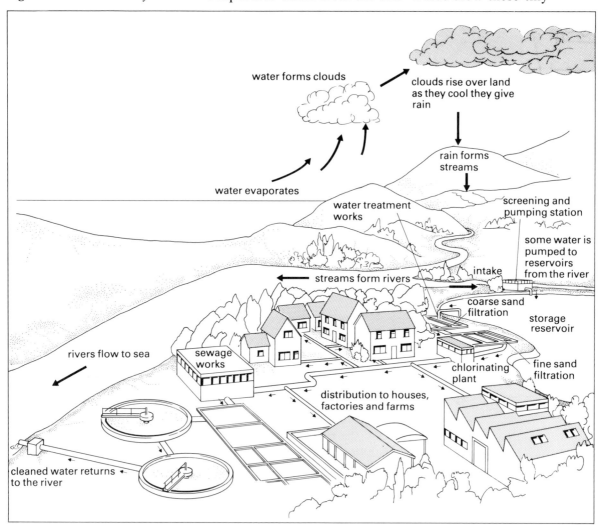

droplets of water vapour together to form clouds. As the clouds go inland, they cool down and rain falls. The rain runs into streams and rivers. Some water is taken from the rivers to go into reservoirs. From reservoirs the water is purified for people and factories to use. After it has been used, it is dirty. So it goes to the sewage works to be partially cleaned. Then it is pumped back into the rivers and out to sea. In this way, all the water we use returns to the sea, and the cycle is complete.

### Tap water

Most of our tap water comes from rivers. Some comes from underground wells called bore-holes. In figure 10.4 you can see a typical waterworks. Sending water through a waterworks does not make the water pure in the chemists' sense. It simply makes sure that it is safe to drink.

River water is first pumped into a reservoir. Floating sticks and other debris are removed by meshes called screens.

The water is then filtered through coarse sand. This removes most of the small bits of solid in it. A second filtration through much finer sand removes any smaller particles of solid. It also takes out many of the bacteria.

The bacteria that get past the filtering stages are now killed by adding a small amount of chlorine to the water. The water is then safe to pipe to homes and factories.

## 10.3 Hard and soft water

### What is hard water?

Hard water is water in which soap will not easily form a lather.

The hardness is caused by calcium or magnesium compounds dissolved in the water. These react with soap to form a scum. So to obtain a lather you have to add extra soap to overcome the effects of these compounds.

| Advantages | Disadvantages |
|---|---|
| has taste many people prefer | leaves scum on baths |
| provides calcium compounds for teeth and bones | may leave clothes with a 'hard' feeling after washing |
| less likely to dissolve lead pipes (lead is poisonous) | wastes soap |
| good for brewing beer | causes kettles to 'fur up' and pipes to become blocked |

Table 10.2 *Some advantages and disadvantages of hard water*

## Hard water – good or bad?

Hard water was once a great nuisance to laundries. Nowadays it is not a serious problem, however, because many modern washing powders are not affected by hardness. Table 10.2 (opposite) lists some benefits and disadvantages of hard water.

## Different types of hardness

There are two different types of hard water. These are shown in table 10.3.

| Types of hardness | Causes of hardness |
| --- | --- |
| temporary hardness | calcium hydrogencarbonate |
| permanent hardness | calcium sulphate |

**Table 10.3** *The types and causes of hard water*

Limestone is a common form of calcium carbonate. Limestone rocks are attacked by water from the rain and carbon dioxide from the air. The water and carbon dioxide react with the calcium carbonate forming calcium hydrogencarbonate. This then dissolves in the water.

calcium carbonate  +   water   + carbon dioxide ⟶ calcium hydrogencarbonate

$$CaCO_3(s) \quad + \ H_2O(l) + \quad CO_2(g) \quad \longrightarrow \quad Ca(HCO_3)_2(aq)$$

Hard water containing calcium hydrogencarbonate becomes soft if it is boiled. This sort of hardness is called **temporary hardness**.

During heating, the dissolved calcium hydrogencarbonate decomposes to give solid limestone again. This reaction is the exact opposite of the one just described.

$$Ca(HCO_3)_2(aq) \longrightarrow \underset{\text{limestone}}{CaCO_3(s)} + H_2O(l) + CO_2(g)$$

The limestone is now a solid. It is no longer dissolved in the water. Because it is not dissolved, it cannot react with the soap to form a scum.

**Figure 10.5** *The inside of this kettle is covered in 'kettle fur' after being used to boil hard water*

**Figure 10.6** *Limestone has built up inside this hot water cylinder. It was used in a house in a hard water area for only a few years*

In areas with temporary hardness, a lot of limestone builds up inside kettles. This is usually called 'kettle fur' (see figure 10.5).

Unfortunately this limestone is also formed inside pipes, especially hot water pipes (see figure 10.6). In time, it can completely block them.

## Measuring the hardness of water

We can compare the hardness of different samples of water. This is done by adding soap solution to the water a small amount at a time. The soap is added until the water gives a good lather. The amount of soap used is then noted. Hard water will need more soap than soft water.

## Softening hard water

There are several ways hard water can be softened. Some of these methods are shown in table 10.4.

| Method | Comments |
|---|---|
| boiling | expensive; only removes temporary hardness |
| add calcium hydroxide | quite cheap; only removes temporary hardness |
| add sodium carbonate (washing soda) | fairly cheap; removes all hardness |
| use an ion-exchange resin | the resin is expensive but can be used over and over again; removes all hardness |
| distillation | very expensive; removes all hardness |

**Table 10.4** *Methods of softening water*

We have already seen how boiling the water can remove temporary hardness. Another way of removing temporary hardness is to add calcium hydroxide. This converts the calcium hydrogencarbonate into calcium carbonate. But calcium carbonate does not dissolve in water. So when it is formed in the water it **precipitates** out. This means that it appears as particles of white solid calcium carbonate. In this way all the dissolved calcium compounds are removed from the water.

calcium hydroxide  +  calcium hydrogencarbonate  ⟶  calcium carbonate  +  water

$Ca(OH)_2(aq)$     +     $Ca(HCO_3)_2(aq)$     ⟶     $2CaCO_3(s)$     +  $2H_2O(l)$

Sodium carbonate can be used to remove both temporary and permanent hardness. As before, it removes dissolved calcium compounds by changing them into insoluble calcium carbonate.

calcium sulphate  +  sodium carbonate  ⟶  calcium carbonate  +  sodium sulphate

$CaSO_4(aq)$     +     $Na_2CO_3(aq)$     ⟶     $CaCO_3(s)$     +     $Na_2SO_4(aq)$

Another method of softening water is to use an **ion-exchange resin**. This is a compound made of sodium and a resin loosely combined together (resin–sodium).

This compound is placed inside a column and the hard water is poured through it. The resin works by exchanging ('swapping') its sodium for the calcium present in the water. So the water coming out of the column contains dissolved sodium compounds instead of dissolved calcium compounds. You may find this easier to understand by looking at the word equation below.

resin–sodium     +     calcium sulphate     ⟶     resin–calcium     +     sodium sulphate

⎡ ion-
 exchange
 resin at
 start ⎤     ⎡ causes
 hardness ⎤     ⎡ ion-
 exchange
 resin at
 end ⎤     ⎡ doesn't cause
 hardness ⎤

Dissolved sodium compounds do not react with soap. So water containing such compounds is soft.

Ion-exchange resins are expensive to buy, but can be used over and over again. When all the sodium in the resin has been replaced by calcium, a strong solution of sodium chloride (common salt) is poured through the used resin. The sodium chloride converts the resin back into the sodium form. It is then ready to use again.

$$\text{resin–calcium} \quad + \quad \text{sodium chloride} \quad \longrightarrow \quad \text{resin–sodium} \quad + \quad \text{calcium chloride}$$

resin–calcium ['used-up' resin]    +    sodium chloride [common salt]    ⟶    resin–sodium [resin ready for softening]    +    calcium chloride [waste solution]

Distillation is another way of softening water. The water is boiled and passed through a condenser where it turns back into a liquid (see section 4.1). Solids do not boil and so these are left in the flask. To use this method on a large scale is very expensive.

The main uses of distilled water, outside of laboratories and factories, are for topping up car batteries and in steam irons.

## 10.4 Solutions, solubility and hydration

This section is about the solutions that are produced when water is used to dissolve substances.

### Some definitions

Solutions in which water is the solvent are sometimes called **aqueous solutions**. The word aqueous means 'in water'.

| Word | Meaning |
|------|---------|
| solute | a substance that is to be dissolved |
| solvent | the liquid used to dissolve a solute |
| solution | a mixture of solute and solvent |

**Table 10.5** *Some definitions*

### Solubility

Some substances dissolve more easily than others.

A substance that does not dissolve at all is **insoluble**.

A substance that does not dissolve very well has a **low solubility**.

A substance that dissolves easily has a **high solubility**, that is, it is very soluble.

### Solubility and temperature

The amount of a solute that will dissolve in a solvent depends on the temperature. Usually solids are more soluble in hot solvents than in cold ones. In contrast, gases tend to dissolve better in cold solvents than in hot ones. In figure

10.2, the warm water returned to the river from the cooling towers contains less dissolved oxygen than normal river water.

## Saturated solutions

A **saturated solution** is one that contains as much dissolved solute as it can at a given temperature. But a solution that is saturated at 20 °C may not be saturated at 80 °C. This is because solubility is different at different temperatures (see figure 10.7).

**Figure 10.7** *How solubility can vary with temperature*

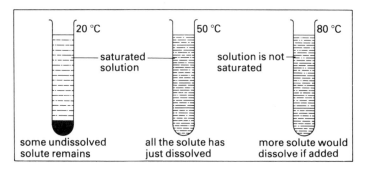

## Calculating solubilities

To calculate the solubility of any solute you have to know three things:

1 the temperature,
2 the mass of solute that has dissolved,
3 the mass of solvent that the solute dissolved in.

For example, at 20 °C the solubility of sodium chloride in water is 36 g of sodium chloride for every 100 g of water.

Usually we say how much of the solute would dissolve in 100 g of solvent, even if some other quantity was actually used.

So, if 5 g of copper sulphate dissolve in 25 g of water at 20 °C, then the solubility is said to be 20 g of copper sulphate per 100 g of water at 20 °C.

## Measuring solubility

Figure 10.8 shows how you can measure the solubility of a substance. The method used is as follows.

1 Heat the solvent to the required temperature in a water bath.
2 Then add solute to the stirred solvent until no more will dissolve.
3 Let the undissolved solute settle.
4 While the solute is settling, weigh an evaporating basin.
5 Then transfer some of the clear saturated solution to the evaporating basin using a warm pipette.

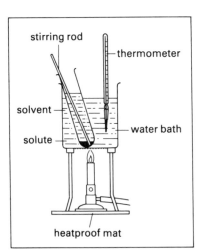

**Figure 10.8** *Finding the solubility of a solute*

**6** Weigh the basin again.
**7** Now carefully evaporate away the solvent.
**8** Finally, re-weigh the evaporating basin.

Here are some results that were obtained from an experiment like this. The solubility of potassium nitrate in water was being investigated.

| | |
|---|---|
| Temperature of solvent (water) | = 70 °C |
| Mass of basin + solution | = 63.5 g |
| Mass of basin + solute (at end) | = 53.5 g |
| Mass of empty basin (at start) | = 40.0 g |
| Name of solute: potassium nitrate | |

From these results we can tell:

mass of solute in the basin = 53.5 g − 40.0 g = 13.5 g

mass of solvent that was in the basin = 63.5 g − 53.5 g
= 10.0 g

This tells us that 13.5 g of potassium nitrate dissolved in 10.0 g of water at 70 °C.

So the solubility of potassium nitrate is 135 g of potassium nitrate per 100 g of water at 70 °C.

## Solubility curves

You could carry out the above experiment at several temperatures, and then plot a graph of the results. Figure 10.9 shows such a graph. It is called a **solubility curve**.

**Figure 10.9** *Solubility curve for potassium nitrate*

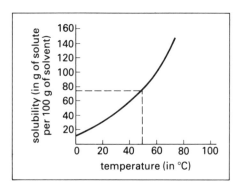

From the curve you can find the solubility of potassium nitrate at any temperature between 0 °C and 100 °C. For example, the solubility of potassium nitrate at 50 °C is about 75 g per 100 g of water.

The effect of temperature on the solubilities of some solids is shown in table 10.6.

| Temperature (°C) | Ammonium chloride | Calcium hydroxide | Copper(II) sulphate | Lithium carbonate | Potassium nitrate | Sodium chloride |
|---|---|---|---|---|---|---|
| 0 | 29.4 | 0.185 | 14.0 | 1.54 | 13.3 | 35.6 |
| 20 | 37.2 | 0.165 | 20.2 | 1.33 | 31.6 | 35.8 |
| 40 | 45.8 | 0.141 | 28.7 | 1.17 | 63.9 | 36.3 |
| 60 | 55.2 | 0.116 | 39.8 | 1.01 | 110 | 37.1 |
| 80 | 65.6 | 0.094 | 56.0 | 0.85 | 169 | 38.0 |
| 100 | 77.3 | 0.077 | 76.9 | 0.72 | 246 | 39.4 |

**Table 10.6** *The solubilities of some substances at various temperatures, in grams of anhydrous substance per 100 g of water*

## Hydrated compounds

Many reactions are carried out in aqueous solutions. The solids that form in these reactions often contain water molecules trapped inside their crystals. These solids are called **hydrated** ('with water') solids. This water inside the solid affects the shape of the crystals and may even change their colour. Hydrated copper sulphate, which is bright blue, is an example.

Often the water in hydrated compounds can be easily driven off by gentle heating. This gives the **anhydrous** compound. Anhydrous means 'contains no water'. Anhydrous copper sulphate is white.

Some hydrated compounds turn into the anhydrous compounds even without being heated. They lose their water to the air even at room temperature. These compounds are called **efflorescent** compounds. Efflorescent means 'gives up water'.

Some anhydrous compounds absorb water from the air. These compounds are called **hygroscopic** compounds (see figure 10.10). Hygroscopic means 'absorbs water'.

Some hygroscopic compounds can absorb so much water that if they are left in the air they actually liquefy. This is because they have dissolved in the absorbed water. These compounds are called **deliquescent** compounds. Deliquescent means 'becomes liquid'.

**Figure 10.10** *A hygroscopic sachet is often packed in with new radios. This stops the radio's parts becoming damaged by damp – the sachet absorbs any extra water in the atmosphere*

## 10.5 Hydrogen

Most of the hydrogen on Earth is combined with oxygen in the form of water.

## How is hydrogen made industrially?

Hydrogen is made from natural gas and water. The process is called **steam reforming**. Natural gas (mainly methane) and steam are passed over heated catalysts. Hydrogen and carbon dioxide are formed in the reaction.

$$\text{methane} + \text{water} \longrightarrow \text{hydrogen} + \text{carbon dioxide}$$
$$CH_4 + 2H_2O \longrightarrow 4H_2 + CO_2$$

## How is hydrogen made in the laboratory?

Hydrogen is made by allowing zinc metal to react with hydrochloric acid. The reaction is:

$$\text{zinc} + \text{hydrochloric acid} \longrightarrow \text{hydrogen} + \text{zinc chloride}$$
$$Zn + 2HCl \longrightarrow H_2 + ZnCl_2$$

One type of apparatus for collecting the gas is shown in figure 10.11.

**Figure 10.11** *Preparing hydrogen in the laboratory*

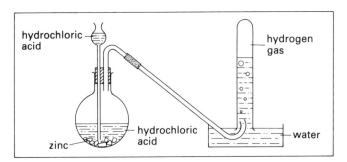

The gas that first collects in the test tube will be a mixture of hydrogen and air. Air and hydrogen mixtures like this are very explosive (figure 10.12). So the first test tube of gas should not be used.

**Figure 10.12** *The airship Hindenburg exploded because it was filled with hydrogen. Mixtures of hydrogen and air are highly explosive*

Table 10.7 shows some of the properties of hydrogen gas.

| |
|---|
| least dense (lightest) of all gases |
| colourless and has no smell |
| almost insoluble in water |
| neither acidic nor alkaline; it is neutral |
| burns in oxygen to form water |
| removes oxygen from some metal oxides |

**Table 10.7** *Some properties of hydrogen gas*

## The reaction of hydrogen with oxygen

If a lighted splint is placed in a test tube of hydrogen, there is a squeaky 'pop'. The pop is a small explosion that takes place when the hydrogen reacts with oxygen in the air around the tube. The product of the reaction is water. This 'popping' reaction is often used as a test for hydrogen.

The water formed when hydrogen reacts with oxygen can be collected in the apparatus shown in figure 10.13.

**Figure 10.13** *Hydrogen burns to form water*

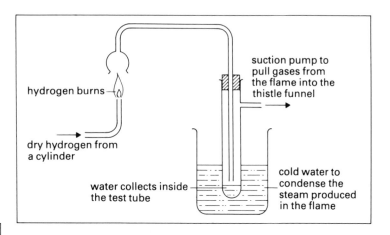

## The reaction of hydrogen with some metal oxides

Hydrogen is able to remove oxygen from some metal oxides. For example, it will remove the oxygen from hot copper oxide. Copper metal and steam are produced.

copper oxide + hydrogen ⟶ copper metal + steam

$$CuO \quad + \quad H_2 \quad \longrightarrow \quad Cu \quad + \quad H_2O$$

Hydrogen will react with lead oxide in the same way to produce lead metal. A suitable apparatus for this reaction is shown in figure 10.14.

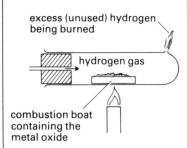

**Figure 10.14** *The reaction of a metal oxide with hydrogen*

## Summary: Water and hydrogen

* Water is the commonest chemical compound.
* Water has the formula $H_2O$.
* Cobalt chloride paper can be used to detect water. The paper turns from blue to pink.
* Water takes part in a continuous cycle: sea $\longrightarrow$ rain $\longrightarrow$ land $\longrightarrow$ rivers $\longrightarrow$ sea. This is called the water cycle.
* Tap water is made safe by filtering and chlorinating it.
* Calcium compounds dissolved in water make the water hard and prevent soap from lathering.
* There are two types of hardness of water: temporary hardness and permanent hardness.
* There are several ways of softening hard water including:
    1 boiling,
    2 adding calcium hydroxide,
    3 adding sodium carbonate,
    4 using an ion-exchange resin,
    5 distillation.
* A saturated solution is one which contains as much solute as it can at that temperature.
* Most substances are more soluble at higher temperatures.
* Solubility is expressed in grams of solute per 100 grams of solvent.
* Hydrated compounds contain water.
* Anhydrous compounds do not contain water.
* Efflorescent compounds lose their water of hydration.
* Hygroscopic compounds absorb moisture from the air.
* Deliquescent compounds can absorb so much water from the air that they form solutions.
* Hydrogen is a very light, highly flammable gas.
* A test for hydrogen is to put a lighted splint in a test tube of the gas. It gives a squeaky 'pop'.
* Hydrogen burns to form water.
* Hydrogen can convert some metal oxides into metals.

## Questions

1 Choose words from the list below to help you complete the following sentences:
    anhydrous
    calcium hydrogencarbonate
    cobalt chloride
    ion-exchange resin
    chlorine

(a) Water can be softened using an _____ .

(b) Substances that contain no water are said to be _____ .

(c) Any bacteria present in tap water are killed by adding _____ .

(d) Limestone rocks react with rain water forming

_____ .

(e) You can test to see if water is present by using

_____ .

2 This question is about an experiment in which the hardness of different samples of water was investigated. A definite amount of water was measured out, and then soap was added until a good lather was obtained. In some of the experiments the water was boiled for a few minutes before adding the soap. The results are shown below.

| Type of water | Amount of water used (cm$^3$) | Amount of soap needed (cm$^3$) |
|---|---|---|
| X | 10 | 2 |
| Y | 10 | 8 |
| Z | 20 | 12 |
| boiled X | 10 | 2 |
| boiled Y | 10 | 3 |
| boiled Z | 20 | 12 |

(a) One of the samples of water was distilled water. Was X, Y or Z the distilled water?
(b) Which of the unboiled samples of water was hardest? Explain how you arrived at this answer.
(c) Which sample of water has temporary hardness? Explain your answer.
(d) Which sample of water has permanent hardness? Explain your answer.
(e) What chemical could you have used to remove both temporary and permanent hardness from the water?

3 A class of pupils was asked to investigate the solubility of potassium chloride. Each group used the method described in section 10.4 to find the solubility at some particular temperature. One group of pupils recorded their results as follows:

| | | |
|---|---|---|
| Mass of basin + solution | = | 76°C |
| Mass of empty basin | = | 36.0 g |
| Mass of basin + solution | = | 70.8 g |
| Mass of basin + solute | = | 45.8 g |

(a) What mass of potassium chloride was left in the basin?

(b) What mass of water was boiled away in the experiment?

(c) Find the solubility of potassium chloride at 76 °C, in grams of KCl per 100 g of water.

4 (a) Using data from table 10.6, plot the solubility curve for ammonium chloride.

(b) What is the maximum mass of ammonium chloride that can dissolve in 50 cm³ of water at (i) 100 °C, (ii) 20 °C?

(c) What mass of ammonium chloride will be precipitated when 50 cm³ of a saturated solution at 100 °C is cooled to 20 °C?

(d) What is the solubility of ammonium chloride at 50 °C?

---

# Water pollution

In the past waterborne diseases such as cholera, typhoid and dysentery were quite common. They were usually caused by seepage of sewage into wells used for drinking water. For example, in 1854 over 600 people using the Broad Street well in London died as a result of a sewage seepage. Fortunately, improvements to sewers and chlorination of drinking water have largely removed such risks.

But although our water is free from infectious disease, it is not free from pollutants. Treated sewage and industrial and agricultural pollutants all affect the quality of our water.

Good water contains about 10 p.p.m. (parts per million) of dissolved oxygen. If this oxygen level drops then the water will no longer support aquatic life such as fish.

Adding sewage or certain industrial wastes to water usually results in a drop in oxygen content. What happens is that certain organisms in the water feed upon the waste. At the same time they consume oxygen from the water. Such water is said to have a high **biochemical oxygen demand** (BOD). The oxygen may be used up faster than fresh oxygen can dissolve. The BOD of different waters can be compared by adding a few drops of potassium manganate(VII) to a full, stoppered bottle of the water. Once the water's dissolved oxygen has been used up the pinkish colour of the manganate(VII) fades.

Another pollutant that gets into our water is nitrate, $NO_3^-$. In the stomach nitrates are converted into nitrites, which can combine with the oxygen-carrying part of the blood (haemoglobin). The body may become short of oxygen for respiration. This is especially dangerous for young babies.

It is estimated that about 30 per cent of the nitrate in our water comes from artificial fertilizers being washed by rain into the rivers. This represents about 5 to 10 per cent of the total amount of fertilizer used. Most of the remaining nitrate is thought to come from the decay of natural organic materials. The graph reveals an increasing use of artificial fertilizers containing nitrogen.

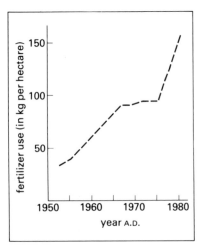

*In the United Kingdom, nitrogen fertilizers are being applied to the soil at increasing rates*

Insecticides are another source of pollution. DDT a
similar compounds seemed a huge benefit when they wer
developed during the Second World War. Many of them
now banned, however. One problem is that they dissolve in
fats better than they do in water. As a result they tend to
build up along food chains. For instance, concentrations of the
insecticides DDT have been studied in the water and the
wildlife of Clear Lake, California. Some results are as follows:

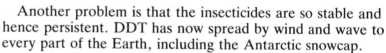

water (0.02 p.p.m.)

↓

plankton (50 p.p.m.)

↓

non-predatory fish (40–1000 p.p.m.)

↓

predatory fish (80–2500 p.p.m.)

Another problem is that the insecticides are so stable and
hence persistent. DDT has now spread by wind and wave to
every part of the Earth, including the Antarctic snowcap.

Metal pollutants show similar concentration effects along
food chains. Following a serious leak of mercury compounds
into the sea at Minamata, Japan, it was found that shrimps
contained about 0.05 p.p.m. mercury whereas sharks and
swordfish contained from 2000 to 16000 p.p.m. mercury.
Clearly care is needed when considering what levels of
pollution are acceptable.

*Skin damage to a fish living in
chemically polluted water*

## Questions

**5** What does the word *pollutant* mean?

**6** Is water with a high BOD likely to be pure or polluted?
Explain your answer.

**7** Why are full stoppered bottles of water used when finding
BOD values?

**8** Fertilizer manufacturers recommend an optimum (best)
quantity of nitrogen fertilizers. What problems can arise if
too much is used?

**9** Fertilizers cause nitrate pollution. Look carefully at the
data provided and decide whether you think these artificial
fertilizers should be banned. Explain your reasons.

**10** From the concentration of mercury in shrimps and
swordfish at Minamata, comment upon the position of
these creatures in the food chain.

**11** Why is care needed in deciding upon acceptable levels of
pollution?

# 11 The periodic table

The world contains many millions of different substances. To remember what each individual substance is like would be impossible. What we can do is group similar substances together. This can make remembering much easier. This chapter is about how chemists divide up the elements into groups.

## 11.1 Metals and non-metals

To start with, we can divide the elements into **metals** and **non-metals**. Most metals have a lot in common with one another. Just knowing that an element is a metal tells us a lot about it. Some of the properties of metals are shown in table 11.1.

At first glance, non-metals seem to be much less alike. For example, bromine is a red liquid, sulphur a yellow solid

**Table 11.1** *Properties of metals and non-metals*

| Some properties of metals | Some properties of non-metals |
|---|---|
| Metals are lustrous (shiny). They usually have a silvery colour, although other colours are possible | Non-metals may be solids, liquids or gases. Some are black, others are colourless. Their appearances have little in common |
| They are good conductors of electricity | They do not usually conduct electricity (the only exception is a form of carbon known as graphite) |
| They are not brittle (they will not shatter into pieces if hit with a hammer). They are malleable (can be hammered into a different shape) | They are brittle when solid. They shatter when struck with a hammer |

| Some properties of metals | Some properties of non-metals |
|---|---|
| They are strong. They can usually hold heavy loads without breaking | Non-metals are not usually very strong |
| They usually have high melting points | Some non-metals have very high melting points. Others have very low melting points |
| Many are very dense. Even a small lump of metal can be quite heavy | They are much less dense than metals |
| Many react with dilute acids to give hydrogen gas. This burns with a squeaky 'pop' | They do not react with dilute acids |
| They usually react with oxygen in the air. Some, like magnesium, burn with a bright flame. Others, like iron, only react very slowly. Such reactions produce metal oxides. Metal oxides are solids. Those metal oxides that dissolve in water form alkaline solutions | They may react with oxygen in the air. Many non-metal oxides dissolve in water to give an acidic solution |

and oxygen a colourless gas. But despite their different appearances, they do have much in common.

You should try to remember as much of the information in table 11.1 as you can. You will then be able to say a lot about an element simply by knowing if it is a metal or not.

## The names of metals

There are over one hundred elements. How can you remember which ones are metals? Fortunately, this is easier than you might think. You can tell most of them from their names.

The names of most metals end in **-ium**. Examples are aluminium, magnesium and sodium. There are actually sixty-two elements that have names ending in -ium. Sixty of these are metals. The two exceptions are helium and selenium, both of which are non-metals. There are, of

| cobalt | mercury |
|---|---|
| copper | nickel |
| gold | platinum |
| iron | silver |
| lead | tin |
| manganese | zinc |

Table 11.2 *Some metals with names that don't end in -ium*

course, other metals with names that do not end in -ium. The most common ones are listed in table 11.2.

## 11.2 Why we need the periodic table

We know that metals have many properties in common. Does this mean that they are all exactly alike? The answer to this question is clearly 'no'. Iron rusts, but gold doesn't. Magnesium burns, but platinum doesn't. In table 11.3 we can see how four different metals react with some chemicals.

But how can you remember the reactions of these four metals? One way is to divide them into groups. Copper and nickel have similar reactions. Magnesium and calcium are also rather like one another.

### Attempts to group elements together

The German chemist Döbereiner was one of the first to notice that there are **groups of elements** with very similar properties. Some of his groups, or triads as he called them, are shown in figure 11.1. If you know what sodium is like, you would also have a very good idea of what lithium and potassium are like.

|  | Copper (Cu) | Nickel (Ni) | Calcium (Ca) | Magnesium (Mg) |
|---|---|---|---|---|
| **Effect of hydrochloric acid** | no reaction | no visible reaction | gets hot and gives off hydrogen gas | gets hot and gives off hydrogen gas |
| **Effect of heating in air** | blackens | darkens | burns with a bright red flash | burns with a brilliant white flame |
| **Effect of nitric acid** | gives off a gas and leaves a blue solution | gives off a gas and leaves a green solution | gives off a gas and leaves a colourless solution | gives off a gas and leaves a colourless solution |

Table 11.3 *The reactions of some metals*

**Figure 11.1** *Döbereiner's triads*

| | | |
|---|---|---|
| lithium | calcium | chlorine |
| sodium | strontium | bromine |
| potassium | barium | iodine |

At the time Döbereiner was working, chemists knew that elements contained only one sort of atom. They also knew that some atoms were heavier than others. This led them to arrange the elements in a list according to the **masses** of their atoms (as in figure 11.2). When this is done an interesting pattern emerges. (We have to swap the positions of just a few elements to get a perfect pattern.)

**Figure 11.2** *Elements listed according to the masses of their atoms*

⟵ light atoms    H He Li Be B C N O F Ne Na Mg Al Si P S Cl Ar K Ca Sc Ti V   heavier atoms ⟶

**Figure 11.3** *The Russian chemist Mendeleev first arranged the elements in order in the periodic table*

**Figure 11.4** *The arrangement of the elements in the periodic table*

If we begin at lithium and count eight elements along the list, we come to sodium. Now count another eight elements along. We arrive at potassium. We know that lithium, sodium and potassium all have similar properties.

In the same way, counting eight elements along from helium brings us to neon. Eight elements further along brings us to argon. Helium, neon and argon are elements that are very much alike.

An English chemist, Newlands, was the first to notice that similar elements occur periodically (at regular intervals) along this list.

The Russian chemist Mendeleev 'chopped up' the list and arranged it so that similar elements came underneath each other. Mendeleev's arrangement became known as the **periodic table**. It is shown in figure 11.4. The number shown

| 1<br>H | | | | | | | | | | | | | | | | | 2<br>He |
|---|---|---|---|---|---|---|---|---|---|---|---|---|---|---|---|---|---|
| 3<br>Li | 4<br>Be | | | | | | | | | | | 5<br>B | 6<br>C | 7<br>N | 8<br>O | 9<br>F | 10<br>Ne |
| 11<br>Na | 12<br>Mg | | | | | | | | | | | 13<br>Al | 14<br>Si | 15<br>P | 16<br>S | 17<br>Cl | 18<br>Ar |
| 19<br>K | 20<br>Ca | 21<br>Sc | 22<br>Ti | 23<br>V | 24<br>Cr | 25<br>Mn | 26<br>Fe | 27<br>Co | 28<br>Ni | 29<br>Cu | 30<br>Zn | 31<br>Ga | 32<br>Ge | 33<br>As | 34<br>Se | 35<br>Br | 36<br>Kr |
| 37<br>Rb | 38<br>Sr | 39<br>Y | 40<br>Zr | 41<br>Nb | 42<br>Mo | 43<br>Tc | 44<br>Ru | 45<br>Rh | 46<br>Pd | 47<br>Ag | 48<br>Cd | 49<br>In | 50<br>Sn | 51<br>Sb | 52<br>Te | 53<br>I | 54<br>Xe |
| 55<br>Cs | 56<br>Ba | 57<br>La | 58 to 71 | 72<br>Hf | 73<br>Ta | 74<br>W | 75<br>Re | 76<br>Os | 77<br>Ir | 78<br>Pt | 79<br>Au | 80<br>Hg | 81<br>Tl | 82<br>Pb | 83<br>Bi | 84<br>Po | 85<br>At | 86<br>Rn |
| 87<br>Fr | 88<br>Ra | 89<br>Ac | | | | | | | | | | | | | | | |

above each element is its position in the list of elements arranged according to the masses of their atoms. This is called its **atomic number**.

## 11.3 How to use the periodic table

### What is the use of the periodic table?

Elements that are similar appear in the same part of the periodic table. The elements most alike are in the same downward column. There are also definite trends (gradual changes) across the table and down the columns. Once you know these trends you can predict the properties of almost any element.

### How is the periodic table labelled?

Sometimes you need to be able to say where an element is in the periodic table. To help you do this the rows and downward columns are numbered, as shown in figure 11.5.

| | groups | | | | | | | | | | | | | | | | groups | | | | | |
|---|---|---|---|---|---|---|---|---|---|---|---|---|---|---|---|---|---|---|---|---|---|---|
| | 1 | 2 | | | | | | | | | | | | | 3 | 4 | 5 | 6 | 7 | 0 |
| period 1 | H | | | | | | | | | | | | | | | | | | | | He |
| period 2 | Li | Be | | | | | | | | | | | | | B | C | N | O | F | Ne |
| period 3 | Na | Mg | | | | transition metals | | | | | | | | | Al | Si | P | S | Cl | Ar |
| period 4 | K | Ca | Sc | Ti | V | Cr | Mn | Fe | Co | Ni | Cu | Zn | Ga | Ge | As | Se | Br | Kr |
| period 5 | Rb | Sr | Y | Zr | Nb | Mo | Tc | Ru | Rh | Pd | Ag | Cd | In | Sn | Sb | Te | I | Xe |
| period 6 | Cs | Ba | La | Hf | Ta | W | Re | Os | Ir | Pt | Au | Hg | Tl | Pb | Bi | Po | At | Rn |
| period 7 | Fr | Ra | Ac | | | | | | | | | | | | | | | |

**Figure 11.5** *How the periodic table is labelled*

The downward columns are known as **groups**. The first column is group 1, the second is group 2 and so on as shown. Between groups 2 and 3 there are some elements known as the **transition metals**. Transition means 'in between'. The last group in the table is called group 0. The rows across the table are known as **periods**. These are numbered from 1 to 7 going down the table.

## What are the other numbers on the periodic table?

Figure 11.6 is another copy of the periodic table. This has some extra information on it.

The number below the symbol for each element is its **relative atomic mass** (see section 7.1). Relative atomic mass is a measure of how heavy an atom is.

**Figure 11.6** *The periodic table*

| 1 | | | | | | | | | | | | | | | | | 2 |
|---|---|---|---|---|---|---|---|---|---|---|---|---|---|---|---|---|---|
| H | | | | | | | | | | | | | | | | | He |
| 1 | | | | | | | | | | | | | | | | | 4 |
| 3 | 4 | | | | | | | | | | | 5 | 6 | 7 | 8 | 9 | 10 |
| Li | Be | | | | | | | | | | | B | C | N | O | F | Ne |
| 7 | 9 | | | | | | | | | | | 11 | 12 | 14 | 16 | 19 | 20 |
| 11 | 12 | | | | | | | | | | | 13 | 14 | 15 | 16 | 17 | 18 |
| Na | Mg | | | | | | | | | | | Al | Si | P | S | Cl | Ar |
| 23 | 24 | | | | | | | | | | | 27 | 28 | 31 | 32 | 35.5 | 40 |
| 19 | 20 | 21 | 22 | 23 | 24 | 25 | 26 | 27 | 28 | 29 | 30 | 31 | 32 | 33 | 34 | 35 | 36 |
| K | Ca | Sc | Ti | V | Cr | Mn | Fe | Co | Ni | Cu | Zn | Ga | Ge | As | Se | Br | Kr |
| 39 | 40 | 45 | 48 | 51 | 52 | 55 | 56 | 59 | 59 | 63.5 | 65.4 | 70 | 72.6 | 75 | 79 | 80 | 84 |
| 37 | 38 | 39 | 40 | 41 | 42 | 43 | 44 | 45 | 46 | 47 | 48 | 49 | 50 | 51 | 52 | 53 | 54 |
| Rb | Sr | Y | Zr | Nb | Mo | Tc | Ru | Rh | Pd | Ag | Cd | In | Sn | Sb | Te | I | Xe |
| 85.5 | 87.6 | 89 | 91 | 93 | 96 | 99 | 101 | 103 | 106.4 | 108 | 112 | 115 | 119 | 122 | 127.6 | 127 | 131 |
| 55 | 56 | 57 | 58 | 72 | 73 | 74 | 75 | 76 | 77 | 78 | 79 | 80 | 81 | 82 | 83 | 84 | 85 | 86 |
| Cs | Ba | La | to | Hf | Ta | W | Re | Os | Ir | Pt | Au | Hg | Tl | Pb | Bi | Po | At | Rn |
| 133 | 137 | | 71 | | 181 | 184 | 186 | 190 | 192 | 195 | 197 | 201 | 204 | 207 | 209 | 210 | 210 | 222 |
| 87 | 88 | 89 | | | | | | | | | | | | | | | |
| Fr | Ra | Ac | | | | | | | | | | | | | | | |
| 223 | 226 | 228 | | | | | | | | | | | | | | | |

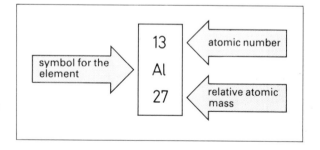

**Figure 11.7** *Where atomic number and relative atomic mass appear in the periodic table in figure 11.6*

## Where do metals and non-metals come in the periodic table?

Non-metals are on the right of the periodic table. Metals are found on the left and in the middle.

A dividing line between metals and non-metals is drawn on many copies of the periodic table (see figure 11.8, over the page). It is towards the right-hand side and runs down the table rather like a staircase. Elements on the right of this line are non-metals. Elements on the left are metals.

| | | | | | | | | | | | | | | | | | |
|---|---|---|---|---|---|---|---|---|---|---|---|---|---|---|---|---|---|
| 1<br>H | | | | | | | | | | | | | | | | | 2<br>He |
| 3<br>Li | 4<br>Be | | | | | | | | | | | 5<br>B | 6<br>C | 7<br>N | 8<br>O | 9<br>F | 10<br>Ne |
| 11<br>Na | 12<br>Mg | | | | | | | | | | | 13<br>Al | 14<br>Si | 15<br>P | 16<br>S | 17<br>Cl | 18<br>Ar |
| 19<br>K | 20<br>Ca | 21<br>Sc | 22<br>Ti | 23<br>V | 24<br>Cr | 25<br>Mn | 26<br>Fe | 27<br>Co | 28<br>Ni | 29<br>Cu | 30<br>Zn | 31<br>Ga | 32<br>Ge | 33<br>As | 34<br>Se | 35<br>Br | 36<br>Kr |
| 37<br>Rb | 38<br>Sr | 39<br>Y | 40<br>Zr | 41<br>Nb | 42<br>Mo | 43<br>Tc | 44<br>Ru | 45<br>Rh | 46<br>Pd | 47<br>Ag | 48<br>Cd | 49<br>In | 50<br>Sn | 51<br>Sb | 52<br>Te | 53<br>I | 54<br>Xe |
| 55<br>Cs | 56<br>Ba | 57<br>La | 58<br>to<br>71 | 72<br>Hf | 73<br>Ta | 74<br>W | 75<br>Re | 76<br>Os | 77<br>Ir | 78<br>Pt | 79<br>Au | 80<br>Hg | 81<br>Tl | 82<br>Pb | 83<br>Bi | 84<br>Po | 85<br>At | 86<br>Rn |
| 87<br>Fr | 88<br>Ra | 89<br>Ac | | | | | | | | | | | | | | | |

☐ metal

■ non-metal

**Figure 11.8** *Metals and non-metals in the periodic table*

Elements near this dividing line are often called **metalloids**. They are like metals in some ways and like non-metals in others.

### Where are the most reactive elements in the table?

Metals get more reactive going down the groups. They get less reactive going across periods from left to right. So the most reactive metal of all is francium (symbol Fr).

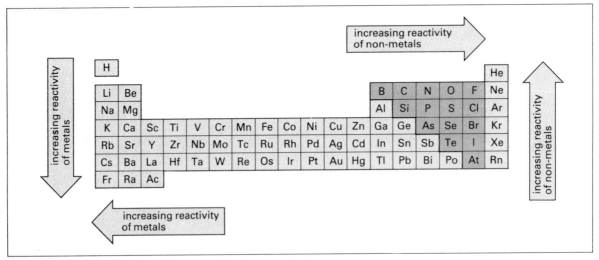

**Figure 11.9** *The reactivities of the elements*

Non-metals get less reactive going down the groups. They get more reactive going across the periods from left to right, except for group 0. (The group 0 elements do not normally react at all.) So the most reactive non-metal is fluorine (symbol F).

## 11.4 Similarities and trends within groups of the periodic table

We have already seen that there are many similarities between the elements in any group. There are also differences. Where such differences appear, they follow a pattern or trend (gradual change). This idea of trends within groups will be clearer if we look at some examples.

### Group 0 – the noble gases

The elements at the extreme right of the periodic table are in group 0. This whole group of elements is very unreactive. It was 1963 before anyone managed to get any of these elements to take part in a chemical reaction. Many chemists call this group the **noble gases**. They are called noble because they do not react easily with the other 'common' elements.

Some information about these gases is given in table 11.4.

| Name | Symbol | Boiling point (°C) | Density (g/dm$^3$) |
|---|---|---|---|
| helium | He | −269 | 0.17 |
| neon | Ne | −246 | 0.84 |
| argon | Ar | −186 | 1.66 |
| krypton | Kr | −152 | 3.46 |
| xenon | Xe | −108 | 5.45 |
| radon | Ra | −62 | 8.90 |

**Table 11.4** *Some properties of the noble gases*

All the noble gases have boiling points well below room temperature. So they are all gases. If you look closer at the boiling points, you should be able to see a regular trend. Imagine that the boiling point of radon had not been given. Would you have been able to predict that it would be about −60 °C?

We can also see a trend for density. If the density of the gas xenon had been left out, could you have estimated it?

You might think that because the noble gases are so unreactive, they are not of any use. But it is their lack of reactivity that makes them useful (see figures 11.10 and 11.11).

**Figure 11.10** *Electric light bulbs are filled with noble gases, usually argon. Even when the filament is white-hot, argon will not react with it*

**Figure 11.11** *Helium is a very light gas. Because it is so unreactive, it is safe to use in airships*

## Group 1 – the alkali metals

These are found on the extreme left of the periodic table. The members of this family of elements are:

lithium      Li
sodium       Na
potassium    K
rubidium     Rb
caesium      Cs
francium     Fr

Sodium and potassium are the most common. Francium is very rare. They are all metals and so they all conduct heat and electricity well.

They are lustrous when freshly cut, but soon lose their shine. This is because they corrode quickly by reacting with the air. Like the noble gases, they show similar patterns in their physical properties.

Their melting points and boiling points are shown in figure 11.12.

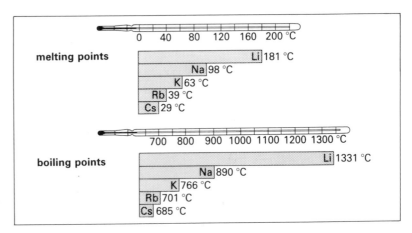

**Figure 11.12** *The melting points and boiling points of the group I elements*

These elements also have similar chemical properties. Their properties show trends.

The elements of this group are called the 'alkali metals' because they all react with water, giving off hydrogen gas and forming an alkaline solution. The diagram equation below shows what happens when sodium reacts with water.

sodium  +  water  $\longrightarrow$  sodium hydroxide  +  hydrogen

2Na  +  2H$_2$O  $\longrightarrow$  2NaOH  +  H$_2$

Similarly, lithium reacts with water to give lithium hydroxide solution. Potassium gives potassium hydroxide solution.

Although their reactions are very similar, they are not identical. This is shown in table 11.5.

| Element | How it reacts |
|---------|---------------|
| lithium | floats, producing a steady stream of hydrogen gas |
| sodium | darts across the water, rapidly producing hydrogen gas |
| potassium | reacts violently, producing hydrogen gas which catches fire |
| rubidium | reacts more violently than potassium does |
| caesium | reacts more violently than rubidium does |

**Table 11.5** *Reactions of the group 1 elements with water*

As you go down the group, each element reacts more strongly with water. This increase in reactivity is found in other reactions with these elements.

## Group 7 – the halogens

This group comes last but one in the periodic table. The elements in the group are shown in table 11.6 (over the page).

Astatine is extremely rare. It has to be made artificially. So little is made at any one time that no one knows what it looks like.

The other halogens are quite common. They are all non-metals and their physical properties show regular trends. Their melting points and boiling points are shown in figure 11.13.

| Element | Symbol | Appearance |
|---------|--------|------------|
| fluorine | F | yellow gas |
| chlorine | Cl | yellowy-green gas |
| bromine | Br | dark red liquid |
| iodine | I | shiny very dark solid |
| astatine | At | unknown |

**Table 11.6** *The group 7 elements – the halogens*

**Figure 11.13** *The melting points and boiling points of the halogens*

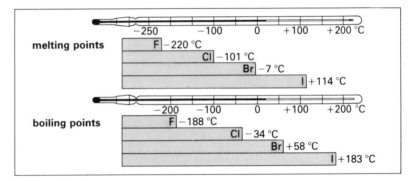

The halogens are the most reactive non-metals. Their reactions get less strong going down the group. The equation below shows what happens when the group 1 element sodium reacts with the group 7 element chlorine.

$$\text{sodium} + \text{chlorine} \longrightarrow \text{sodium chloride}$$
$$2\text{Na} + \text{Cl}_2 \longrightarrow 2\text{NaCl}$$

The sodium and chlorine join together to form the compound sodium chloride (common salt). We can write similar equations for the reaction of sodium with the other group 7 elements. Sodium reacts with bromine to give sodium bromide and with iodine to give sodium iodide.

The compounds formed when metals react with these elements are all **salts** (see section 19.3). The group 7 elements are called 'the halogens' because they form salts so easily. 'Halogen' means 'salt maker'.

The halogens also react with many non-metals. The reaction between hydrogen and chlorine to give hydrogen chloride is shown below:

$$\text{hydrogen} + \text{chlorine} \longrightarrow \text{hydrogen chloride}$$

$$\text{H}_2 + \text{Cl}_2 \longrightarrow 2\text{HCl}$$

We could draw similar diagrams to represent the reactions between hydrogen and fluorine, bromine or iodine. The reactions of the halogens with hydrogen are compared in table 11.7.

| Element | Reaction with hydrogen | |
|---------|------------------------|---|
| fluorine | reacts instantly even at $-200\,°C$ | reactivity decreases |
| chlorine | reacts fairly slowly in the dark at room temperature; explodes in sunlight | |
| bromine | needs heating to $+200\,°C$ in order to react | |
| iodine | does not react completely even at $500\,°C$ | |

**Table 11.7** *The reactions of the halogens with hydrogen*

From this table, we can see that fluorine is the most reactive of the halogens. This is true in all reactions. The reactivity of the halogens gets less as we go down the group.

## The transition elements

The **transition elements** are rather like a special 'group' in the periodic table. But they are not a proper group, because they do not form a downward column in the table. They are similar in many ways and also very different from the metals of groups 1 and 2. Some of these differences are:

1 they are much less reactive,
2 they have higher densities,
3 their compounds are usually coloured,
4 they have higher melting points and boiling points,
5 the elements and their compounds often act as catalysts (see section 27.1),
6 they are harder and stronger.

## How similar are elements within a period?

This section has been about elements in groups. You have seen that the elements in a group show similarities and trends.

The elements going across a period may be very different from each other. But there are still general patterns in their behaviour. These patterns are often more complicated than those found in a group.

One example of a fairly simple pattern is the reaction between hydrochloric acid and some elements from period 4. It is shown in table 11.8 (over the page).

| Element | Potassium | Calcium | Chromium | Iron | Copper, arsenic, bromine, krypton |
|---|---|---|---|---|---|
| Reaction | explodes | reacts very violently | reacts slowly | reacts if heated | do not react |

**Table 11.8** *How some elements from period 4 react with dilute hydrochloric acid*

## Summary: The periodic table

★ The elements can be divided into metals and non-metals.
★ All metals are shiny and conduct electricity.
★ Most metals have names ending in -ium.
★ The elements in the periodic table are arranged in order of increasing atomic number.
★ Columns going down the periodic table are called groups.
★ Rows going across the periodic table are called periods.
★ Metals are on the left and in the middle of the periodic table.
★ Non-metals are on the right of the periodic table.
★ Metals get more reactive going down groups and less reactive going across periods from left to right.
★ Non-metals get less reactive going down groups and more reactive going across periods. The noble gases do not follow this rule.
★ Elements in the same group of the periodic table have similarities and trends in their physical and chemical properties.
★ The properties of the elements in a group change in a regular way.

## Questions

1 Rewrite the following, choosing the correct words from each of the pairs given in brackets:

Metals are all (good/poor) conductors of electricity. They are usually (brittle/malleable) and often have (high/low) melting points. Many react with (acids/alkalis) to give (hydrogen/oxygen) gas. Many will react with (oxygen/nitrogen) from the air. If a metal oxide dissolves in water, then it usually gives an (acid/alkaline) solution.

**2** Below you will find a 'skeleton' periodic table with ten elements in it. Choose from the elements given here to answer these questions.

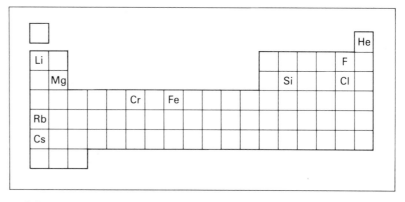

(a) Name three elements in the same period.
(b) Name three elements in the same group.
(c) Name two elements likely to form coloured compounds.
(d) Name the most reactive metal shown in this table.
(e) Name the most reactive non-metal shown in this table.
(f) Name a metalloid.
(g) Name the least reactive element.
(h) Name an element likely to act as a catalyst.
(i) Name a metal that burns with a brilliant white flame.

**3** The tables below show some data for elements in group 1 and group 4.

| Element | Melting point (°C) | Boiling point (°C) | Density (g/cm³) |
|---|---|---|---|
| lithium | 181 | 1331 | 0.53 |
| sodium | 98 | 890 | 0.97 |
| potassium | 63 | 766 | 0.86 |
| rubidium | 39 | 701 | 1.53 |
| caesium | 29 | 685 | 1.90 |
| francium | | | |

*Group 1 elements*

| Element | Melting point (°C) | Boiling point (°C) | Density (g/cm³) |
|---|---|---|---|
| carbon | 3570 | – | 2.22 |
| silicon | 1414 | 2355 | 2.33 |
| germanium | 958 | 2700 | 5.36 |
| tin | 232 | 2362 | 7.31 |
| lead | 328 | 1755 | 11.34 |

*Group 4 elements*

(a) Predict the melting point, boiling point and density of the group 1 element francium. Justify your predictions.
(b) Predict the chemical behaviour of francium with water. Justify your prediction and write an equation for the reaction.

(c) The elements in group 4 are much less alike than the elements in group 1. To what extent would you say the physical properties of these elements follow any patterns? (You may wish to draw graphs to help here.)

(d) Carbon is a non-metal. Lead is a metal. Try and find information about these and other elements in the group and their reactions. Use this to discuss whether or not there is a gradual change from non-metal to metal going down the group.

## The search for anti-knock

Inside the cylinder of a car engine, a mixture of petrol vapour and air burns. There are different grades of petrol. Good petrol burns smoothly inside the engine. Low-grade petrol tends to explode rather than burn. This gives jerky engine movements and causes a 'knocking' noise. Besides giving less power, petrol that 'knocks' can also damage the engine.

Early in the history of the car an American named Midgely headed a research team looking for an 'anti-knock' additive for petrol. For almost five years they tested hundreds of different substances but with no real success. Then they gave up their trial and error methods and began matching their results to the periodic table. They had already found that iodine reduced knocking whereas bromine increased it. Selenium compounds also reduced knocking. They tried tellurium compounds and found that these were better than selenium. Very soon they were achieving remarkable success using other compounds of elements at the bottom of groups 4 and 5.

A compound called tetraethyl-lead (TEL) was particularly effective. Just one part of TEL in 2000 parts of petrol was enough to prevent knocking.

At that time there was no way of manufacturing TEL on a large scale. A method was developed in which chloroethane was made to react with a lead–sodium alloy. What happens is this:

chloroethane + lead–sodium ⟶ TEL + lead metal + sodium chloride
a gas ... a solid ... a liquid ... a solid ... a solid

The manufacture of TEL and its addition to petrol thus solved the problem of achieving smooth combustion. It did, however, bring a new problem: that of lead pollution of the atmosphere. Lead is now recognized as being dangerously poisonous. It seems to have especially serious effects on children.

*It is hoped that as lead-free petrol comes into wider use, lead contamination of street dust and air will diminish*

Fortunately chemists have now developed lead-free petrol that burns smoothly without TEL. The use of TEL is therefore being phased out. It is already banned in several countries.

### Questions

**4** Why were anti-knock additives used in petrol?

**5** Using elements from period 3, illustrate that elements get less metallic going across periods from left to right.

**6** Using elements from group 4, illustrate that elements get more metallic going down groups.

**7** Selenium compounds reduced knocking in petrol.
(a) Why did Midgely's team then investigate tellurium compounds rather than sulphur compounds?
(b) Explain why they were soon investigating compounds of the elements from the bottom of groups 4 and 5.

**8** The manufacture of TEL (a volatile liquid) produces a mixture of TEL, lead metal and salt. (Salt does not dissolve in TEL.)
(a) Explain with words and diagrams how you might prepare and separate pure TEL in the laboratory.
(b) What safety precautions would you need to take when handling (i) TEL, and (ii) lead–sodium alloy?

**9** Try and change the following paragraph into simple English:

TEL and other organolead compounds are now known to affect the central nervous system. At quite low levels they can cause behavioural problems in children and at higher levels they cause insomnia, nausea and hallucinations. Still higher concentrations can be lethal.

# 12 Solids, liquids and gases

**Figure 12.1** *Tiny particles leave the hot loaves and spread into the air. This is why you can smell the bread even while you are outside the shop*

Scientists believe that all substances are made of very tiny particles. Not bits of solid like grains of sand, but particles that are much, much smaller. These are atoms, ions and molecules. We cannot see them even with the most powerful microscopes. When these particles move about, they make substances change from solids to liquids or from liquids to gases. This chapter is about these particles and their movement.

## 12.1 Why believe in particles?

Here are three pieces of evidence for the existence of particles.

### Diffusion

If you walk past a baker's shop you may be able to smell the fresh bread inside. Tiny particles must be leaving the loaves and spreading out in the air. It is difficult to think of any other explanation.

Many people add a little salt to vegetables before boiling them. How can so little salt spread so evenly through the food? It must be that tiny salt particles **diffuse** (spread out) through the water and food in the pan.

If you use a coloured solid, you can actually watch this diffusion happening (see figure 12.2).

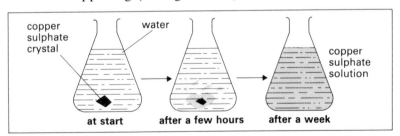

**Figure 12.2** *Diffusion of a coloured solid into a liquid*

A coloured food dye will spread out, or diffuse, into a liquid. Try adding some orange food dye to some lemon squash. Before long it looks like orange squash. You can even try this on someone. Few people notice that the taste is not orange!

We can explain all these processes if we believe that substances are made of small particles that can move around. The smaller they are the faster they move. Increasing the temperature also speeds up particles.

## Crystals and crystallizing

Copper sulphate is a substance that can be grown in large crystals. You just hang a small crystal in a saturated copper sulphate solution (see figure 12.3). Over a few weeks, the crystal gets bigger and bigger.

**Figure 12.3** *Growing a large copper sulphate crystal*

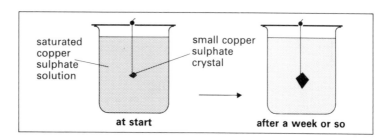

Particles of copper sulphate which were dissolved in the solution must be leaving it and joining the crystal.

Many other substances, including sodium chloride, also form crystals (see figure 12.4).

**Figure 12.4** *(a) Crystals of sodium chloride, common salt. The crystals always grow in a cube shape. (b) The structure of salt*

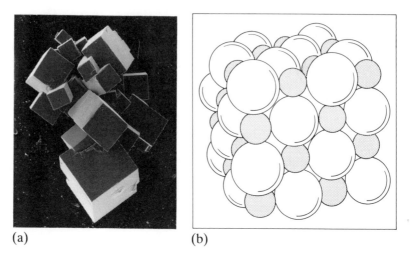

(a)  (b)

## Brownian motion

In 1827 Robert Brown was looking at pollen grains suspended in water. Through his microscope he noticed that the grains were moving about in a random (haphazard) way. This movement is now called **Brownian motion** (see figure 12.5).

The pollen grains move because water particles are constantly bumping into them. The pollen grains are much bigger than the water particles. If the same number of water molecules hit opposite sides of a grain, it will not move. But

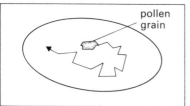

**Figure 12.5** *Brownian motion. The arrow shows the way the pollen grain moves*

**Figure 12.6** *Why Brownian motion happens*

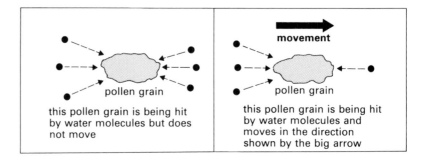

this pollen grain is being hit by water molecules but does not move

this pollen grain is being hit by water molecules and moves in the direction shown by the big arrow

if more molecules hit one side than the other, the pollen grain will move (figure 12.6).

## 12.2  Changes of state

### Melting

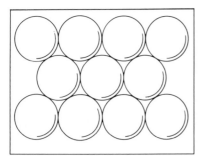

**Figure 12.7** *Particles are arranged in a regular pattern in a solid at −273 °C. A regular arrangement of atoms is called a lattice*

The coldest temperature that can ever be reached is −273 °C. At this temperature even substances like petrol and oxygen are solids. Suppose there was some way of seeing the particles in such a cold solid. What would they look like? You would see that they were arranged in a regular pattern with not much space in between them. They would also be completely still (see figure 12.7).

If you warmed the solid up from −273 °C, the particles would start to move. They would vibrate, slowly at first, and then more violently as the temperature went up.

At a certain temperature, the particles vibrate so fast that the regular pattern of the structure breaks down. This temperature is the **melting point**. It takes some time before the entire structure collapses and the solid becomes a liquid.

### Boiling

**Figure 12.8** *How a solid changes on heating first into a liquid and then into a gas*

The particles in a liquid move about quickly from place to place. There is no regular pattern, but the particles are still quite close together. As the temperature is increased, the

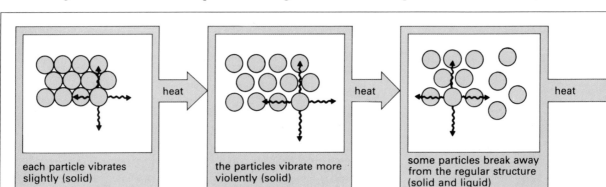

each particle vibrates slightly (solid)

the particles vibrate more violently (solid)

some particles break away from the regular structure (solid and liquid) particles move faster and faster. Some have enough energy to escape from the liquid and form a gas. At a certain temperature, large numbers of particles have enough energy to escape. This temperature is the **boiling point**. The particles in the gas move very quickly. There are large spaces in between them.

## Condensing and solidifying

When a gas is cooled down, the exact opposite of the above changes take place. The particles move more and more slowly. At the boiling point they come together to form a liquid. When they do this, the gas has **condensed**.

When you cool a liquid, the particles move even more slowly. At the melting point they arrange themselves in a regular pattern. This is when the liquid has **solidified**.

## Sublimation

Most solids melt when heated and solidify again when cooled. A few behave differently. When heated they change straight from a solid to a gas. This rapid spreading out of the particles is called **sublimation**. Cooling brings about the

**Figure 12.9** *Solid carbon dioxide, 'dry ice', sublimes: when warmed it changes straight into a cloudy gas*

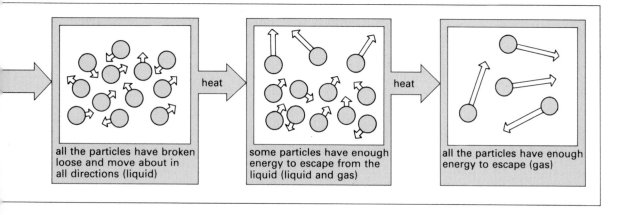

all the particles have broken loose and move about in all directions (liquid)

*heat*

some particles have enough energy to escape from the liquid (liquid and gas)

*heat*

all the particles have enough energy to escape (gas)

opposite process. The gas changes directly to a solid that is called a **sublimate**. (You read about this in section 4.3.) Iodine behaves in this way. When heated it changes straight from a solid to a purple gas. When you cool this gas, you get the dark solid again.

Another common example of a substance that sublimes is carbon dioxide. At temperatures below $-78\,°C$, carbon dioxide is a white solid called 'dry ice'. When allowed to warm up to $-78\,°C$, it turns into a colourless gas (see figure 12.9). This makes it convenient to use in refrigeration.

### Summary of changes of state

Figure 12.10 summarizes what happens during changes of state.

**Figure 12.10** *Changes of state*

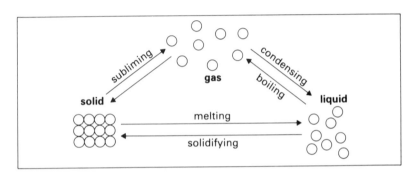

### Heating and cooling curves

When you heat a beaker of water, the temperature of the water rises fairly steadily up to $100\,°C$. Then it stops rising even if you carry on heating it (see figure 12.11). Where is the heat energy going?

In a liquid the particles are fairly close together. Somehow, the particles must be attracted to one another. When a liquid boils, the particles move far apart. This can only happen if enough energy is provided to overcome the attractions.

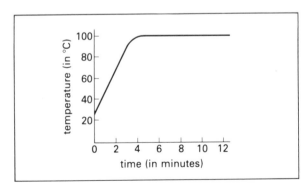

**Figure 12.11** *A heating curve for water*

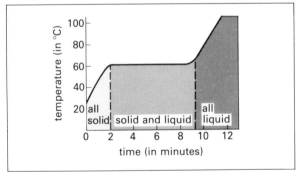

**Figure 12.12** *A heating curve for a solid near its melting point*

This is where the heat energy is going. The amount of heat needed to boil away a given amount of liquid is called the **latent heat of vaporization**. Its value tells you a lot about the forces between the particles. The larger the latent heat of vaporization, the more strongly the particles must be attracted to each other.

In the same way, if you want to melt a solid, you have to give it some extra energy. This energy is called the **latent heat of fusion**. The heating curve for a solid always stops rising at the melting point (see figure 12.12). This is because the latent heat of fusion is needed to separate the particles in the solid to form a liquid.

## 12.3 The properties of solids, liquids and gases

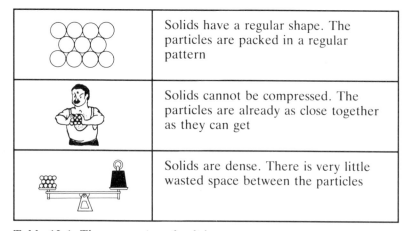

| | |
|---|---|
| | Solids have a regular shape. The particles are packed in a regular pattern |
| | Solids cannot be compressed. The particles are already as close together as they can get |
| | Solids are dense. There is very little wasted space between the particles |

**Table 12.1** *The properties of solids*

| | |
|---|---|
| | Liquids take up the shape of the bottom of whatever container they are put into. The particles are not in any fixed pattern |
| | It is easy to push solid objects through liquids. The particles of the liquid are free to move out of the way |
| | Liquids are very difficult to compress. The particles cannot get much closer |
| | Liquids are quite dense. There is not much empty space between the particles |

**Table 12.2** *The properties of liquids*

| | |
|---|---|
| | Gases have no shape. They fill up whatever space is available. The particles move rapidly about in all directions |
| | Gas particles are so easy to push out of the way that you often don't realize they are there |
| | Gases are easily compressed. The particles can easily be pushed much closer together |
| | Gases have very low densities. There are huge amounts of empty space between the particles |

**Table 12.3** *The properties of gases*

## 12.4 Calculations involving volumes of gases

The volume of a gas is usually measured more easily than its mass. Because of this we sometimes need to work in volumes.

> At room temperature and pressure a mole of any gas occupies a volume of $24 \, dm^3$ ($24\,000 \, cm^3$).

### Example 1

What volume of carbon dioxide is formed when 42 g of magnesium carbonate, $MgCO_3$, is heated?

1  $MgCO_3 \longrightarrow MgO + CO_2$

2  1 mole $\longrightarrow$ 1 mole + 1 mole

3  84 g $\longrightarrow$ 40 g + $24 \, dm^3$

4  | scale-down factor = 42/84                        | multiply by 42/84

   42 g                                              $12 \, dm^3$

i.e. $\frac{42}{84} \times 24 = 12 \, dm^3$

**Answer:** 42 g of $MgCO_3$ gives $12 \, dm^3$ of carbon dioxide gas.

## Example 2

Heating potassium chlorate(v), $KClO_3$, makes it decompose into oxygen and potassium chloride. What volume of oxygen, at room temperature and pressure, will be formed when 49 g of the chlorate is decomposed?

**1** $2KClO_3 \longrightarrow 2KCl + 3O_2$

**2** 2 moles $\longrightarrow$ 2 moles + 3 moles

**3** 245 g $\longrightarrow$ 149 g + $3 \times 24 = 72\,dm^3$

**4** $\downarrow$ scale-down factor = 49/245          $\downarrow$ multiply by 49/245

    49 g                                    $12\,dm^3$

**Answer:** 49 g of $KClO_3$ gives $12\,dm^3$ of oxygen gas.

## Example 3

What volume of hydrogen, at room temperature and pressure, is formed when 120 g of calcium reacts with water to form calcium hydroxide?

**1** $Ca + 2H_2O \longrightarrow H_2 + Ca(OH)_2$

**2** 1 mole + 2 moles $\longrightarrow$ 1 mole + 1 mole

**3** 40 g      36 g      $24\,dm^3$      74 g

**4** $\downarrow$ scale-up factor = 120/40      $\downarrow$ multiply by 120/40

    120 g                  $72\,dm^3$

**Answer:** 120 g of Ca reacts to give $72\,dm^3$ of hydrogen gas.

## Changes in volume

In many reactions involving gases, considerable changes in volume are observed. For instance, when 1 mole of ammonia completely decomposes the volume doubles. The equation for the reaction is:

| ammonia | | nitrogen | + | hydrogen |
|---|---|---|---|---|
| $2NH_3(g)$ | $\longrightarrow$ | $N_2(g)$ | + | $3H_2(g)$ |
| 2 moles | gives | 1 mole | + | 3 moles |
| $2 \times 24\,dm^3$ | gives | $24\,dm^3$ | + | $3 \times 24\,dm^3$ |
| So   $48\,dm^3$ | gives | | $96\,dm^3$ | |

that is, the volume will have doubled.

Of course, in this reaction the volume will double no matter how much we start off with. If $10\,cm^3$ of ammonia decomposes we will get $5\,cm^3$ of nitrogen and $15\,cm^3$ of hydrogen (total = $20\,cm^3$).

Indeed, in any reaction involving gases, the volumes of the gases concerned are always directly proportional to the numbers shown in the equation. In the decomposition of ammonia, for example, these are 2:1:3.

Some more examples will make this clear.

### *Example 4*

What volume of carbon dioxide is formed when $20 \, cm^3$ of carbon monoxide reacts with oxygen?

|  | $2CO(g)$ | $+$ | $O_2(g)$ | $\longrightarrow$ | $2CO_2(g)$ |
|---|---|---|---|---|---|
|  | 2 moles | $+$ | 1 mole | gives | 2 moles |
| So | 2 volumes | $+$ | 1 volume | gives | 2 volumes |
| and | $20 \, cm^3$ | $+$ | $10 \, cm^3$ | gives | $20 \, cm^3$ |

**Answer:** $20 \, cm^3$ of carbon dioxide is formed.

### *Example 5*

What volume of hydrogen chloride is produced when $100 \, cm^3$ of thionyl chloride, $SOCl_2$, reacts with water vapour according to the equation shown:

|  | $SOCl_2(l)$ | $+$ | $H_2O(g)$ | $\longrightarrow$ | $SO_2(g)$ | $+$ | $2HCl(g)$ |
|---|---|---|---|---|---|---|---|
|  | 1 mole | $+$ | 1 mole | gives | 1 mole | $+$ | 2 moles |
| So | 1 volume | $+$ | 1 volume | gives | 1 volume | $+$ | 2 volumes |
| and | $100 \, cm^3$ | $+$ | $100 \, cm^3$ | gives | $100 \, cm^3$ | $+$ | $200 \, cm^3$ |

**Answer:** $200 \, cm^3$ of hydrogen chloride is formed.

## Summary: Solids, liquids and gases

* ★ All substances are made of particles.
* ★ Diffusion and Brownian motion are evidence that particles exist.
* ★ The particles in substances are moving.
* ★ The particles in solids vibrate but are in a regular arrangement.
* ★ When a solid melts, the regular arrangement disappears.
* ★ When a liquid boils, the particles spread out into the air.
* ★ Some solids turn straight into gases when heated. This process is called sublimation.
* ★ Some gases turn straight into solids when cooled. The solid is called a sublimate.
* ★ Energy is needed to separate the particles both in melting and in boiling.
* ★ The properties of solids, liquids and gases are easily understood in terms of particles.
* ★ At room temperature and pressure a mole of any gas occupies a volume of $24\,000 \, cm^3$.

# Questions

1 Below are five diagrams labelled A to E. For each of the questions, choose the diagram which best fits the description.

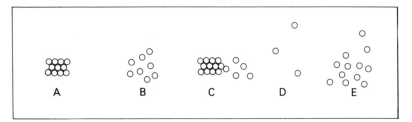

(a) Which diagram could represent a gas?
(b) Which diagram could represent a solid?
(c) Which diagram represents a solid melting?
(d) Which diagram represents a liquid evaporating?
(e) Which diagram represents a liquid?

2 Choose your answers from the substances described in the table below. Assume normal room temperature to be 20 °C.

| Substance | Melting point (°C) | Boiling point (°C) |
|-----------|--------------------|--------------------|
| sulphur trioxide | +17 | +45 |
| tungsten carbide | +2870 | +6000 |
| ethane-1,2-diol | −13 | +197 |
| ammonia | −78 | −33 |
| ethanoic acid | +17 | +118 |

(a) Which substance is a gas at room temperature?
(b) Which substance is a solid at 0 °C but a gas at 80 °C?
(c) Which substance would be a liquid in a normal refrigerator (about −2 °C)?
(d) Which substances might freeze on a cold day?
(e) Which substance is a liquid over the greatest range of temperature?

3 Calculate the volume of gas produced in each of the following reactions:
(a) from 17 g of hydrogen peroxide, $H_2O_2$, in the reaction

$$2H_2O_2(l) \longrightarrow 2H_2O(l) + O_2(g)$$

(b) from 4.0 g of calcium in the reaction

$$Ca(s) + 2H_2O(l) \longrightarrow Ca(OH)_2(aq) + H_2(g)$$

(c) from 8 g of sulphur in the reaction

$$S(s) + O_2(g) \longrightarrow SO_2(g)$$

(d) from 3.2 g of calcium carbide, $CaC_2$, in the reaction

$$CaC_2(s) + 2H_2O(l) \longrightarrow Ca(OH)_2(aq) + C_2H_2(g)$$

(e) from 400 g of calcium carbonate, $CaCO_3$, in the reaction

$$CaCO_3(s) \longrightarrow CaO(s) + CO_2(g)$$

Use values of the relative atomic masses from appendix 1.

**4** Calculate the volume of gas formed in each of these reactions:

(a) when 20 cm³ of hydrogen reacts with 20 cm³ of chlorine:

$$H_2(g) + Cl_2(g) \longrightarrow 2HCl(g)$$

(b) when 10 cm³ of methane, $CH_4$, reacts with 20 cm³ of oxygen:

$$CH_4(g) + 2O_2(g) \longrightarrow CO_2(g) + 2H_2O(g)$$

(c) when 40 cm³ of sulphur dioxide, $SO_2$, reacts with 20 cm³ of oxygen:

$$2SO_2(g) + O_2(g) \longrightarrow 2SO_3(g)$$

(d) when 1 dm³ of nitrogen dioxide, $NO_2$, dimerizes into $N_2O_4$:

$$2NO_2 \longrightarrow N_2O_4$$

(e) when 50 cm³ of nitrogen reacts completely with 150 cm³ of hydrogen:

$$N_2(g) + 3H_2(g) \longrightarrow 2NH_3(g)$$

**5** Read the passage below. Then try and answer the questions.

Both ammonia solution and hydrochloric acid have dangerous and unpleasant smells. When beakers containing these solutions are placed near each other, clouds of smoke form above the beakers.

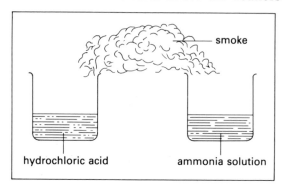

hydrochloric acid                ammonia solution

A similar experiment can be carried out using the apparatus shown below. Plugs of cotton wool are soaked in ammonia solution and in hydrochloric acid. These are placed at either end of a long glass tube. Nothing seems to happen for a few minutes, but then a ring of smoke forms nearer to the plug soaked in hydrochloric acid.

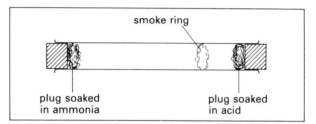

smoke ring

plug soaked in ammonia

plug soaked in acid

(a) Both these solutions have a strong smell. How does the smell get from the solution to your nose?

(b) When a solid melts its volume usually increases by about 5 to 10 per cent. When a liquid boils its volume increases by 10 000 per cent. Using both words and diagrams, explain this huge difference.

(c) In the second experiment a few minutes passed before the smoke ring formed, yet we know gas molecules travel at high speeds. Suggest a reason for this delay.

(d) Both plugs were put into the tube at the same time but the smoke ring appeared nearer the plug soaked in the acid. What does this tell you about the ammonia molecules and the acid molecules?

## A new type of storage heater

Over the last fifteen or twenty years energy costs have risen sharply. People have become more conscious of the costs of heating their homes, and new ways of saving fuel – and therefore money – have been developed. One result is that many homes are now insulated against heat loss to a fairly high standard. Loft insulation, cavity wall insulation and double glazing are now quite common. These all rely on the fact that trapped pockets of air are bad conductors of heat.

Some houses even have solar panels on the roof. These use the Sun's rays (free!) to warm up water. The trouble is that we don't usually want the radiators on when the Sun is shining. What is needed is a way to store such **solar energy**.

One method is to use the heat of fusion of a solid. A large tank of solid is kept in the roof-space. During the day this absorbs heat from the Sun and melts. At night it solidifies again and gives out the heat of fusion that it absorbed in the day. This keeps the house warm at night.

*Even in a cloudy country like Britain, worthwhile amounts of energy can be obtained from the Sun's radiation by using solar panels*

(a)                    (b)

*(a) A test tube containing supersaturated sodium thiosulphate solution. (b) When the tube is shaken, a solid mass of crystals forms*

A similar idea uses dissolving instead of melting. Sodium thiosulphate absorbs energy as it dissolves. This energy is known as the **heat of solution**. Although very soluble in hot water, sodium thiosulphate dissolves much less well in cold water. You might expect that if a hot saturated solution was cooled then crystals of sodium thiosulphate would form. In practice this only happens if the cool (and now supersaturated) solution is shaken, or if it is 'seeded' with a crystal or perhaps just with dust. It is thus possible to cool a hot saturated solution of sodium thiosulphate right down to room temperature and then to make it crystallize whenever is convenient simply by shaking it.

Energy is taken in when the sodium thiosulphate dissolves and given back when it crystallizes. Here is a system that can, in theory at least, store energy indefinitely.

## Questions

6 The figure below shows a heating curve for a solid to go into a roof-space to store energy.

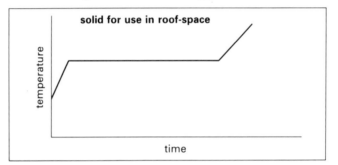

(a)  Sketch this graph.
(b)  Decide on a suitable melting point and draw in a scale for the vertical axis.
(c)  Draw in a scale indicating the time of day (or night) on the horizontal axis.
7 Look up and explain the following words or terms:
   (a)  saturated solution,
   (b)  supersaturated solution,
   (c)  heat of fusion.
8 Why is energy needed to melt a solid even if the temperature does not increase?
9 Why is energy needed to enable sodium thiosulphate to dissolve in water?
10 Explain in what way a supersaturated solution of sodium thiosulphate can be regarded as storing energy.
11 In what way is dissolving and crystallizing sodium thiosulphate a better way of storing energy than melting and solidifying a solid?

# 13 Atoms and bonding

**Figure 13.1 (above)** *Lord Rutherford put forward the idea of atoms having a central nucleus with electrons moving round it*

**Figure 13.2 (right)** *The particles in atoms*

## 13.1 What are atoms made of?

Earlier this century, scientists discovered that atoms are made of three types of even smaller particles. These are **protons, neutrons and electrons**.

The protons and neutrons are in the **nucleus** at the centre of the atom. The electrons move very quickly around the outside of the nucleus (see figure 13.2. This diagram and similar diagrams in this chapter are not drawn to scale, however. The electrons are much, much smaller than the protons and neutrons.)

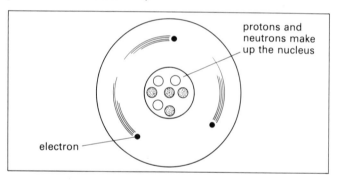

### The properties of the particles in atoms

Table 13.1 shows two important properties of these particles.

| Particle | Mass | Electric charge |
|----------|------|-----------------|
| proton | 1 | +1 |
| neutron | 1 | 0 |
| electron | $\dfrac{1}{1840}$ | −1 |

**Table 13.1** *Properties of the particles in atoms*

The **masses** of these particles are incredibly small. So we cannot use units like grams to weigh them. Instead, we compare the masses of the particles to the mass of the lightest atom, hydrogen. The proton and neutron both have about the same mass as a hydrogen atom. So they are given a mass of 1. The electron weighs very much less than a hydrogen atom. In chemistry we often treat it as though it weighed nothing at all.

The **electric charges** of these particles are very important. The proton and electron have equal but opposite charges. The neutron has no charge.

## No charge for atoms!

Although atoms contain electrically charged particles, the atoms themselves have no charge. This is because atoms contain equal numbers of protons and electrons. The charges on the protons and electrons balance and make each atom electrically neutral.

## Atomic number

The number of protons in the nucleus of an atom is called the **atomic number**. Each element has its own number of protons in its atoms (see figure 13.3). The atomic numbers of the elements are given above their symbols in the periodic table in figure 11.4 on page 113.

**Figure 13.3** *The structures of atoms*

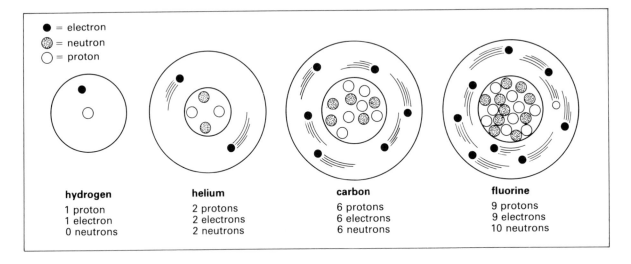

● = electron
◉ = neutron
○ = proton

| hydrogen | helium | carbon | fluorine |
|---|---|---|---|
| 1 proton | 2 protons | 6 protons | 9 protons |
| 1 electron | 2 electrons | 6 electrons | 9 electrons |
| 0 neutrons | 2 neutrons | 6 neutrons | 10 neutrons |

## Mass number

Protons weigh the same as neutrons. Electrons weigh almost nothing at all. So the mass of any atom depends on the numbers of protons and neutrons in its nucleus. The sum of the numbers of protons and neutrons is called the **mass number**. Here are some examples.

## Example 1

Iron has 26 protons, 30 neutrons and 26 electrons, so

mass number of iron = number of protons + neutrons

= 26 + 30

So the mass number of iron = 56.

## Example 2

Lithium has 3 protons, 4 neutrons and 3 electrons, so

mass number of lithium = number of protons + neutrons

= 3 + 4

So the mass number of lithium = 7.

### Isotopes

All atoms of a particular element contain the same number of protons. But atoms of the same element can contain different numbers of neutrons. Atoms of the same element that have different numbers of neutrons are called **isotopes**. The isotopes of carbon are shown in figure 13.4.

**Figure 13.4** *The isotopes of carbon*

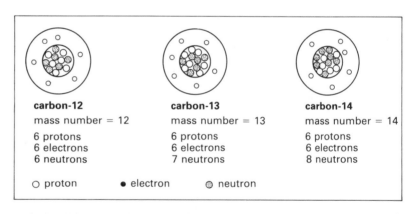

**carbon-12**
mass number = 12
6 protons
6 electrons
6 neutrons

**carbon-13**
mass number = 13
6 protons
6 electrons
7 neutrons

**carbon-14**
mass number = 14
6 protons
6 electrons
8 neutrons

○ proton     ● electron     ◉ neutron

**Figure 13.5** *It is difficult to separate the isotopes of an element, because their chemical reactions are alike. Their densities are slightly different, however, and in this factory the difference is used to separate the isotopes of uranium*

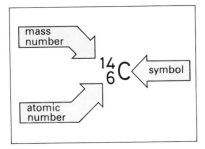

**Figure 13.6** *A shorthand way of writing the isotope carbon-14*

The three isotopes of carbon all behave in the same way during chemical reactions. But the isotope carbon-14 is different from the other two in one way. Carbon-14 is radioactive. Radioactivity is looked at later, in sections 26.2 and 26.3.

You can distinguish between different isotopes by writing the mass number and atomic number at the side of the symbol. Remember, the atomic number is the same for all the isotopes of an element, but the mass number is different for each isotope. Figure 13.6 shows how this works with carbon-14.

### Relative atomic mass

The approximate average mass of the large collection of atoms of an element compared with the mass of a hydrogen atom is called the **relative atomic mass**. Some elements have only one isotope. So for these elements, the relative atomic mass is equal to the mass number of that isotope.

Other elements contain mixtures of isotopes. The relative atomic masses of these elements depend upon two factors:

**1** the mass numbers of the isotopes present,
**2** the relative amounts of the different isotopes present.

Chlorine contains a mixture of two isotopes: chlorine-35 and chlorine-37. There are always three times as many atoms of mass number 35 as there are of mass number 37. So the average mass of a collection of these atoms will be somewhere between 35 and 37. As there are more of the lighter atoms, it will be nearer 35 than 37. It is in fact 35.5.

## 13.2  How are electrons arranged in atoms?

The electrons that move around the outside of the atom are not all the same distance from the nucleus. They are arranged in definite layers. These layers are called **shells** (see figure 13.7). Each shell can only hold a certain number of electrons.

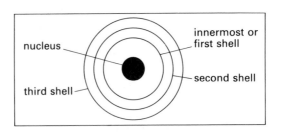

**Figure 13.7** *The electron shells around the nucleus of an atom*

The most the first shell can hold is: 2 electrons
The most the second shell can hold is: 8 electrons

**Figure 13.8** *The arrangement of electrons in hydrogen and carbon atoms*

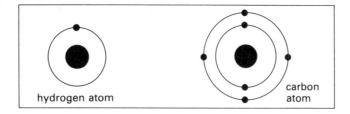

Electrons always get as close to the nucleus as they can. The hydrogen atom has only one electron. So this goes in the first (innermost) shell (figure 13.8). Carbon atoms have six electrons. These cannot all go into the first shell, since this shell can only hold two electrons. Four electrons have to go into the next shell. Table 13.2 shows which shells the electrons fit into for the elements with atomic numbers 1 to 20.

| Element | Symbol | Number of electrons | Number of electrons in | | | |
|---|---|---|---|---|---|---|
| | | | 1st shell | 2nd shell | 3rd shell | 4th shell |
| hydrogen | H | 1 | 1 | | | |
| helium | He | 2 | 2 (full) | | | |
| lithium | Li | 3 | 2 (full) | 1 | | |
| beryllium | Be | 4 | 2 (full) | 2 | | |
| boron | B | 5 | 2 (full) | 3 | | |
| carbon | C | 6 | 2 (full) | 4 | | |
| nitrogen | N | 7 | 2 (full) | 5 | | |
| oxygen | O | 8 | 2 (full) | 6 | | |
| fluorine | F | 9 | 2 (full) | 7 | | |
| neon | Ne | 10 | 2 (full) | 8 (full) | | |
| sodium | Na | 11 | 2 (full) | 8 (full) | 1 | |
| magnesium | Mg | 12 | 2 (full) | 8 (full) | 2 | |
| aluminium | Al | 13 | 2 (full) | 8 (full) | 3 | |
| silicon | Si | 14 | 2 (full) | 8 (full) | 4 | |
| phosphorus | P | 15 | 2 (full) | 8 (full) | 5 | |
| sulphur | S | 16 | 2 (full) | 8 (full) | 6 | |
| chlorine | Cl | 17 | 2 (full) | 8 (full) | 7 | |
| argon | Ar | 18 | 2 (full) | 8 (full) | 8 (full) | |
| potassium | K | 19 | 2 (full) | 8 (full) | 8 (full) | 1 |
| calcium | Ca | 20 | 2 (full) | 8 (full) | 8 (full) | 2 |

**Table 13.2** *How the electrons fit into shells for the first twenty elements in the periodic table (sometimes the third shell can hold more than eight electrons)*

## What's special about the noble gases?

The **noble gases** are unusual elements. Their atoms seem to like being alone! They will not join up with each other to form molecules. Nor will they react easily with other elements to form compounds.

We can see why these elements do not react if we look at their electron structures, shown in table 13.3 and figure 13.9.

| Noble gas | Number of electrons in | | |
|---|---|---|---|
| | 1st shell | 2nd shell | 3rd shell |
| helium | 2 (full) | | |
| neon | 2 (full) | 8 (full) | |
| argon | 2 (full) | 8 (full) | 8 (full) |

**Table 13.3** *The noble gases: number of electrons per shell*

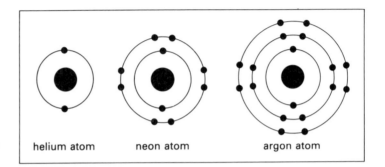

helium atom          neon atom              argon atom

**Figure 13.9** *The atoms of noble gases have full electron shells*

You can see that each shell is either completely full or completely empty. Having a completely full or empty shell seems to make atoms very stable (unreactive).

## Have any other elements got full electron shells?

The noble gases are the only elements whose shells are all completely full.

Let us look at the first twenty elements in the periodic table. From the atomic number we can tell how many electrons each element has in its outer shell. The arrangement of electrons for the group 1 elements are given in table 13.4.

| Group 1 element | Number of electrons in | | | |
|---|---|---|---|---|
| | 1st shell | 2nd shell | 3rd shell | 4th shell |
| lithium | 2 (full) | 1 | | |
| sodium | 2 (full) | 8 (full) | 1 | |
| potassium | 2 (full) | 8 (full) | 8 (full) | 1 |

**Table 13.4** *Group 1 elements: number of electrons per shell*

Each of these elements contains a single electron in its outer shell (see figure 13.10).

**Figure 13.10** *The atoms of group 1 elements all have one electron in their outer shells*

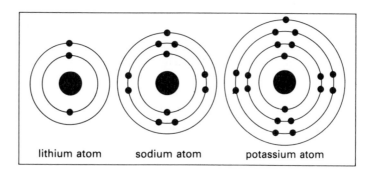

lithium atom     sodium atom     potassium atom

There is a similar pattern for the group 7 elements, shown in table 13.5.

| Group 7 element | Number of electrons in | | | |
|---|---|---|---|---|
| | **1st shell** | **2nd shell** | **3rd shell** | **4th shell** |
| fluorine | 2 (full) | 7 | | |
| chlorine | 2 (full) | 8 (full) | 7 | |

**Table 13.5** *Group 7 elements: number of electrons per shell*

These elements each contain seven electrons in the outer shell.

### How to describe the electrons in an atom

It is important to know how many electrons an atom has in each shell. Chemists have a short way of doing this. The numbers of electrons in each shell are written down, using dots to separate the numbers. For example: He 2, Na 2.8.1 and Cl 2.8.7.

### Chemical bonding

We have already seen that the noble gases are stable (unreactive) because they have only full electron shells. We could just as well have said that the other elements are reactive because they do not have full shells. It seems that elements react to try and get only full shells.

There are three ways in which an element can fill a shell:

**1** it can lose electrons,
**2** it can gain electrons,
**3** it can share electrons.

## 13.3 Electrovalent bonding

**Electrovalent bonds** are found in most compounds that contain a metal and a non-metal. This is best explained by looking at some examples.

## Example 3 – Lithium fluoride

Lithium has just one electron in its outer shell. If it could lose this it would have only a full inner shell.

Fluorine has seven electrons in its outer shell. It needs just one more electron to fill this shell up.

So the lithium atom gives its outermost electron to the fluorine atom (see figure 13.11). This way both atoms become like noble gases. They both have shells that are all either entirely full or entirely empty.

**Figure 13.11** *Lithium gives an electron to fluorine to form lithium fluoride*

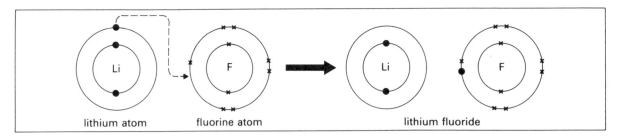

lithium atom         fluorine atom                    lithium fluoride

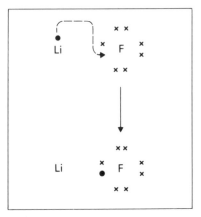

**Figure 13.12** *Simplified diagram of lithium fluoride bonding*

In figure 13.11, the lithium electrons are shown as circles and the fluorine electrons as crosses. This is just to make the diagram easier to follow. All electrons are alike. It doesn't matter what kind of atom they are part of.

Diagrams like figure 13.11 can be made simpler. Since the electrons in the full inner shells do not take part in this reaction, they can be left out of the diagram. The reaction of lithium with fluorine would then look like figure 13.12.

## Ions and bonding

You have already seen that the atoms of an element contain equal numbers of protons (with a positive charge) and electrons (with a negative charge). As a result the atoms have no overall electric charge. Now look at what has happened in figure 13.12. The lithium atom has lost an electron. So it now has an overall charge of 1+. The fluorine atom has gained an electron. Its overall charge becomes 1−. Figure 13.13 shows what has happened.

**Figure 13.13** *How atoms become charged*

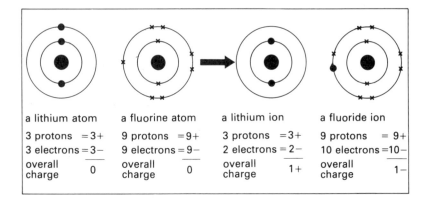

| a lithium atom | a fluorine atom | a lithium ion | a fluoride ion |
|---|---|---|---|
| 3 protons  =3+ | 9 protons  =9+ | 3 protons  =3+ | 9 protons  = 9+ |
| 3 electrons =3− | 9 electrons =9− | 2 electrons = 2− | 10 electrons =10− |
| overall charge   0 | overall charge   0 | overall charge   1+ | overall charge   1− |

Atoms that have an electric charge are called **ions**. Lithium fluoride contains $Li^+$ ions and $F^-$ ions. (The + and − stand for the overall charges on the ions.) Particles with opposite charges attract each other. Because of this, the $Li^+$ ion and $F^-$ ion are pulled or 'bonded' together. This type of bonding is called electrovalent bonding because the electric charges on the atoms help to form the bond. Some chemists simply call this **ionic bonding** (because ions take part).

### Example 4 – Magnesium oxide

The arrangement of electrons in these elements are: Mg 2.8.2 and O 2.6.

Magnesium can have a full shell if it loses two electrons. Oxygen can also have a complete shell, but it needs to gain two electrons. Figure 13.14 shows what happens.

**Figure 13.14** *The bonding in magnesium oxide*

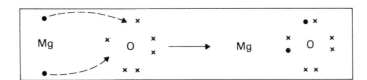

The $Mg^{2+}$ and $O^{2-}$ ions produced are then attracted to each other because they are oppositely charged.

### Example 5 – Magnesium chloride

The electron structures of the atoms are: Mg 2.8.2 and Cl 2.8.7.

The magnesium atom needs to lose two electrons. The chlorine atom only needs to gain one electron. Here the magnesium atom can only lose two electrons if it reacts with two chlorine atoms, as shown in figure 13.15.

**Figure 13.15** *The bonding in magnesium chloride*

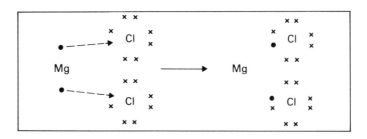

So the compound formed between these elements contains two chlorine ions for each magnesium ion.

## 13.4 Covalent bonding

**Covalent bonds** are found in elements and compounds that contain only non-metals. Here are some examples.

### Example 6 – The hydrogen molecule

The hydrogen atom has just one electron. To have a noble gas structure it must have two electrons.

This problem is neatly solved by the hydrogen molecule. In the molecule two hydrogen atoms both share their electron (see figure 13.16).

**Figure 13.16** *The bonding in a hydrogen molecule*

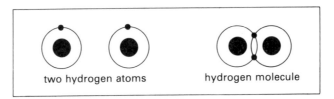

To carry on sharing in this way, the two atoms must stay together. They are said to be **bonded**. The shared pair of electrons form a bond called a covalent bond.

### Example 7 – The hydrogen chloride molecule

This is a very similar example. The electron arrangements are: H 1 and Cl 2.8.7.

Both the hydrogen atom and the chlorine atom need one more electron in order to have full shells. They do this by sharing one pair of electrons (see figure 13.17).

**Figure 13.17** *The bonding in a hydrogen chloride molecule*

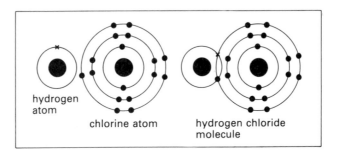

The inner electron shells of the chlorine atom do not take part in the reaction. So figure 13.17 can be made simpler by leaving these out (see figure 13.18).

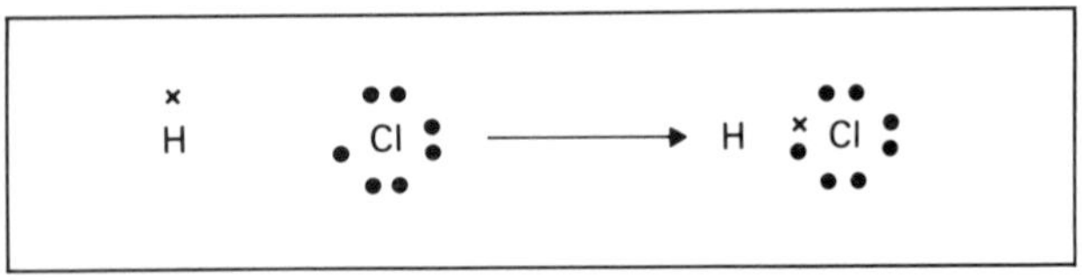

**Figure 13.18** *Simplified diagram of hydrogen chloride bonding*

### Example 8 – Methane

Methane (natural gas) contains one carbon atom joined to four hydrogen atoms. The electron arrangements for these atoms are: C 2.4 and H 1.

The carbon atom needs four more electrons to have a full shell. It does this by sharing its outer four electrons with four hydrogen atoms. This also gives each hydrogen atom two electrons. So all the atoms have full outer shells (see figure 13.19).

**Figure 13.19** *The bonding in a methane molecule*

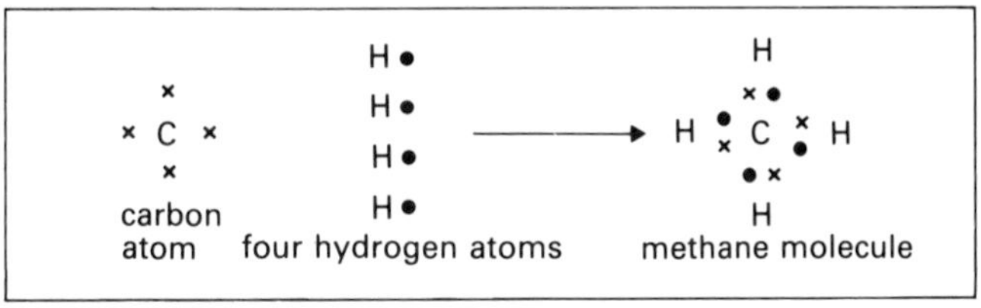

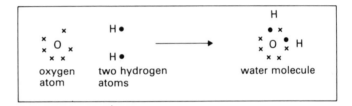

## Example 9 – Water

A water molecule contains two hydrogen atoms and one oxygen atom. The electron arrangements are: H 1 and O 2.6.

The oxygen atom needs to share two electrons. Each hydrogen atom needs to share one electron. Together they form a water molecule (see figure 13.20).

**Figure 13.20** *The bonding in a water molecule*

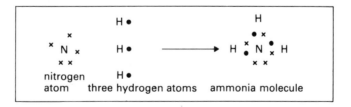

## Example 10 – Ammonia

Ammonia has the formula $NH_3$. The electron arrangements in the atoms are: H 1 and N 2.5.

The nitrogen atom needs to share three electrons. Each hydrogen atom needs to share one. Together they make an ammonia molecule (see figure 13.21).

**Figure 13.21** *The bonding in an ammonia molecule*

## Covalent double bonds

All of the compounds we have looked at so far have contained single covalent bonds. The atoms in these compounds were held together by sharing one pair of electrons.

Some substances contain **covalent double bonds**. Here the atoms are held together by sharing two pairs of electrons.

An oxygen molecule contains this sort of double bond. Oxygen has the electron arrangement O 2.6.

Each oxygen atom needs two more electrons to have a noble gas structure. So each atom shares two of its electrons in a covalent double bond (see figure 13.22).

Ethene is another important substance that contains a double bond. Its structure is shown in figure 13.23.

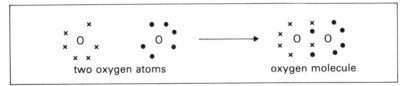

two oxygen atoms          oxygen molecule

**Figure 13.22** *The bonding in an oxygen molecule*

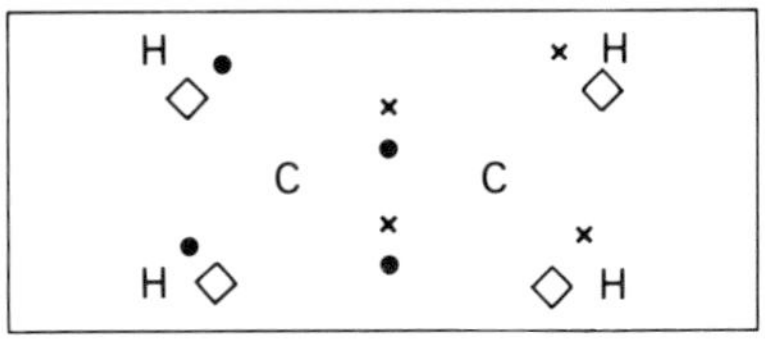

**Figure 13.23** *The double covalent bond in ethene*

## Simplified drawings of molecules

A group of atoms joined by covalent bonds is called a **molecule**. Chemists often use diagrams to represent such molecules. In these diagrams a single covalent bond is represented by a single line. A double bond is represented by a pair of lines. Some examples are given in figure 13.24.

**Figure 13.24** *Simplified diagrams of single and double covalent bonds*

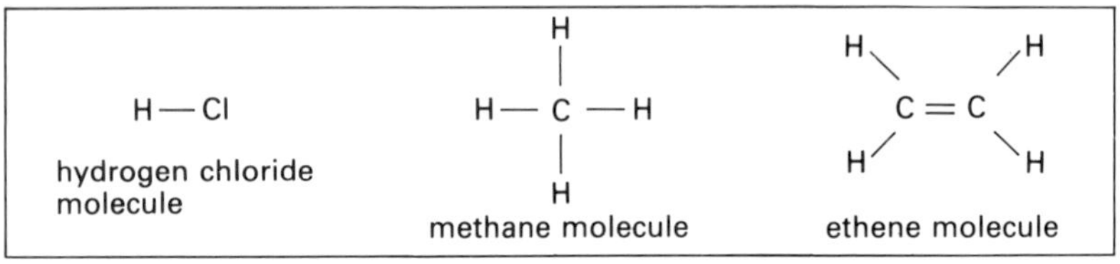

$$H-Cl$$

hydrogen chloride molecule

$$H-\overset{\displaystyle H}{\underset{\displaystyle H}{C}}-H$$

methane molecule

ethene molecule

## Big and small molecules

The molecules that we have already looked at all contained relatively few atoms. Not all molecules are so simple. Figure 13.25 shows two elements. Diamond is a form of carbon. It has many millions of atoms in each molecule. It is described as a giant molecule.

**Figure 13.25** *A giant molecule and a small molecule*

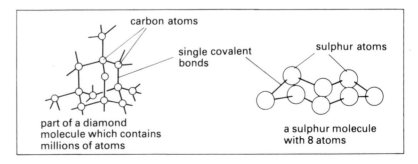

carbon atoms

single covalent bonds

sulphur atoms

part of a diamond molecule which contains millions of atoms

a sulphur molecule with 8 atoms

A sulphur molecule contains only eight atoms. This is considered to be a small molecule.

## 13.5 Chemical formulae

We have already met some chemical formulae, especially in section 8.1. For example, the formula for water is $H_2O$. These formulae are not only shorter than the names of the chemicals. They also tell us what is in the substance. This section is about the formulae of compounds.

### The formulae of covalent compounds

When two or more non-metals react together a covalent compound is formed. We can 'work out' what formulae some of these have. To do this we have to draw dot-and-cross diagrams of the type shown in section 13.4.

The formulae of some of the most common covalent compounds are given in table 13.5. Try and memorize as many of these as you can.

| Name of compound | Formula of compound |
|---|---|
| ammonia | $NH_3$ |
| carbon dioxide | $CO_2$ |
| carbon monoxide | $CO$ |
| ethane | $C_2H_6$ |
| ethanol | $C_2H_6O$ |
| hydrogen chloride | $HCl$ |
| methane | $CH_4$ |
| nitrogen dioxide | $NO_2$ |
| sulphur dioxide | $SO_2$ |
| sulphur trioxide | $SO_3$ |
| sulphuric acid | $H_2SO_4$ |
| water | $H_2O$ |

**Table 13.5** *The formulae of some common covalent compounds*

### What does the formula of a covalent compound tell us?

The formula of a covalent compound tells us two things:

1 the elements that are present in the compound,
2 the number of each type of atom in a molecule.

For example, a molecule of water, $H_2O$, contains two hydrogen atoms and one oxygen atom.

A problem does arise when we come to write the formula of a giant molecule. The formula for a grain of sand might be $Si_{100\,000\,000}O_{200\,000\,000}$.

This looks rather clumsy! It is also only correct for one particular size of grain. What we do know is that any grain of sand contains twice as many oxygen atoms as silicon atoms. Because of this we simplify the formula to $SiO_2$.

## The formulae of simple electrovalent compounds

Section 13.3 explained electrovalent (ionic) bonding. When such compounds are formed, electrons move from the metal to the non-metal. As a result ions are produced.

The charges on some ions are given in table 13.6.

| Metal ions | | Non-metal ions | |
|---|---|---|---|
| **Name** | **Charge** | **Name** | **Charge** |
| lithium | $Li^+$ | bromide | $Br^-$ |
| potassium | $K^+$ | chloride | $Cl^-$ |
| silver | $Ag^+$ | fluoride | $F^-$ |
| sodium | $Na^+$ | iodide | $I^-$ |
| calcium | $Ca^{2+}$ | oxide | $O^{2-}$ |
| copper | $Cu^{2+}$ | sulphide | $S^{2-}$ |
| iron(II) | $Fe^{2+}$ | | |
| magnesium | $Mg^{2+}$ | | |
| zinc | $Zn^{2+}$ | | |
| aluminium | $Al^{3+}$ | nitride | $N^{3-}$ |
| chromium | $Cr^{3+}$ | | |
| iron(III) | $Fe^{3+}$ | | |

**Table 13.6** *Simple ions and their charges*

The raised number in the table tells you how many electrons were lost or gained when the atom became an ion. A plus sign means electrons were lost. A minus sign means electrons were gained. Some elements can form more than one type of ion. For example, iron can form iron(II) compounds. These contain $Fe^{2+}$ ions. Iron can also form iron(III) compounds. These contain $Fe^{3+}$ ions.

We shall now look at the formation of calcium chloride as an example.

Each Ca atom must lose 2 electrons to give $Ca^{2+}$.
Each Cl atom must gain 1 electron to give $Cl^-$.

So we need more than one chlorine atom to accept the two electrons from the calcium atom. Two are needed. The formula of the compound is $CaCl_2$ (see figure 13.26).

**Figure 13.26** *The formation of calcium chloride*

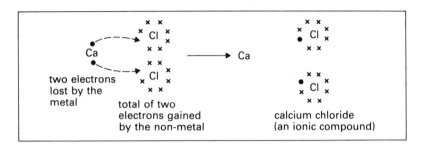

You can see that the total number of plus and minus charges in the compound are equal. This is always true for any electrovalent compound.

We can use this idea to predict the formula of any electrovalent compound. This is done in five stages, shown in the examples that follow.

### Example 11 – Zinc sulphide

| | |
|---|---|
| **1** What is the name of the compound? | zinc sulphide |
| **2** What ions are present? | $Zn^{2+}$ and $S^{2-}$ |
| **3** Which ions do you need more of? | Neither. The plus and minus charges are already equal |
| **4** How many of each ion do you need? | one $Zn^{2+}$ and one $S^{2-}$ |
| **5** What is the formula? | ZnS |

### Example 12 – Sodium oxide

| | |
|---|---|
| **1** What is the name of the compound? | sodium oxide |
| **2** What ions are present? | $Na^+$ and $O^{2-}$ |
| **3** Which ions do you need more of? | $Na^+$ |
| **4** How many of each ion do you need? | two $Na^+$ and one $O^{2-}$ |
| **5** What is the formula? | $Na_2O$ |

## Example 13 – Aluminium oxide

1  What is the name of the compound?    aluminium oxide

2  What ions are present?    $Al^{3+}$ and $O^{2-}$

3  Which ions do you need more of?    $O^{2-}$

4  How many of each ion do you need?    two $Al^{3+}$ and three $O^{2-}$

5  What is the formula?    $Al_2O_3$

## The formulae of more complicated electrovalent compounds

The ions that we have mentioned so far have been charged atoms. An ion may also be a group of atoms that has a charge. Such ions are often formed in reactions with acids. Table 13.7 lists some of these ions and also explains the meaning of each ion's formula.

| Name of ion | Formula of ion | Atoms present in each ion | Charge on the ion |
|---|---|---|---|
| hydroxide | $OH^-$ | one oxygen and one hydrogen | 1− |
| nitrate | $NO_3^-$ | one nitrogen and three oxygens | 1− |
| hydrogen-carbonate | $HCO_3^-$ | one hydrogen, one carbon and three oxygens | 1− |
| sulphate | $SO_4^{2-}$ | one sulphur and four oxygens | 2− |
| carbonate | $CO_3^{2-}$ | one carbon and three oxygens | 2− |
| phosphate | $PO_4^{3-}$ | one phosphorus and four oxygens | 3− |

Table 13.7 *Some ions and their formulae*

The formulae of compounds containing these ions are worked out in five stages as before. Here are two examples.

## Example 14 – Anhydrous copper sulphate

1  What is the name of the compound?    copper sulphate

| | |
|---|---|
| **2** What ions are present? | $Cu^{2+}$ and $SO_4^{2-}$ |
| **3** Which ions do you need more of? | Neither. The electric charges are already equal |
| **4** How many of each ion do you need? | one $Cu^{2+}$ and one $SO_4^{2-}$ |
| **5** What is the formula? | $CuSO_4$ |

### *Example 15 – Anhydrous copper nitrate*

| | |
|---|---|
| **1** What is the name of the compound? | copper nitrate |
| **2** What ions are present? | $Cu^{2+}$ and $NO_3^{-}$ |
| **3** Which ions do you need more of? | $NO_3^{-}$ |
| **4** How many of each ion do you need? | one $Cu^{2+}$ and two $NO_3^{-}$ |
| **5** What is the formula? | $Cu(NO_3)_2$ |

Look at the brackets around the $NO_3^{-}$ ion in the final formula. They are there to show that there are two complete $NO_3$ groups.

## 13.6 Metallic bonding

When metal atoms bond together they do not form a noble gas structure. Electrons are still shared in the bonds, but metallic bonds are different from covalent bonds in one important way. In a covalent bond the electrons spend their time between the two atoms that are bonded. When metal atoms share their electrons, these electrons become free to move anywhere within the metal structure. In figure 13.27 the atom X will share some of its electrons with all of its neighbouring atoms (A to F). In the same way, the atoms A to F share their electrons with several neighbouring atoms. This results in giant structures containing billions of atoms.

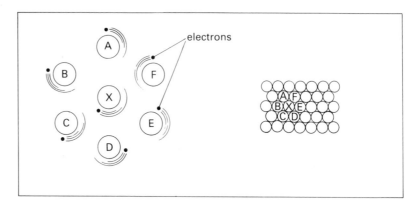

**Figure 13.27** *Each metal atom shares electrons with several other atoms*

## Summary: Atoms and bonding

* ★ Atoms contain protons, neutrons and electrons.
* ★ The charges on these particles are: protons +1, electrons −1, neutrons 0.
* ★ The relative masses of these particles are: protons 1, neutrons 1, electrons 0.
* ★ The atoms of an element have equal numbers of protons and electrons.
* ★ The atomic number is the number of protons in the atom.
* ★ The mass number is the number of protons + the number of neutrons.
* ★ Isotopes are atoms with the same number of protons but different numbers of neutrons.
* ★ The electrons in atoms are arranged in shells.
* ★ The first shell can hold up to 2 electrons, the second shell can hold up to 8 electrons, the third shell can hold up to 8 electrons.
* ★ The noble gas atoms have only full electron shells. This makes them very stable.
* ★ Other elements often try to copy the electron arrangements of the noble gases.
* ★ There are three types of bonding:
    1 covalent,
    2 electrovalent,
    3 metallic.
* ★ Covalent bonds contain shared electrons.
* ★ Electrovalent bonds are formed by transferring electrons.
* ★ When electrovalent bonds form, the atoms do not stay electrically neutral. They gain or lose charge and become ions.
* ★ Formulae can be worked out from the charges on the ions involved.

## Questions

1 Choose from the list below the answer which best fits the description given.

    nucleus
    isotopes
    shell
    neutrons
    electrons

(a) Atoms having equal numbers of protons but different numbers of neutrons.
(b) A 'layer' of electrons.
(c) Particles having no electric charge.
(d) The lightest particles in atoms.
(e) The central part of an atom.

**2** Choose from the list below the answer which fits each of the following descriptions.

    0   2   8   16   31

(a) The number of electrons that can fit into the first electron shell.

(b) The atomic number of $^{16}_{8}O$ atoms.

(c) The number of neutrons in a $^{31}_{15}P$ atom.

(d) The overall electric charge on a $^{4}_{2}He$ atom.

(e) The number of protons in a $^{70}_{31}Ga$ atom.

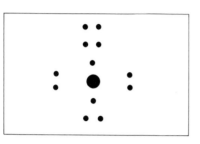

**3** Magnesium atoms have a 2.8.2 arrangement of electrons. A shell diagram of this arrangement is shown on the left. Use first numbers and then shell diagrams to show the arrangement of electrons in each of the following atoms.

(a) $^{32}_{16}S$    (b) $^{27}_{13}Al$    (c) $^{40}_{20}Ca$    (d) $^{20}_{10}Ne$

**4** (a) Draw a shell diagram of a $^{23}_{11}Na$ atom.

(b) Draw a shell diagram of a $^{35}_{17}Cl$ atom.

(c) What is the electric charge on a sodium atom?

(d) What is the electric charge on a chlorine atom?

(e) What must a sodium atom do to get the same arrangement of electrons as a neon atom?

(f) What must a chlorine atom do to get the same arrangement of electrons as an argon atom?

(g) Explain how a reaction between sodium atoms and chlorine atoms could change each of these atoms into ions with noble gas structures.

(h) Why should such ions be attracted to each other?

**5** (a) Aluminium nitride has the formula AlN. Draw shell diagrams for:

    an aluminium atom,      a nitrogen atom,
    an aluminium ion ($Al^{3+}$),   a nitride ion ($N^{3-}$).

(b) Explain what happens to the electrons when aluminium and nitrogen atoms combine to form aluminium nitride.

**6** Write out the correct chemical formulae for the chemicals in these questions. (Use table 13.6 on page 152.)

(a) sodium bromide        (b) potassium oxide

(c) calcium sulphide       (d) copper iodide

(e) aluminium chloride     (f) aluminium sulphide

(g) potassium nitrate      (h) sodium sulphate

(i) zinc nitrate          (j) aluminium sulphate

**7** (a) Draw a shell diagram of a chlorine atom.

(b) How many electrons does a chlorine atom need to have the same electron arrangement as an argon atom?

(c) Draw a shell diagram of a chlorine molecule ($Cl_2$).

(d) Show clearly how sharing electrons gives each chlorine atom an argon electron structure.

**8** This question is about the carbon dioxide molecule. Its structure can be drawn as $O=C=O$. You can see that it contains two double bonds.

(a) Draw a shell diagram for a carbon atom.

(b) How many extra electrons does it need to have the same electron arrangement as a neon atom?

(c) Draw a shell diagram for an oxygen atom.

(d) How many extra electrons does it need to have the same electron arrangement as a neon atom?

(e) Draw a shell diagram for the carbon dioxide molecule.

(f) Show clearly how each atom achieves a neon electron arrangement by sharing electrons.

9 Explain the difference between ionic and covalent bonding. In what way are the noble gases important to both these theories of bonding?

10 The table lists some properties of the group 1 elements.

| Element | Relative atomic mass | Atomic radius $(m \times 10^{-9})$ | Ionic radius $(m \times 10^{-9})$ |
|---|---|---|---|
| lithium | 6.9 | 1.55 | 0.60 |
| sodium | 23.0 | 1.86 | 0.95 |
| potassium | 39.1 | 2.31 | 1.33 |
| rubidium | 85.5 | 2.44 | 1.48 |
| caesium | 132.9 | 2.62 | 1.69 |
| francium | 223.0 | | |

(a) Estimate likely values for the atomic radius and the ionic radius of the element francium.

(b) What is similar about the arrangement of electrons in all of the group 1 elements?

(c) Why do the radii of the atoms get bigger going down a group of the periodic table?

(d) What charge will ions of the group 1 elements have?

(e) Why is the radius of the ion much smaller than the radius of the corresponding atom for all these elements?

(f) Which of the elements listed is most clearly made up of a mixture of isotopes? (Explain your answer.)

(g) When metals react they form positive ions. Can you suggest why caesium should form positive ions more readily than sodium does?

## Chemistry can surprise you!

Many chemical discoveries are the result of systematic scientific research by chemists working in universities or industry. For example, a great deal of carefully controlled research went into developing a satisfactory method of manufacturing ammonia by the reaction of nitrogen with hydrogen. (This method, which is known as the Haber process, is described in section 20.3.)

Other discoveries are the result of research that has taken an unexpected direction. The first preparation of polythene by a team of ICI chemists comes into this category. In 1931 they had started studies of reactions at high pressures in connection with the Haber process but decided to try using similar conditions to make ethene react with another organic chemical. Instead the ethene molecules reacted with each other, and the result was the plastic we call polythene. Only about half a gram of the plastic was obtained and it was 1936 before further work produced sufficient of the plastic for the importance of the discovery to be recognized.

The plastic PTFE (or Teflon) was also discovered in a chance manner. A chemist called Dr Plunkett noticed that a cylinder of tetrafluoroethene gas seemed to have lost all its pressure. But when he weighed the cylinder he noticed that its mass was still the same as that of a full cylinder. He cut the cylinder open to find the white plastic that is now used to coat our non-stick frying pans.

Yet another chance discovery was the method of preparing calcium carbide. This was made by an American named John Morehead, whose father had built a $50 000 furnace in the hope of extracting aluminium. Morehead hoped that heating aluminium oxide and carbon to 2000 °C would give aluminium and carbon dioxide. Pure aluminium was never obtained by this method, however. One day Morehead decided to try and make calcium by heating lime (calcium oxide) with carbon. Instead of getting calcium he got calcium carbide. At first he thought this was calcium, as it gave off a gas when added to water. When he found that the gas burned with a bright, smoky flame it was clear that the gas was not hydrogen and that the solid was not calcium. In fact the gas was ethyne (acetylene) and the method became of great industrial importance.

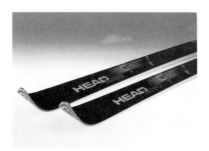

*The PTFE coating reduces friction between the skis and the snow*

## Questions

**11** Draw dot-and-cross diagrams of the structures of a nitrogen atom, a hydrogen atom and an ammonia molecule.

**12** Ethene has the formula $C_2H_4$. Draw a dot-and-cross diagram of the ethene molecule.

**13** Tetrafluoroethene has the formula $C_2F_4$. Draw a dot-and-cross diagram of this molecule.

**14** Aluminium oxide is an ionic compound. Explain the charges present upon aluminium ions and on oxide ions and hence account for the formula of aluminium oxide.

**15** Ethyne has the formula $C_2H_2$. Draw a dot-and-cross diagram of the ethyne molecule.

**16** Calcium carbide is an ionic compound with the formula $CaC_2$. What is the charge on the carbide ion, $C_2$?

**17** The chemist Kekulé is said to have discovered the structure of the benzene molecule in a dream. Find out about the structure of benzene.

# 14 Structure

Only the noble gases exist as separate atoms. The atoms of all other substances become bonded together in some way. In this chapter we shall look at how the arrangement of atoms can affect the properties of a substance.

## 14.1 Types of structure

There are three main types of substance:

1 metallic,
2 ionic (or electrovalent),
3 covalent.

### Metallic substances

In metals the bonding between atoms is often strong. The structures that are formed contain millions of atoms packed together very tightly. The ways in which the atoms are arranged in these structures are very simple. The same patterns can be obtained by dropping a lot of marbles into a small box.

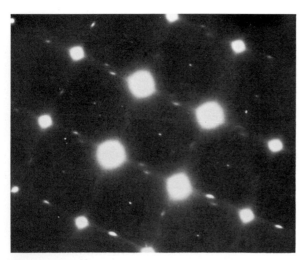

**Figure 14.1** *An electron diffraction pattern of brass. You can see that the atoms in this material are arranged in a regular pattern*

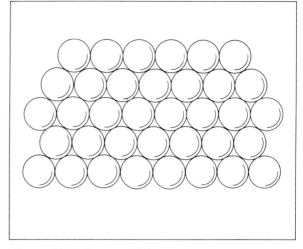

**Figure 14.2** *One possible arrangement of atoms in a metal*

## Ionic compounds

Ionic compounds are formed by electron transfer. The ions that are formed are attracted to each other by the opposite electric charges. These ions quickly arrange themselves into giant structures in which each positive ion is surrounded by negative ions and vice versa (see figure 14.3).

**Figure 14.3** *How ions form into giant structures*

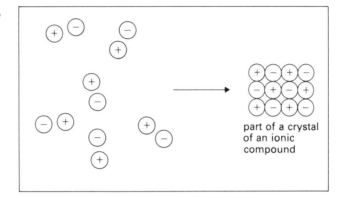

The oppositely charged ions are very strongly attracted to one another. This keeps the structure tightly held together.

## Covalent substances

Substances that contain only non-metals consist of molecules. We shall divide them into two groups:

**1** small molecules,
**2** giant molecules.

In both small and giant molecules the atoms are bonded to each other by strong covalent bonds. Most small molecules contain between two and twenty atoms. Giant molecules have many millions of atoms per molecule.

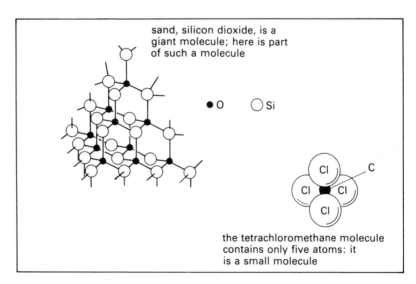

**Figure 14.4** *A giant molecule and a small molecule*

### Summary of the types of structure

|  | **Small molecular** | **Giant molecular** | **Metallic** | **Ionic** |
|---|---|---|---|---|
| **Size of structure** | small | giant | giant | giant |
| **Type of elements** | non-metal only | non-metal only | metal only | both a metal and a non-metal |

**Table 14.1** *Types of structure*

## 14.2 Structure and properties

### Melting points

The melting points of four different substances are given in table 14.2.

|  | **Methane** | **Salt** | **Iron** | **Diamond** |
|---|---|---|---|---|
| **Melting point (°C)** | −183 | 801 | 1540 | 3550 |
| **Type of structure** | small molecules | ionic | metallic | giant molecules |
| **Size of structure** | small | giant | giant | giant |

**Table 14.2** *Melting points*

From this table you can see that substances with giant structures have high melting points (see figure 14.5).

**Figure 14.5** *Tapping iron from a blast furnace. Iron has a giant structure, so it has a high melting point*

Methane, which consists of small molecules, melts well below 0 °C.

The reasons for the very much lower melting point of methane are explained in figure 14.6.

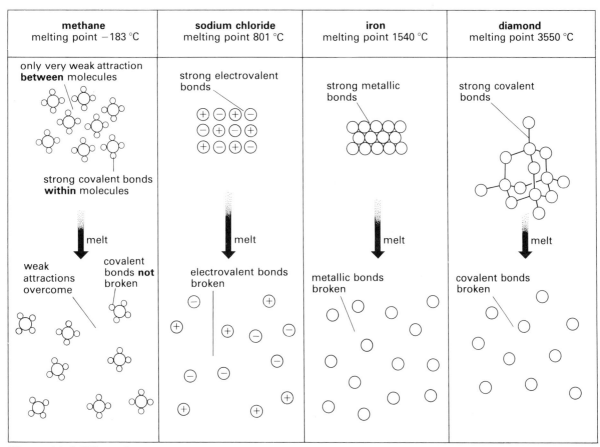

**Figure 14.6** *Why substances have different melting points*

The particles of a solid must move away from each other for the solid to be able to melt. The regular pattern then collapses. To separate the particles in any giant structure means breaking the strong bonds between the atoms. This is difficult and so substances with giant structures have high melting points.

It is much easier to melt a substance with a small molecular structure. There is no need to break the strong covalent bonds inside the molecules. You only have to separate the molecules from each other. The molecules are only weakly attracted to each other.

## Density

How dense (or heavy) a substance is depends on h
its atoms are. It also depends on how much emp
between its particles. Many metal atoms are h
closely together and such metals are dense (

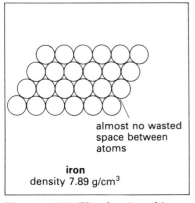

**Figure 14.7** *The density of iron*

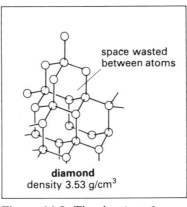

**Figure 14.8** *The density of diamond*

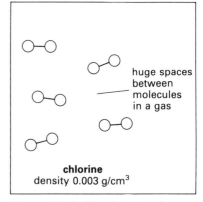

**Figure 14.9** *The density of chlorine*

All other types of substance have more empty space between their particles. So they are less dense (see figure 14.8).

Some substances with small molecular structures are gases at room temperature. These have huge spaces between the molecules. So they have very low densities (see figure 14.9).

## Solubility in petrol

Dissolving, like melting, means separating the particles that make up the structure. We have already seen that small molecules are only weakly attracted to each other. This means that they separate easily when the substance is added to solvents such as petrol (see figure 14.10).

Substances with giant structures do not dissolve in solvents like petrol. The strong bonds between the atoms or ions prevent them from separating.

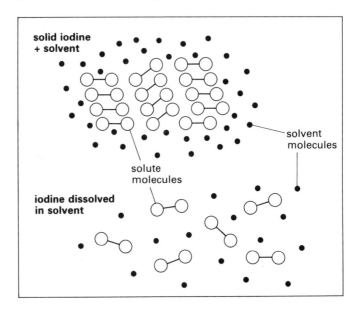

## Solubility in water

Water is a very unusual solvent. It is quite good at dissolving ionic compounds and poor at dissolving most other substances.

## Brittleness

When a metal is struck with a hammer it normally changes shape. When you hit the metal, the layers of atoms can slip over each other as shown in figure 14.12.

**Figure 14.11** *In a metal the layers of atoms can slip over one another easily when the metal is beaten with a hammer. These metal bowls have been beaten into shape. You can see the hammer marks on them*

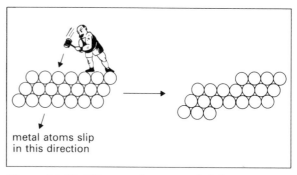

**Figure 14.12** *Why metals are not brittle*

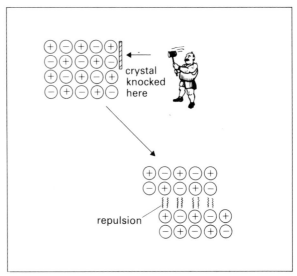

**Figure 14.13** *Why ionic crystals are brittle*

Most molecular substances have more complicated structures than metals. This prevents the atoms slipping past each other in this way.

Ionic compounds seem to be very brittle, even though they have fairly simple structures. The reason for this can be seen in figure 14.13.

When the ions slide over each other, ions with the same charge come together. Particles with the same charge repel each other. As a result the crystal splits into two pieces.

## Electrical conductivity

An electric current is a flow of electrons. In metals the outer electrons are free to move from atom to atom (see section 13.6). Normally the electrons in a piece of metal move in a disorganized way. Some move left. Just as many move right. So there is no overall electron flow. But when you connect the ends of the metal to a battery, this changes. Electrons are repelled from the negative terminal of the battery and attracted to the positive terminal. It is this movement of electrons through the metal that is the electric current. We say that the metal is conducting (see figure 14.14).

**Figure 14.14** *Why metals conduct electricity*

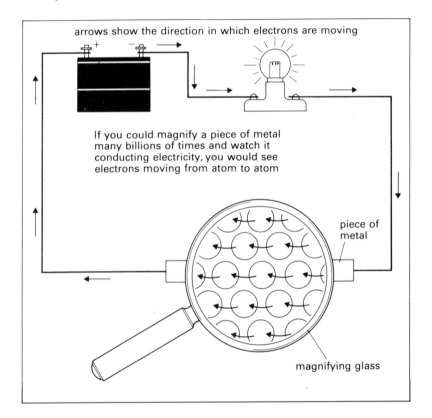

arrows show the direction in which electrons are moving

If you could magnify a piece of metal many billions of times and watch it conducting electricity, you would see electrons moving from atom to atom

piece of metal

magnifying glass

In covalent substances all the outer electrons are needed to give a noble gas structure (see section 13.4). As a result, these electrons are not free to move through the substance. So covalent substances are usually electrical **insulators** (they do not conduct electricity).

In most ionic compounds the outer electrons are used to give the ions noble gas structures. These electrons cannot move away from their ions. So you might expect ionic compounds to be insulators. This is true of ionic solids. It is not true if the ionic compound is in solution or has been melted.

The reasons for this are explained in the next chapter.

| Type of structure | Size of structure | Type of substance | Physical state at room temperature | Melting point | Electrical conductivity of | | Solubility in | |
|---|---|---|---|---|---|---|---|---|
| | | | | | solid | liquid | petrol | water |
| small molecules | small | elements or compounds that contain non-metals *only* | gas, liquid or solid | low | poor | poor | sol. | insol. |
| giant molecules | giant | elements or compounds that contain non-metals *only* | solid | very high | poor | poor | insol. | insol. |
| metals | giant | elements or alloys that contain metals *only* | solid | very high | good | good | insol. | insol. |
| ionic compounds | giant | compounds that contain *both* a metal and a non-metal | solid | very high | poor | good | insol. | most are sol. |

**Table 14.3** *The properties of different structures ('sol.' means soluble and 'insol.' means insoluble)*

## 14.3 Elements that have more than one structure

In section 14.2 we looked at the reason for the high melting point of diamond. You may have been surprised to see that diamond consists entirely of carbon atoms. Surely carbon is that black stuff! But diamonds are transparent. The explanation for this is rather interesting. The atoms in carbon can be arranged in two possible ways. One arrangement gives rise to diamond and the other to graphite.

When an element can form more than one type of structure, this is called **allotropy**. Diamond and graphite are **allotropes** of carbon. The structures and properties of these substances are shown in figure 14.16 (over the page).

They are both giant structures made of millions of atoms joined by covalent bonds. You need very high temperatures to break down either of these structures. Diamond is hard because all the atoms are part of one giant structure. Graphite is flaky and slippery. Although the atoms within a layer of graphite are strongly joined by covalent bonds, there are much weaker forces between the layers. This

**Figure 14.15** *Diamond is a valuable allotrope of carbon*

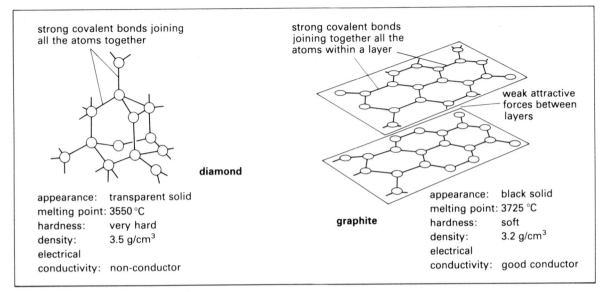

strong covalent bonds joining all the atoms together

strong covalent bonds joining together all the atoms within a layer

weak attractive forces between layers

**diamond**

**graphite**

| appearance: | transparent solid |
| melting point: | 3550 °C |
| hardness: | very hard |
| density: | 3.5 g/cm$^3$ |
| electrical conductivity: | non-conductor |

| appearance: | black solid |
| melting point: | 3725 °C |
| hardness: | soft |
| density: | 3.2 g/cm$^3$ |
| electrical conductivity: | good conductor |

**Figure 14.16** *The structure and properties of the allotropes of carbon*

means that the layers can slide easily over each other. Graphite has a lower density than diamond because of the rather large distance (in atomic terms) between the layers.

Graphite is the only non-metallic element that will conduct electricity. This makes it unique. The reason is that the carbon atoms in graphite have an electron that is free to move from one atom to the next.

Sulphur is another element that exists as allotropes. We shall be looking at it in more detail in section 21.3.

## Summary: Structure

★ Substances with giant structures usually have high melting points.
★ Substances with small molecular structures have low melting points.
★ Giant structures do not dissolve in petrol or similar solvents.
★ Small molecular substances usually dissolve in petrol.
★ Most ionic substances dissolve in water.
★ Metals are denser and more malleable than other substances.
★ Liquid and solid metals conduct electricity.
★ Ionic compounds conduct electricity when molten or dissolved in water.
★ Molecular substances do not conduct electricity.
★ Different forms of the same element are called allotropes.
★ The properties of allotropes depend on how the atoms or molecules are arranged.

# Questions

**1** These questions are about the substances A to E described in the table.

| Substance | Does it conduct when solid? | Does it conduct when liquid? | Does it have a high melting point? |
|---|---|---|---|
| A | no | no | yes |
| B | yes | yes | yes |
| C | no | no | no |
| D | no | yes | yes |
| E | yes | yes | no |

(a) Which substance is an ionic compound?
(b) Which substance is made of small molecules?
(c) Which substance is a typical metal?
(d) Which substance consists of giant molecules?

**2** Explain, with the help of diagrams, each of the following facts.
(a) Ionic compounds are brittle.
(b) Metals are malleable.
(c) Diamond has a very high melting point.
(d) Iodine dissolves in petrol.
(e) Metals conduct electricity.
(f) Graphite is flaky.

**3** Four types of structure are mentioned in section 14.1. They are: ionic, small molecular, giant molecular and metallic. What type of structure has each of the following substances?
(a) aluminium
(b) nitrogen
(c) magnesium oxide
(d) rust
(e) water
(f) brass
(g) chromium chloride
(h) sand
(i) candle wax
(j) limestone (calcium carbonate)

**4** The table below gives some data upon the chlorides of compounds in period 3 of the periodic table.

| Compound | Melting point (°C) | Boiling point (°C) | Solubility in water (mol/dm³) |
|---|---|---|---|
| sodium chloride | 1081 | 1738 | high |
| magnesium chloride | 987 | 1691 | high |
| aluminium chloride | 190 | unknown | moderate |
| silicon chloride | −70 | 57 | reacts |
| phosphorus chloride | −112 | 76 | reacts |
| sulphur chloride | −80 | 136 | reacts |

(a) Which two substances have giant structures?

(b) Explain, with words and diagrams, why sodium chloride will only melt at a high temperature.

(c) Silicon is a metalloid. Its chloride might, therefore, be either ionic or molecular. Which of these structures does it have? Justify your answer.

(d) Comment upon the type of structure that aluminium chloride must have. How does this fit with the position of aluminium in the periodic table?

(e) What type of structure does sulphur chloride have?

(f) At normal atmospheric pressure aluminium chloride sublimes. Its melting point was found by taking the measurement at a pressure 2.5 times atmospheric. What does 'to sublime' mean? Why do you think that increasing the pressure above the chloride might prevent this?

(g) Write a brief summary of any trends you can see in the structures of the chlorides across period 3.

## It's the way they join together

Several elements exist in more than one form. These different forms are called **allotropes** (or **polymorphs**).

One element that exists as allotropes is phosphorus. At least three allotropes are known, and the table below contains some information upon them.

| Allotrope | Melting point (°C) | Boiling point (°C) | Density (g/cm³) |
|---|---|---|---|
| white | 44 | 200 | 1.82 |
| red | 590 | unknown | 2.20 |
| black | high | high | 2.70 |

*In the nineteenth century, matches were made from the highly poisonous white phosphorus. Many of the factory girls developed a disease known as 'phossy-jaw' in which the lower jaw crumbles away*

*A 'planet-friendly' aerosol – some aerosols contain chlorofluoromethanes that could damage the ozone shield, but this one does not*

White phosphorus has a structure in which the atoms join in groups of four. It is a rather waxy solid. It changes into red phosphorus on heating, or if it is left to stand in the light for a long time. Black phosphorus is a flaky grey-black solid that is a surprisingly good conductor of electricity.

Oxygen is another element that exists as allotropes. Normal oxygen consists of $O_2$ molecules but $O_3$ molecules, called ozone, also exist. The upper atmosphere contains significant amounts of ozone. This ozone is particularly important because it absorbs damaging ultra-violet rays from the Sun. Too much of this radiation reaching the Earth would give us sun-burn or even skin cancer.

There has been concern that this 'ozone shield' might be damaged by chlorofluoromethanes. These compounds are sometimes used inside aerosol cans and they are known to catalyse the reaction in which ozone changes into ordinary oxygen. The USA has banned the use of this chemical in aerosols. Other countries, however, continue to use it.

## Questions

5  What are *allotropes*?
6  In which allotrope of phosphorus are the atoms most closely packed?
7  Does red or white phosphorus contain the biggest molecules?
8  What allotrope of carbon has properties similar to those of black phosphorus? What kind of structure do you think black phosphorus has?
9  Find out what the word *catalyst* means.
10  What is meant by the 'ozone shield'?
11  Write a balanced equation for the change from ozone into normal oxygen.
12  Chlorofluoromethanes from aerosols are not likely to be present in the atmosphere in large amounts. Why has the USA banned their use?

# 15 Electricity and chemicals

## 15.1 Not just insulators and conductors!

**Figure 15.1** *Lightning being conducted through the air during a thunderstorm*

### Insulators

Something very unusual is happening in the photograph in figure 15.1. Air is conducting electricity. Air does not normally do this. Air is usually an **insulator** or **non-conductor**.

Almost all non-metals are insulators. This is true whether the substance is an element or a compound. The only common exception is graphite.

### Conductors

Metals are good conductors of electricity. Silver, copper and aluminium are the three best conductors. We can find out whether a substance is a conductor or an insulator by using the circuit shown in figure 15.2. If the substance is a conductor, the bulb will light up. If it is an insulator the bulb remains unlit.

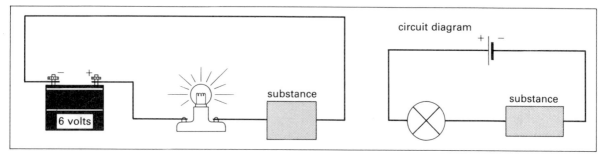

**Figure 15.2** *Testing to see if a substance conducts electricity*

When metals conduct electricity, no chemical reaction takes place. This is because no new substance is formed. The metals just allow electrons to move through their structure. This is what an electric current is, a flow of electrons. So far we have seen that:

1 metals are good conductors,
2 non-metals (except graphite) are insulators.

There are some substances that are neither simple conductors nor simple insulators.

### Semiconductors

These are substances that are normally poor conductors. But under the right conditions they can become quite good conductors. They are called **semiconductors**. Some metalloids are semiconductors. (A metalloid is an element that is like a metal in some ways and like a non-metal in others.) Metalloids are used to make transistors. Most transistors are made from the elements silicon or germanium.

### Electrolytes

**Electrolytes** are liquids or solutions that conduct electricity but which are chemically changed in the process. We do not usually talk of an electrolyte conducting electricity. We say that it is undergoing **electrolysis** (electrical splitting up). The rest of this chapter is about electrolysis.

## 15.2 Electrolytes, anions and cations

### How can you tell if a substance is an electrolyte?

**Figure 15.3** *The semiconductor silicon is the basis of the silicon chips used in computers*

Electrolytes conduct electricity. When they conduct there is some chemical change. The apparatus over the page can be used to test this. Two conductors, called **electrodes**, are dipped into the liquid or solution. These are then connected to the battery and bulb as shown in figure 15.4 (over the page). The electrode connected to the (+) side (or terminal) of the battery is the **positive electrode** or **anode**. The electrode connected to the (−) terminal of the battery is the **negative**

**Figure 15.4** *Testing for an electrolyte*

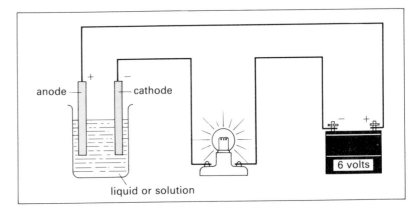

**electrode** or **cathode**. Two things will happen if the substance is an electrolyte:

1 the bulb will glow,
2 some new substances will form at the electrodes.

You might see some or all of the following:

1 effervescence (fizzing),
2 a new substance (perhaps a coloured substance) being formed,
3 an electrode being coated with a metal.

## What sort of substances are electrolytes?

An electrolyte is a liquid or a solution that contains ions. Substances that melt or dissolve to form electrolytes usually contain a metal and a non-metal. These substances are called ionic compounds. Solutions of acids in water are also electrolytes. These also contain ions.

## Where do ions go during electrolysis?

Ions have an electric charge. We know that opposite charges attract each other. This happens during electrolysis. Ions are attracted to the electrode that has the opposite charge (see figure 15.5).

Ions with a positive charge move towards the cathode (−). Because of this, positive ions are often called **cations**. Ions

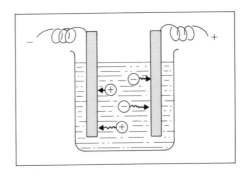

**Figure 15.5** *The movement of ions in electrolysis*

that have a negative charge move towards the anode (+). These ions are called **anions**.

Table 15.1 shows the electric charges on some common ions.

| Cations | | Anions | |
|---|---|---|---|
| silver | $Ag^+$ | hydroxide | $OH^-$ |
| copper | $Cu^{2+}$ | iodide | $I^-$ |
| hydrogen | $H^+$ | bromide | $Br^-$ |
| lead | $Pb^{2+}$ | chloride | $Cl^-$ |
| zinc | $Zn^{2+}$ | nitrate | $NO_3^-$ |
| magnesium | $Mg^{2+}$ | sulphate | $SO_4^{2-}$ |
| aluminium | $Al^{3+}$ | carbonate | $CO_3^{2-}$ |
| sodium | $Na^+$ | phosphate | $PO_4^{3-}$ |

**Table 15.1** *Some common ions and their electric charges*

You can see that metal ions are all cations. Non-metal ions, except hydrogen, are all anions.

When these ions reach the electrode some new substance is formed. Just what is formed does not only depend on the compound used. It also depends on whether we are using a molten (melted) electrolyte or a solution. We shall look at the electrolysis of molten electrolytes first.

## 15.3 The electrolysis of molten electrolytes

Table 15.2 gives some examples of what happens at the electrodes when molten compounds are electrolysed.

| Compound | Cations | Anions | Product at cathode | Product at anode |
|---|---|---|---|---|
| copper bromide | $Cu^{2+}$ | $Br^-$ | copper metal | bromine gas |
| lead iodide | $Pb^{2+}$ | $I^-$ | lead metal | iodine vapour |
| magnesium chloride | $Mg^{2+}$ | $Cl^-$ | magnesium metal | chlorine gas |
| silver bromide | $Ag^+$ | $Br^-$ | silver metal | bromine gas |

**Table 15.2** *The electrolysis of molten compounds*

You can see that every time:

**1** metals are formed at the cathode,
**2** non-metals are formed at the anode.

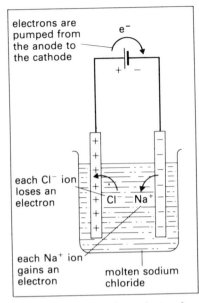

electrons are pumped from the anode to the cathode

$e^-$

each $Cl^-$ ion loses an electron

$Cl^-$    $Na^+$

each $Na^+$ ion gains an electron

molten sodium chloride

**Figure 15.6** *The electrolysis of molten sodium chloride*

The positive metal ions move towards the cathode. When they get there they lose their charge and change from metal ions into metal atoms. This means that a layer of metal builds up on the electrode.

The negatively charged non-metal ions go to the anode. When they get there they lose their charge and change into the non-metallic element.

## Where do the charges go?

Ions are atoms or groups of atoms that have an electric charge. They have this charge because the atoms have either gained or lost electrons. As we saw in section 13.3, the $Na^+$ ion is a sodium atom that has lost one of its electrons. The $Mg^{2+}$ ion is a magnesium atom that has lost two electrons. Negative ions are atoms that have gained electrons. The $Cl^-$ ion is a chlorine atom that has gained an extra electron.

Let us take the electrolysis of molten sodium chloride as an example. The apparatus needed is shown in figure 15.6. The battery is really an electron pump. Electrons are pulled in through the positive terminal and pushed out through the negative one (see figure 15.7).

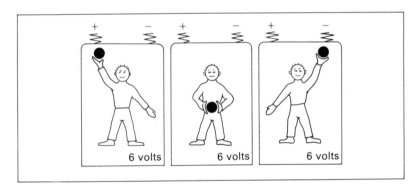

6 volts    6 volts    6 volts

**Figure 15.7** *Electrons are pulled in from the anode and pushed out to the cathode*

Electrons are negatively charged and so the cathode becomes negatively charged. The anode becomes positively charged because the battery is pulling electrons away from it.

Opposite charges attract each other, and so the $Cl^-$ ions move to the positive anode. Here the extra electrons are pulled off. So the chlorine ions becomes chlorine atoms:

a chloride ion loses an electron $\longrightarrow$ a chlorine atom

or         $Cl^-$        $-$        $e^-$        $\longrightarrow$        $Cl$

These chlorine atoms soon join up into ordinary chlorine molecules:

two chlorine atoms join $\longrightarrow$ a chlorine molecule

or                $2Cl$                $\longrightarrow$                $Cl_2$

The electrons that have been pulled off the chlorine ions travel through the wiring. The battery pumps them around the circuit to the cathode. When each $Na^+$ ion arrives at the cathode it gains an electron. As a result it becomes a sodium atom:

a sodium ion gains an electron $\longrightarrow$ a sodium atom

$$Na^+ \quad + \quad e^- \quad \longrightarrow \quad Na$$

In the end these sodium atoms join together and form a metallic coating of sodium on the cathode.

So we can see that in electrolysis, electrons are taken from anions and given to cations.

### Why don't solid ionic compounds conduct?

In electrolysis the ions must be free to move to the electrodes. In solid ionic compounds the ions are held firmly in one place. They are not free to move (see figure 15.8). So solid ionic compounds cannot undergo electrolysis.

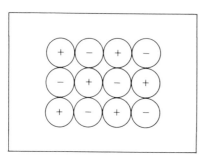

**Figure 15.8** *The structure of a solid ionic compound*

## 15.4 The electrolysis of aqueous electrolytes

Most ionic compounds have very high melting points. This makes the electrolysis of their molten compounds difficult. We can overcome this by using an aqueous solution (a solution in water). When ionic compounds dissolve in water, the ions separate into the solution.

| Compound | Cations present | Anions present | Product at cathode | Product at anode |
|---|---|---|---|---|
| copper bromide | $Cu^{2+}$ | $Br^-$ | copper | bromine |
| nickel iodide | $Ni^{2+}$ | $I^-$ | nickel | iodine |
| hydrochloric acid | $H^+$ | $Cl^-$ | hydrogen | chlorine |
| silver nitrate | $Ag^+$ | $NO_3^-$ | silver | oxygen |
| sodium chloride | $Na^+$ | $Cl^-$ | hydrogen | chlorine |
| magnesium sulphate | $Mg^{2+}$ | $SO_4^{2-}$ | hydrogen | oxygen |

**Table 15.3** *The products of electrolysis of an aqueous electrolyte*

These aqueous solutions often behave in just the same way as the molten ionic compound. This is so for each of the first three compounds in table 15.3. The fourth compound, silver nitrate, seems to be more complicated. Do we expect to get nitrogen, oxygen or a mixture of both at the anode? There is a simple rule that will give us the answer to this question:

During the electrolysis of an aqueous electrolyte chlorine is formed at the anode from chlorides, bromine from bromides, iodine from iodides and oxygen with all other anions.

The fifth compound in table 15.3 is sodium chloride. You might think you would get sodium at the cathode. Instead, hydrogen is formed. Hydrogen is also formed in place of other very reactive metals, such as sodium, potassium and calcium. This gives us another general rule:

Very reactive metals are not formed at the cathode during the electrolysis of aqueous electrolytes. Hydrogen gas is given off instead.

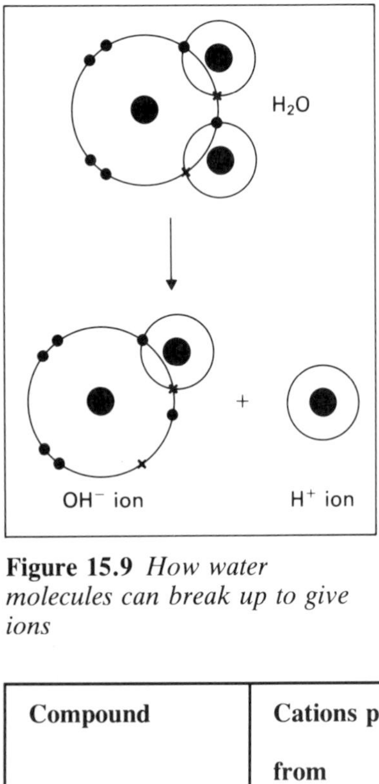

Figure 15.9 *How water molecules can break up to give ions*

### Why does electrolysing aqueous electrolytes sometimes give unexpected results?

Water is one of the very few solvents that will dissolve many ionic compounds. This is why water is used as a solvent. (It is also cheap!) Water consists mainly of molecules. These have no overall electric charge and so they are not affected by electrolysis. But some of these molecules break up to form ions, as shown in figure 15.9.

Because of this, aqueous solutions of electrolytes contain *two* sets of ions. There are the ions from the compound dissolved in the water. There are also the $H^+$ and $OH^-$ ions from the water itself. $H^+$ are called hydrogen ions and $OH^-$ hydroxide ions. Table 15.4 shows the products you get when you electrolyse some concentrated aqueous solutions.

| Compound | Cations present | | Anions present | | Product at cathode | Product at anode |
|---|---|---|---|---|---|---|
| | from compound | from water | from compound | from water | | |
| copper bromide | $\underline{Cu^{2+}}$ | $H^+$ | $\underline{Br^-}$ | $OH^-$ | copper | bromine |
| nickel iodide | $\underline{Ni^{2+}}$ | $H^+$ | $\underline{I^-}$ | $OH^-$ | nickel | iodine |
| hydrochloric acid | $\underline{H^+}$ | $H^+$ | $\underline{Cl^-}$ | $OH^-$ | hydrogen | chlorine |
| silver nitrate | $\underline{Ag^+}$ | $H^+$ | $NO_3^-$ | $\underline{OH^-}$ | silver | oxygen |
| sodium chloride | $Na^+$ | $\underline{H^+}$ | $\underline{Cl^-}$ | $OH^-$ | hydrogen | chlorine |

Table 15.4 *The products of electrolysis of some concentrated aqueous solutions (the underlined ions are those which are discharged at the electrodes)*

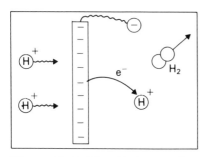

**Figure 15.10** *The discharge of H⁺ ions*

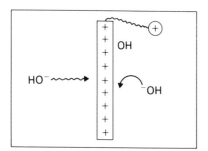

**Figure 15.11** *The discharge of OH⁻ ions*

The first three compounds in the table give the products you would expect. Ions from the water play no part in the electrolysis. But in the last two compounds, one of the ions from the ionic compound has not been discharged. Instead ions from the water have been discharged. Let us look at what happens when these H⁺ and OH⁻ ions reach the electrodes.

The H⁺ ion moves to the cathode. Here it accepts an electron and becomes a hydrogen atom. These atoms soon join up into molecules of hydrogen gas (see figure 15.10):

$$H^+ + e^- \longrightarrow H$$

$$2H \longrightarrow H_2$$

The OH⁻ ions move to the anode. Here they give up their extra electron. So the hydroxide group no longer has any charge (see figure 15.11):

$$4OH^- - 4e^- \longrightarrow 4OH$$

What happens next is shown in figure 15.12. The atoms from four hydroxide groups rearrange themselves, and two water molecules and an oxygen molecule are formed. The oxygen gas can be seen coming from the anode.

This is how it is possible to obtain hydrogen at the cathode and oxygen at the anode when we use aqueous solutions.

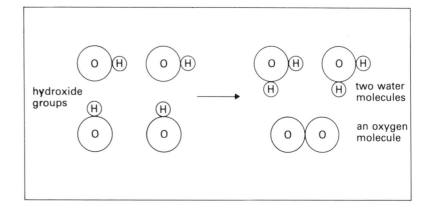

**Figure 15.12** *The formation of oxygen at the anode*

## 15.5 The electrolysis of 'water'

Pure water is a very poor conductor. It consists mainly of molecules. Despite this chemists often talk about the electrolysis of water! What they usually mean is the electrolysis of water containing sulphuric acid. When you add this acid to water, you get a solution which conducts well and which produces hydrogen and oxygen when it is electrolysed. These are just the products that you would expect from water. They are even produced in the right amounts: two parts of hydrogen for every one part of oxygen.

**Figure 15.13** *The electrolysis of 'water'*

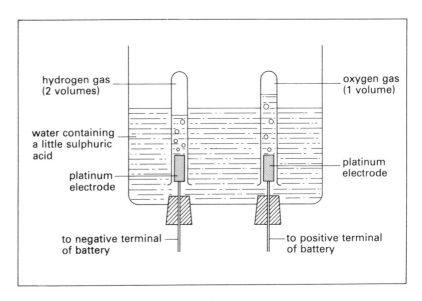

You can use the apparatus in figure 15.13 to test this.

Now let us look at what is happening during this electrolysis. The dilute acid contains three types of ion:

sulphuric acid $\longrightarrow$ hydrogen ions $+$ sulphate ions

$$H_2SO_4 \longrightarrow 2H^+ + SO_4^{2-}$$

Some of the water molecules break up into ions:

water $\longrightarrow$ hydrogen ions $+$ hydroxide ions

$$H_2O \longrightarrow H^+ + OH^-$$

Both the sulphate and the hydroxide ions are attracted to the anode $(+)$. When they reach the anode, it is the hydroxide ions that give up their electrons. Water and oxygen are formed, as we saw in section 15.4:

hydroxide ions $-$ electrons $\longrightarrow$ water $+$ oxygen

$$4OH^- - 4e^- \longrightarrow 2H_2O + O_2$$

These four electrons are pumped through the wiring by the battery. When they arrive at the cathode $(-)$, they are given to four hydrogen ions:

hydrogen ions $+$ electrons $\longrightarrow$ hydrogen

$$4H^+ + 4e^- \longrightarrow 2H_2$$

You can see from the above equations that two hydrogen molecules are produced for each oxygen molecule. Just what you would expect from the electrolysis of water!

## 15.6 How the electrodes can affect electrolysis

### The electrolysis of 'water' using different electrodes

Figure 15.13 showed what happens when water is electrolysed. In this case platinum electrodes were used. Platinum is often used as an electrode material. This is because it is inert (unreactive). It carries the electrons to and from the solution but does not itself react.

Graphite is also widely used as an electrode material. It is much cheaper than platinum. It is not, however, as inert. If dilute sulphuric acid is electrolysed using graphite electrodes, you do not get pure oxygen at the anode. Instead, you get a mixture of oxygen, carbon monoxide and carbon dioxide. The graphite, which is a form of carbon, has reacted with the oxygen.

### The electrolysis of copper sulphate solution

Sometimes we deliberately use an electrode that is not inert. This happens in copper refining.

First, let us look at the electrolysis of copper sulphate using inert platinum electrodes. There are four types of ion present:

$$\text{copper sulphate} \longrightarrow \text{copper ions} \quad + \text{sulphate ions}$$
$$CuSO_4 \longrightarrow Cu^{2+} \quad + \quad SO_4^{2-}$$

and

$$\text{water} \longrightarrow \text{hydrogen ions} + \text{hydroxide ions}$$
$$H_2O \longrightarrow H^+ \quad + \quad OH^-$$

Both the sulphate and the hydroxide ions move to the anode (+). Only the hydroxide ions give up their electrons, and so water and oxygen are formed.

$$\text{hydroxide ions} - \text{electrons} \longrightarrow \text{water} + \text{oxygen}$$
$$4OH^- \quad - \quad 4e^- \longrightarrow 2H_2O + \quad O_2$$

The electrons are pumped around to the cathode (−). Here there are both hydrogen ions and copper ions. The electrons are given to the copper ions:

$$\text{copper ions} \quad + \quad \text{electrons} \longrightarrow \text{copper metal}$$
$$2Cu^{2+} \quad + \quad 4e^- \longrightarrow 2Cu$$

So the cathode gets coated with copper and oxygen gas is formed at the anode.

Now look what happens if you use copper electrodes. You have the same ions in the solution as before. The sulphate and the hydroxide ions go to the anode. But neither loses its electrons. Instead, the copper anode itself loses electrons. As it does so it becomes copper ions which dissolve in the solution:

$$\text{copper} \longrightarrow \text{copper ions} \quad + \quad \text{electrons}$$
$$\text{Cu} \quad \longrightarrow \quad Cu^{2+} \quad + \quad 2e^-$$

The electrons from the copper are pumped around the circuit to the cathode. Here the electrons are given to copper ions:

$$\text{copper ions} \quad + \quad \text{electrons} \longrightarrow \text{copper metal}$$
$$Cu^{2+} \quad + \quad 2e^- \quad \longrightarrow \quad \text{Cu}$$

These results are different from the ones we had when we used a platinum electrode. This time copper has dissolved from the anode and been coated on to the cathode. The end result is just as if copper had been moved from one electrode to the other.

## 15.7  Some uses of electrolysis

### Electroplating

Sometimes it is useful to coat a metal with a thin layer of another metal to improve its appearance or prevent corrosion. We can do this in the laboratory using the apparatus shown in figure 15.15.

**Figure 15.14** *These three soup ladles have been electroplated with silver*

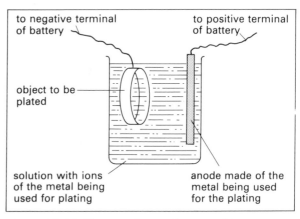

**Figure 15.15** *How to electroplate an object*

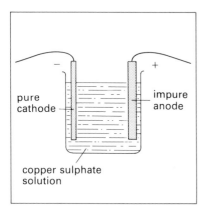

**Figure 15.16** *Purifying copper*

The object to be coated must be the cathode. The electrolyte must contain ions of the metal that are to go to the cathode and plate the object. This is known as **electroplating**.

## Copper purification

Very pure copper is used for electrical wiring. Even small amounts of impurities can cut down the conductivity of the copper noticeably. Figure 15.16 shows diagrammatically the apparatus used for purifying copper; figure 15.17 shows the purification being carried out on a large scale. The process is the one we have already looked at in section 15.6. Copper leaves the anode and is plated on to the cathode. Any impurities in the anode sink to the bottom of the container. The substance left at the bottom is sometimes called 'anode mud'! At the end, the copper cathode is very much larger because it now has a thick coating of pure copper.

**Figure 15.17** *Electrolytic refining of copper*

## The extraction of metals

Electrolysis is the only method that can be used to make some of the more reactive metals. Section 15.3 explained how sodium can be obtained from salt (sodium chloride) by electrolysis. This is in fact how sodium is manufactured from salt. You will find more about the electrolytic extraction of metals in section 17.3.

## The manufacture of sodium hydroxide

Sodium hydroxide is very important in industry. It is made by the electrolysis of a salt solution using a mercury cathode.

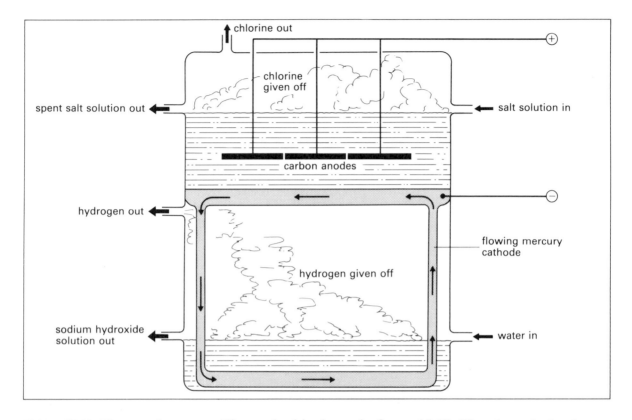

**Figure 15.18** *The manufacture of sodium hydroxide*

**Figure 15.19** *Hydroelectric power provides a relatively cheap source of electricity for industrial electrolyses*

The method is shown in figure 15.18. The electrolysis of aqueous sodium hydroxide would usually produce hydrogen at the cathode. This does not happen if a mercury cathode is used. Instead, sodium is formed and dissolves in the mercury. This solution of sodium in mercury is called an **amalgam**.

The amalgam is pumped into another container. Here it meets a current of water. The sodium in the amalgam reacts with the water and forms sodium hydroxide solution and hydrogen. The mercury is then pumped back into the electrolysis cell. The sodium hydroxide solution is taken out of the container and evaporated. This gives solid sodium hydroxide.

## 15.8 How much electricity?

Industrial processes using electrolysis can be expensive to run because of the electricity costs. Obviously we get more of the chemicals that are produced in an electrolysis if we use more electricity. This section shows you how to calculate how much.

The unit of electrical charge is the coulomb (C). We can calculate the number of coulombs that have passed from this equation:

charge (coulombs) = current (amps) × time (seconds)

Let's take the electrolysis of silver nitrate solution as an example. The $Ag^+$ ions will go to the cathode. Here each 1+ ion will gain an electron, changing the ion into a neutral atom of silver metal:

a silver ion gains an electron $\longrightarrow$ a silver atom

$$Ag^+(aq) \quad + \quad e^- \quad \longrightarrow \quad Ag(s)$$

Now a mole of any ion contains $6 \times 10^{23}$ ions (see section 7.2). Clearly this many $Ag^+$ ions will need a mole of electrons, or $6 \times 10^{23}$ electrons.

Experiments show that it takes 96 500 coulombs to form a mole of silver. So 96 500 C of charge must be $6 \times 10^{23}$ electrons, or a mole of electrons.

| $Ag^+(aq)$ | + | $e^-$ | $\longrightarrow$ | $Ag(s)$ |
|---|---|---|---|---|
| $6 \times 10^{23}$ $Ag^+$ ions | + | $6 \times 10^{23}$ electrons | $\longrightarrow$ | $6 \times 10^{23}$ Ag atoms |
| or 1 mole of silver ions | + | 1 mole of electrons | $\longrightarrow$ | 1 mole of silver atoms |
| that is, | | 96 500 C | gives | 108 g of silver |

Other metals having 1+ ions will need the same amount of electricity, namely 96 500 C or 1 mole of electrons. (This quantity of electricity used to be called the faraday; you may meet this name in older books.) Metals with 2+ ions will obviously need more electricity in order to produce a mole of the metal. They will need two moles of electrons.

| $Cu^{2+}(aq)$ | + | $2e^-$ | $\longrightarrow$ | $Cu(s)$ |
|---|---|---|---|---|
| 1 mole of copper ions | + | 2 moles of electrons | $\longrightarrow$ | 1 mole of copper atoms |
| | | $2 \times 96\,500$ C | gives | 63.5 g of copper |

### Example 1

What mass of copper will be formed when a current of 0.965 A flows for 1000 seconds through copper sulphate solution, $CuSO_4$ (figure 15.20)?

**1** 1 mole of copper = relative atomic mass in grams = 63.5 g

**2** Charge passed = 0.965 A × 1000 s = 965 C

**3** It will take 2 × 96 500 = 193 000 C to change 1 mole of a 2+ ion ($Cu^{2+}$) into the metal.

Electrical charge         Mass of metal
   193 000 C                 63.5 g

scale-down factor        multiply
= 965/193 000 = 0.005     by 0.005
   965 C                   0.3175 g

**Answer:** 965 C of charge gives 0.3175 g of copper.

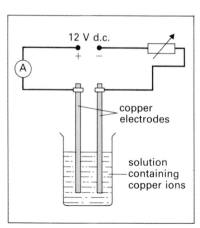

**Figure 15.20** *Example 1 –*
*electrolysis of copper sulphate*

12 V d.c.
+ −
A
copper electrodes
solution containing copper ions

Since we know that 1 mole of electrons is equal to 96 500 C, and that this amount of electricity will deposit 1 mole of a metal from 1+ ions or $\frac{1}{2}$ mole of metal from 2+ ions, we can find out what charge a metal ion has.

### Example 2

Copper is able to form two kinds of compound, in which the copper ions carry either a 1+ or a 2+ charge. Using the apparatus shown in figure 15.20, an electric current of 0.193 amps was passed through a solution of a copper compound for 2000 seconds. The mass of the cathode increased by 0.254 g as a result of a copper coating. What is the charge on the copper ion?

**1** 1 mole of copper = 63.5 g

**2** Charge passed = 0.193 A × 2000 s = 386 C

**3** Mass of metal           Electrical charge

     0.254 g                 386 C

     | scale-up factor         | multiply

     ↓ = 63.5/0.254 = 250     ↓ by 250

     63.5 g                 96 500 C

So it takes 96 500 C to produce 1 mole of copper.
**Answer**: The copper ions must have had a 1+ charge.

## Summary: Electricity and chemicals

- ★ Insulators do not conduct electricity.
- ★ Conductors conduct electricity.
- ★ Electrolytes conduct electricity and change chemically in the process.
- ★ Electrolysis takes place when an electrolyte conducts electricity.
- ★ Electrolytes are solutions or liquids that contain ions.
- ★ Anions are ions that move towards the anode. They are negative ions.
- ★ Cations are ions that move towards the cathode. They are positive ions.
- ★ Cations are usually metal ions.
- ★ Anions are usually non-metal ions.
- ★ Metals (and hydrogen) are formed at the cathode.
- ★ Non-metals are formed at the anode.
- ★ Very reactive metals do not form at the cathode during electrolysis of an aqueous electrolyte. Hydrogen gas is given off instead.
- ★ Oxygen is formed at the anode, except in the electrolysis of chlorides, bromides and iodides.
- ★ The products formed during electrolysis may depend on the electrodes used.

* Electrolysis is used in many industrial processes.
* 1 mole of electrons (96 500 coulombs of charge) will deposit:
  1 mole of metal from ions with a 1+ charge,
  $\frac{1}{2}$ mole of metal from ions with a 2+ charge,
  $\frac{1}{3}$ mole of metal from ions with a 3+ charge.

## Questions

**1** Choose from the list below the substance which best fits each of the descriptions that follow:
>  copper
>  graphite
>  silicon
>  sodium chloride solution
>  polythene

(a) an insulator
(b) a material used in transistors
(c) an electrolyte
(d) one of the best electrical conductors
(e) a non-metal but a conductor

**2** Explain the meaning of each of the following words:
(a) electrolysis
(b) anode
(c) cation
(d) battery
(e) aqueous electrolyte

**3** A pupil did an experiment to find out how electricity affected different substances. He dissolved the substances in water. Then he passed electricity into the solutions using two carbon electrodes. He wrote down his results in a table with the following headings:

| Substance | What was formed at the anode | What was formed at the cathode |
|-----------|------------------------------|--------------------------------|
|           |                              |                                |

Copy out these headings and write into the table what you think happened for each of the following substances:
(a) copper bromide
(b) nickel iodide
(c) copper sulphate
(d) sodium bromide
(e) silver nitrate
(f) lead nitrate
(g) sugar
(h) gold chloride
(i) magnesium nitrate
(j) aluminium chloride

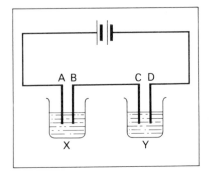

**4** A pupil did an experiment to investigate the effect of using different electrode materials. She set up the circuit shown in the diagram.

Beakers X and Y both contain copper sulphate solution. A and B are carbon electrodes and C and D are copper electrodes.

(a) What would you expect to happen at each of the electrodes A, B, C and D?

(b) What changes would you expect to occur to the solutions in beakers X and Y?

**5** Some substances conduct. Others undergo electrolysis. Give examples of each type of substance. Explain how undergoing electrolysis is different from just conducting.

**6** During electrolysis electrons are taken from anions and are given to cations. Show that this happens in the electrolysis of

(a) molten sodium chloride,

(b) aqueous copper chloride.

Is it true in *all* electrolyses?

**7** The electrolysis of aqueous aluminium sulphate using platinum electrodes gives hydrogen and oxygen at the electrodes. Explain, in as much detail as you can, where these gases come from.

**8** Coating non-metallic objects with metal is sometimes called electroforming. It is important in the manufacture of gramophone records. Imagine you wanted to copper-plate an oak leaf. What problems would you have? Can you think of any ways of overcoming these problems?

**9** A pupil passed an electric current of 0.5 amps through a solution of a cobalt compound for 32 minutes 10 seconds. During this time the mass of the cathode increased by 0.59 g.

(a) What is the mass of 1 mole of cobalt?

(b) How many coulombs of electricity were passed in the experiment?

(c) How many coulombs would have been needed to produce 1 mole of cobalt?

(d) What is the charge on the cobalt ion in this compound?

(e) Draw a diagram of an apparatus suitable for carrying out this experiment.

## Electrochemical polishing

Electroplating is a process in which ions from an electrolyte gain electrons at the cathode and produce a metal coating. In copper-plating, for example, copper ions gain electrons at the cathode:

$$Cu^{2+}(aq) + 2e^- \longrightarrow Cu(s)$$

A less well known technique is **electropolishing**. This process is the reverse of electroplating. A metal object is made the anode in a circuit, and during electrolysis a thin layer of atoms from the metal surface lose electrons and turn into soluble metal ions. For example, if copper is electropolished then:

$$Cu(s) - 2e^- \longrightarrow Cu^{2+}(aq)$$

If currents and electrolyte temperatures are chosen carefully, brightly polished surfaces can be obtained in times as short as half a minute. The method has the added advantage of not contaminating the surface with particles of abrasive. It is suitable for a wide range of metals, including very hard ones such as tungsten and chromium.

  **Electrochemical machining** takes the same idea further. Here the metal to be shaped is made into the anode. The 'cutting tool' is the cathode. To shape the metal the cathode is brought very close to the anode and a low voltage is applied. The metal near to the cathode dissolves away from the anode. By moving the cathode appropriately the anode can be shaped. The method is very good for complex shapes as well as simple ones. It is also good for hard metals as it keeps the anode at a fairly even temperature. (Conventional machining techniques sometimes cause uneven heating which may produce stress and brittleness in the metal.)

  Both electropolishing and electromachining require the use of definite, known amounts of electricity. These can be calculated using the fact that 96 500 coulombs of electricity will remove 1 mole of a metal that forms 1+ ions, $\frac{1}{2}$ mole of a metal that forms 2+ ions, and so on.

## Questions

**10** Write ionic equations for the process occurring at the cathode during electroplating with
  (a) an electrolyte containing $Ag^+(aq)$ ions,
  (b) an electrolyte containing $Ni^{2+}(aq)$ ions.
**11** During a nickel-plating experiment a current of 100 mA was passed for 60 minutes.
  (a) How many coulombs were passed?
  (b) What is the mass of 1 mole of nickel?
  (c) What mass of nickel will be deposited? (Assume the solution contained $Ni^{2+}(aq)$ ions.)
**12** During electropolishing a chromium object it was intended to remove 1.73 g of chromium metal. During the process $Cr^{3+}(aq)$ ions are formed. A current of 1000 amps was used.
  (a) What is the mass of 1 mole of chromium?
  (b) How many coulombs are needed to remove 1 mole of chromium?

(c)  How many moles is 1.73 g of chromium?

(d)  How many coulombs are needed to remove 1.73 g of chromium?

(e)  How long must the 1000 amp current remain switched on for?

13 Suggest possible advantages of using electrochemical machining to shape the following metals:

(a)  tungsten,

(b)  titanium,

(c)  magnesium.

# 16 The activity series

## 16.1 The reactivity of metals

**Figure 16.1** *These cars will rust quickly. The iron they are made of is quite a reactive metal*

Iron is quite a reactive metal. It will rust quickly unless it is protected from the air. Gold is much less reactive. A gold ring will last for many years without tarnishing. This section is about the reactivity of various metals.

### Reactions with oxygen

Sodium is a very reactive metal. If it is heated, it burns with a sudden yellow flash. It is reacting with oxygen in the air to form sodium oxide.

If you heat magnesium, it burns steadily with a brilliant white flame. The ash that remains is magnesium oxide.

When copper is heated in air nothing much seems to happen. But when it is taken out of the flame, it has a black coating of copper oxide on it. The surface of the copper has reacted with oxygen.

Platinum is even less reactive than copper. Platinum keeps its shiny appearance even when strongly heated. It does not react with oxygen.

These reactions are summarized in figure 16.2.

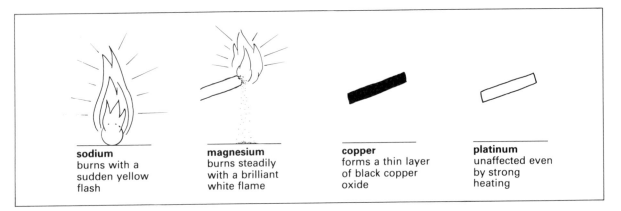

**sodium**
burns with a
sudden yellow
flash

**magnesium**
burns steadily
with a brilliant
white flame

**copper**
forms a thin layer
of black copper
oxide

**platinum**
unaffected even
by strong
heating

**Figure 16.2** *The reactions of some metals with oxygen*

We can make a list of these metals in order of their reactivity. Sodium reacts most vigorously and so it comes at the top. Platinum will come at the bottom.

sodium

magnesium

copper

platinum

decreasing
reactivity

Such a list is called an **activity series**.

## Reactions with water

If a small piece of sodium is put into water, it darts about over the surface giving off hydrogen gas.

Magnesium reacts extremely slowly with cold water. But it reacts vigorously with steam. These experiments are shown in figures 16.3 and 16.4.

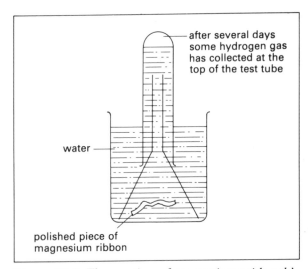

after several days
some hydrogen gas
has collected at the
top of the test tube

water

polished piece of
magnesium ribbon

**Figure 16.3** *The reaction of magnesium with cold water*

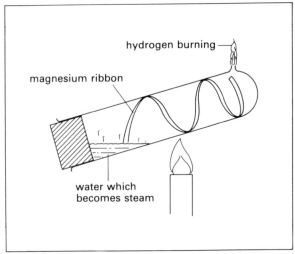

hydrogen burning

magnesium ribbon

water which
becomes steam

**Figure 16.4** *The reaction of magnesium with steam*

Zinc also reacts with steam but less vigorously than magnesium. Neither copper nor platinum reacts with either water or steam. We can now add zinc to our activity series:

sodium
magnesium
zinc
copper
platinum

decreasing reactivity

## The activity series

By carrying out more experiments, we can add other metals to the series. The exact order changes slightly from one reaction to another. But even so, the series still helps us to predict what will happen in many reactions. This is what we get:

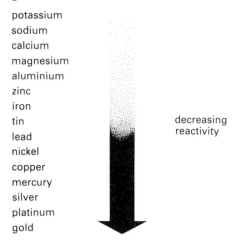

potassium
sodium
calcium
magnesium
aluminium
zinc
iron
tin
lead
nickel
copper
mercury
silver
platinum
gold

decreasing reactivity

## How to remember the order of the common metals

You can remember the order of some of the more common metals in the activity series by learning this 'silly sentence'. The first letter of each word in the sentence is the same as the first letter of the element in the series.

| | |
|---|---|
| Please | Potassium |
| Send | Sodium |
| Charlie's | Calcium |
| Monkeys | Magnesium |
| And | Aluminium |
| Zebras | Zinc |
| In | Iron |
| Large | Lead |
| Cages | Copper |
| Most | Mercury |
| Securely | Silver |
| Guarded | Gold |

## 16.2  The electrochemical series

### How the electrochemical series is found

An electric cell (a battery) can be made by dipping strips of two different metals into salt solution (see figure 16.5).

**Figure 16.5** *Measuring the voltage of a 'home-made' cell*

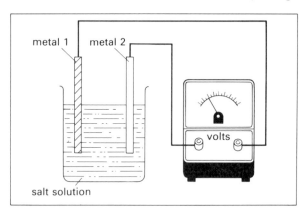

The voltages obtained from several of these cells are shown in table 16.1.

| Metal 1 | Metal 2 | Voltage (volts) |
|---------|---------|-----------------|
| magnesium | copper | 1.5 |
| aluminium | copper | 0.9 |
| zinc | copper | 0.9 |
| iron | copper | 0.5 |
| lead | copper | 0.5 |
| nickel | copper | 0.3 |
| copper | copper | 0.0 |

**Table 16.1** *The voltages obtained from several 'home-made' cells*

You can see that the voltages produced are in the same order as the activity series we looked at in section 16.1. This electrically measured activity series is called the **electrochemical series**.

### How can the electrochemical series tell us about chemical activity?

To understand this, you need to know a bit more about what makes the electric current in these cells.

Take the magnesium–copper cell as an example. If you could magnify the magnesium electrode billions of times,

**Figure 16.6** *A magnesium–copper cell*

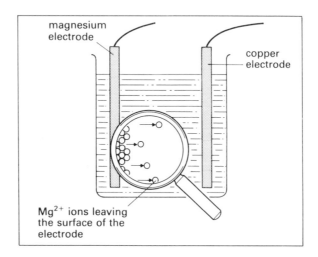

you would see magnesium ions leaving the surface (as in figure 16.6). These are atoms of electrode which have left two of their electrons behind.

magnesium metal atom $\longrightarrow$ magnesium ion + 2 electrons

$$Mg \longrightarrow Mg^{2+} + 2e^-$$

The electrons that are left behind move out of the electrode, along the wire and through the voltmeter. Now magnesium loses electrons very easily. So you get a high reading on the voltmeter.

Zinc does not lose its electrons as easily as the magnesium does. So if you use zinc instead of magnesium you get a lower voltage reading.

So now we can see that the position of a metal in the electrochemical series depends on how easily it loses electrons. This also decides the chemical activity of a metal.

When metals react they form ionic compounds (see section 13.3). For example, zinc reacts with chlorine to form zinc chloride. Magnesium reacts with chlorine to form magnesium chloride.

$$Zn(s) + Cl_2(g) \longrightarrow ZnCl_2(s)$$

$$Mg(s) + Cl_2(g) \longrightarrow MgCl_2(s)$$

In each case the metal loses electrons. The more easily it loses electrons, the more reactive it is.

From this, we can see that the reason why the orders of the electrochemical series and the activity series are the same is that they are both measuring how easily metals lose electrons.

## 16.3 More about the reactions of some metals and their compounds

In section 16.1 we looked at how reactive certain metals were. In this section we are going to look at these reactions

**Figure 16.7** *Aluminium powder burns with an intense flame and so it can be used for distress flares*

again in more detail. We shall also look at the equations for the reactions. You can remind yourself about writing equations by re-reading section 8.1.

## The reactions of metals with oxygen

| Metal | What happens | Chemical equation |
|---|---|---|
| potassium<br>sodium<br>calcium<br>magnesium | all burn brightly; a white, or off-white, ash is left | $K + O_2 \longrightarrow KO_2$<br>$2Na + O_2 \longrightarrow Na_2O_2$<br>$2Ca + O_2 \longrightarrow 2CaO$<br>$2Mg + O_2 \longrightarrow 2MgO$ |
| aluminium<br>zinc<br>iron | all react with oxygen if in powder form; aluminium and zinc leave white ash, iron leaves a black oxide | $4Al + 3O_2 \longrightarrow 2Al_2O_3$<br>$2Zn + O_2 \longrightarrow 2ZnO$<br>$3Fe + 2O_2 \longrightarrow Fe_3O_4$ |
| copper<br>mercury | both react at the surface only; copper forms a black coating; mercury forms a red coating | $2Cu + O_2 \longrightarrow 2CuO$<br>$2Hg + O_2 \longrightarrow 2HgO$ |
| silver<br>gold | neither reacts with oxygen | no reaction<br>no reaction |

**Table 16.2** *How metals react with oxygen*

## The reactions of metals with water

| Metal | What happens | Chemical equation |
|---|---|---|
| potassium | shoots across the surface; hydrogen gas is produced which burns with a lilac flame | $2K + 2H_2O \longrightarrow H_2 + 2KOH$ |
| sodium | similar to potassium, but the hydrogen does not catch fire | $2Na + 2H_2O \longrightarrow H_2 + 2NaOH$ |
| calcium | the grey metal sinks and gives off a slow stream of hydrogen gas | $Ca + 2H_2O \longrightarrow H_2 + Ca(OH)_2$ |
| magnesium | extremely slow reaction, but does react with steam leaving white magnesium oxide | $Mg + H_2O \longrightarrow H_2 + MgO$ |
| aluminium<br>zinc<br>iron | all three react with steam; aluminium is less reactive than magnesium, while zinc and iron are less reactive still | $2Al + 3H_2O \longrightarrow 3H_2 + Al_2O_3$<br>$Zn + H_2O \longrightarrow H_2 + ZnO$<br>$3Fe + 4H_2O \longrightarrow 4H_2 + Fe_3O_4$ |
| copper<br>mercury<br>silver<br>gold | these do not react with either water or steam | no reaction<br>no reaction<br>no reaction<br>no reaction |

**Table 16.3** *How metals react with water*

## The reactions of metals with dilute acids

Many metals react with acids to give hydrogen gas. Nitric acid reacts in a different way from most other acids. This is explained in section 20.5. Table 16.4 describes the reactions of metals with hydrochloric acid.

| Metal | What happens | Chemical equation |
|---|---|---|
| potassium<br>sodium<br>calcium | these three are dangerously reactive and should not be added to acids | $2K + 2HCl \longrightarrow H_2 + 2KCl$<br>$2Na + 2HCl \longrightarrow H_2 + 2NaCl$<br>$Ca + 2HCl \longrightarrow H_2 + CaCl_2$ |
| magnesium<br>aluminium<br>zinc<br>iron | these four react less and less as you go down the activity series | $Mg + 2HCl \longrightarrow H_2 + MgCl_2$<br>$2Al + 6HCl \longrightarrow 3H_2 + 2AlCl_3$<br>$Zn + 2HCl \longrightarrow H_2 + ZnCl_2$<br>$Fe + 2HCl \longrightarrow H_2 + FeCl_2$ |
| copper<br>mercury<br>silver<br>gold | these four do not react with dilute acids | no reaction<br>no reaction<br>no reaction<br>no reaction |

**Table 16.4** *How metals react with hydrochloric acid*

## The reactions of some compounds of metals

The activity series is not only useful for predicting how metals will react. It also tells you about the compounds of metals. Table 16.5 lists how compounds behave depending on the position of the metal in the activity series. You should be able to see how the compounds of metals from the same part of the series have similar properties.

| Metal | Oxide: action of hydrogen | Oxide: effect of heat | Hydroxide: solubility | type of solution | Hydroxide: effect of heat | Nitrate: effect of heat | Carbonate: effect of heat |
|---|---|---|---|---|---|---|---|
| potassium | none | none | soluble | alkaline | none | forms nitrite | no reaction |
| sodium | none | none | soluble | alkaline | none | forms nitrite | no reaction |
| calcium | none | none | slightly soluble | alkaline | forms oxide | forms oxide | forms oxide |
| magnesium | none | none | slightly soluble | alkaline | forms oxide | forms oxide | forms oxide |
| aluminium | none | none | insoluble | – | forms oxide | forms oxide | forms oxide |
| zinc | none | none | insoluble | – | forms oxide | forms oxide | forms oxide |
| iron | slight reaction | none | insoluble | – | forms oxide | forms oxide | forms oxide |
| copper | forms metal | none | insoluble | – | forms oxide | forms oxide | forms oxide |
| mercury | forms metal | forms metal | doesn't exist | | | forms metal | doesn't exist |
| silver | forms metal | forms metal | doesn't exist | | | forms metal | very unstable |
| gold | forms metal | forms metal | doesn't exist | | | forms metal | doesn't exist |

**Table 16.5** *The behaviour of compounds of metals in the activity series*

The chemical equations for some of these reactions are given in table 16.6.

| Metal | Effect of heat upon | | |
|---|---|---|---|
| | hydroxide | carbonate | nitrate |
| potassium<br>sodium | unaffected<br>unaffected | unaffected<br>unaffected | $2KNO_3 \longrightarrow 2KNO_2 + O_2$<br>$2NaNO_3 \longrightarrow 2NaNO_2 + O_2$ |
| calcium<br>magnesium<br>zinc<br>copper | $Ca(OH)_2 \longrightarrow CaO + H_2O$<br>$Mg(OH)_2 \longrightarrow MgO + H_2O$<br>$Zn(OH)_2 \longrightarrow ZnO + H_2O$<br>$Cu(OH)_2 \longrightarrow CuO + H_2O$ | $CaCO_3 \longrightarrow CaO + CO_2$<br>$MgCO_3 \longrightarrow MgO + CO_2$<br>$ZnCO_3 \longrightarrow ZnO + CO_2$<br>$CuCO_3 \longrightarrow CuO + CO_2$ | $2Ca(NO_3)_2 \longrightarrow 2CaO + 4NO_2 + O_2$<br>$2Mg(NO_3)_2 \longrightarrow 2MgO + 4NO_2 + O_2$<br>$2Zn(NO_3)_2 \longrightarrow 2ZnO + 4NO_2 + O_2$<br>$2Cu(NO_3)_2 \longrightarrow 2CuO + 4NO_2 + O_2$ |
| mercury<br>silver<br>gold | hydroxides of these three compounds are too unstable to exist | carbonates of these three compounds are too unstable to exist | $Hg(NO_3)_2 \longrightarrow Hg + 2NO_2 + O_2$<br>silver and gold nitrates both decompose to the metal |

**Table 16.6** *The chemical equations for some of the reactions in table 16.5*

**Figure 16.8** *This bronze incense burner from Cyprus dates back to the 12th century* B.C.

# 16.4 Some uses of metals and alloys

The discovery of how to make metals marked the end of the Stone Age. Ever since that time, metals have been important in our lives. Look around your house and garage. How many items are made out of metal? Why do you think metals were chosen for these items?

## Why choose metals?

Some important properties of most metals are:

1 good thermal and electrical conductivity,
2 good malleability (can be hammered into new shapes),
3 good strength,
4 high melting points,
5 lustrous (shiny) appearance,
6 usually dense (heavy).

Metals are usually chosen because one or more of the above properties are needed.

## Why choose a particular metal?

When you choose a metal for a particular job, you naturally want to choose the most suitable one. For example, all metals conduct but silver, copper and aluminium are some of the best electrical conductors. So copper and aluminium (but not silver – it is too expensive!) are widely used in electrical work.

Magnesium and aluminium are both very light metals. They are used in aircraft.

Titanium has a fairly low density as well as a high melting point. This makes it useful in jet-engine construction.

Lead is very dense. So it is used for diving weights.

As well as the properties listed above, there are two other things that have to be considered when choosing a metal:

**1** its price,
**2** its resistance to corrosion.

## Corrosion

The rate at which metals corrode is usually related to the position of the metal in the activity series. There are some important exceptions to this. Table 16.7 lists how some metals corrode.

| Metal | How easily it corrodes |
|---|---|
| potassium | tarnishes immediately in air |
| sodium | tarnishes immediately in air |
| calcium | corrodes very quickly in air |
| magnesium | corrodes in air |
| aluminium | appears not to corrode |
| zinc | appears not to corrode |
| iron | slowly corrodes to give rust |
| copper | hardly any corrosion |
| mercury | hardly any corrosion |
| silver | hardly any corrosion |
| gold | does not corrode |

**Table 16.7** *How easily some metals corrode*

**Figure 16.9** *Aluminium forms a layer of oxide at its surface. Sometimes this layer is deliberately made thicker by a process called anodizing*

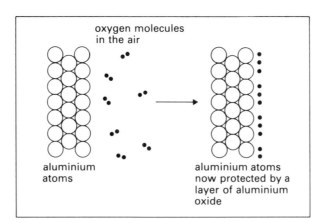

oxygen molecules in the air

aluminium atoms

aluminium atoms now protected by a layer of aluminium oxide

**Figure 16.10** *This piece of jewellery dates back to about 1700 B.C. It is still in perfect condition as the gold does not corrode*

You can see that aluminium and zinc do not seem to corrode. What actually happens is this. These metals do react to form metal oxides. But unlike most metal oxides, these do not flake off the surface. They stick fast. Before long the whole surface is covered in a layer of the oxide (see figure 16.9, on the previous page). This layer prevents further air from attacking the metal, so no further corrosion can take place. Metals low down in the activity series resist corrosion well, and this property often makes them useful.

For thousands of years, gold, silver and copper have been used to make coins. This is partly because they do not corrode. Alloys (mixtures of metals) containing mercury and silver are used for dental fillings. But a filling that corroded would not be much use! Copper is also used in plumbing. It is chosen because it is easy to bend and is not corroded by water passing through it.

Reactive metals such as sodium, potassium and calcium have no everyday uses. They corrode too quickly.

We have already seen that aluminium and magnesium are used in aircraft alloys. Aluminium is also used in many other ways (a few are shown in figure 16.12) because of its good conductivity, low density and high corrosion resistance.

Iron is used even though it corrodes. The rusting of iron is described in section 9.5.

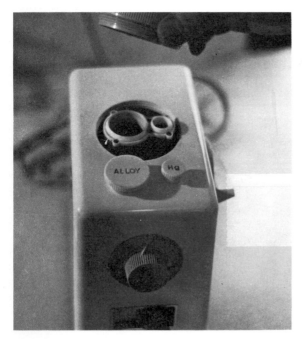

**Figure 16.11** *This machine mixes silver and mercury to make an alloy for filling teeth*

**Figure 16.12** *Aluminium foil is used to protect a variety of foods because it does not corrode easily*

## Alloys

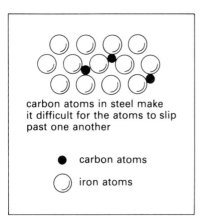

carbon atoms in steel make it difficult for the atoms to slip past one another

● carbon atoms

◯ iron atoms

**Figure 16.13** *Slip is more difficult in alloys than in pure metals*

**Alloys** are mixtures of metals. Steel, which is a mixture of iron and carbon, is also considered an alloy.

One of the main reasons for making alloys is to increase the strength of the metal. Pure iron, for example, can be made to change its shape fairly easily. Wrought iron gates are made from such pure iron. Steel is much stronger. Its shape changes much less readily. So it is a better material to use if strength is an important consideration.

The reason for such increased strength is as follows. In a pure metal the regular arrangement of the atoms allows atoms to slip readily past one another. In an alloy the different sized atoms can make this 'slip' much more difficult (see figure 16.13).

Alloys may also be used for other reasons. Solder is a mixture of lead and tin. It has a melting point lower than that of either pure metal. This makes it useful for joining wires.

An iron alloy containing chromium and nickel does not rust. It is known as stainless steel.

Table 16.8 lists some common alloys and the uses that are made of them.

| Alloy | Approximate composition | Uses |
|---|---|---|
| alnico | iron, aluminium, nickel, cobalt | permanent magnets |
| brass | 60% copper, 40% zinc | machine bearings |
| bronze | 90% copper, 10% tin | machine bearings, some gears |
| constantan | 60% copper, 40% nickel | thermocouple wires |
| coinage bronze | 95% copper, 3.5% tin, 1.5% zinc | 'copper' coins |
| cupronickel | 75% copper, 25% nickel | 'silver' coins |
| duralumin | 95% aluminium, 4% copper | light alloy used in aircraft, racing bikes |
| solder | 63% tin, 38% lead | joining electrical wires |
| steel | 99% iron, 1% carbon | machinery, bridges |
| stainless steel | 74% iron, 18% chromium, 8% nickel | cutlery, tools, bicycle wheels |
| tungsten steel | iron, carbon, tungsten | drills |

**Table 16.8** *Composition and uses of some alloys*

## Summary: The activity series

* Metals can be arranged in order of chemical activity.
* You can remember the order of the more common metals in the activity series:

| Please | Send | Charlie's | Monkeys |
| Potassium | Sodium | Calcium | Magnesium |

| And | Zebras | In | Large | Cages |
| Aluminium | Zinc | Iron | Lead | Copper |

| Most | Securely | Guarded. |
| Mercury | Silver | Gold |

* The activity of a metal can be found from its position in the electrochemical series.
* The activity series can be used for predicting the reactions of the metals. It can also predict the reactions of compounds of metals.
* Alloys (mixtures of metals) are often used instead of pure metals. They are often stronger and may also have other improved properties.

## Questions

1 Five properties of metals are listed below. Which of these properties is the most important reason for choosing the named metal for each of the purposes described?
    good electrical conductors
    high strength
    high malleability
    lustrous
    dense
(a) Diving weights are made of lead.
(b) Bridges are made of steel.
(c) Wire is often made of copper.
(d) Car headlamp reflectors are often chromium-plated.
(e) The filaments in bulbs are made of tungsten.

2 These questions are about the metals A to E in the table below.

| Metal | Action of dilute hydrochloric acid | Effect of heating in air |
|-------|-----------------------------------|--------------------------|
| A | rapid effervescence | burns with a brilliant white flame |
| B | no reaction | no reaction |
| C | no reaction | gets a thin black coating |
| D | violent reaction | burns quickly with a bright yellow flame |
| E | steady effervescence | burns with a greenish flame |

The metals were sodium, gold, magnesium, zinc and copper, but this is not the order in which they are given in the table.

(a) Identify each of the metals A to E.

(b) What gas is produced when metals react with dilute hydrochloric acid?

(c) How can you test for this gas?

(d) Which two of these metals do not react with water?

(e) Zinc reacts quickly with steam. Design an apparatus for carrying out the reaction of zinc with steam and collecting the hydrogen gas that is formed.

**3** Examine the data in the table below.

| Metal | Melting point (°C) | Boiling point (°C) | Density (g/cm$^3$) | Electrical resistance ($10^{-8}$ m) | Strength | Speed at which it corrodes |
|-------|--------------------|--------------------|--------------------|-------------------------------------|----------|----------------------------|
| A | 1083 | 2600 | 8.94 | 1.56 | medium | nil |
| B | 658 | 1800 | 2.70 | 2.45 | medium | nil |
| C | 1725 | 3260 | 4.49 | 43.10 | high | nil |
| D | 1528 | 2735 | 7.86 | 8.90 | high | quite slow |
| E | 3400 | 5700 | 19.10 | 4.90 | high | nil |
| F | 98 | 892 | 0.97 | 4.80 | low | very fast |

Decide which of the metals A to F you would use for the following purposes:

(a) the filament in a light bulb,

(b) electrical house wiring,

(c) overhead electricity cables,

(d) military aircraft and missiles,

(e) a bridge.

In each case give reasons to support your choice.

*The 'flint' in this lighter is really an alloy of cerium*

# Cerium

This is about a silvery-grey metal called cerium. There are many uses of this metal, although the chances are you have never heard of it. Perhaps its most 'everyday' use is as an alloy in cigarette-lighter and gas-lighter 'flints'. Similar alloys are used in 'tracer' bullets and shells. Cerium is also used as a catalyst in the petroleum industry. Small amounts of the metal are often included in television tubes to react with, and thus remove, final traces of oxygen. Adding cerium to steel improves its strength and heat resistance; cerium-containing steel can be used to make surgical instruments.

Cerium–magnesium alloys are light and strong and are used to make jet-engine components.

The metal reacts slowly in moist air forming yellow-green cerium(IV) oxide, $CeO_2$. At temperatures above about 300°C cerium catches fire, forming the same oxide. This oxide is used to make high-quality optical glasses, and as an abrasive for polishing lenses and mirrors.

Cerium reacts slowly in cold water but in hot water it rapidly gives hydrogen gas and cerium hydroxide, $Ce(OH)_3$. Upon heating the hydroxide forms a greyish-green oxide of formula $Ce_2O_3$.

The hydroxide reacts with acids to form cerium salts. With nitric acid cerium(III) nitrate is formed. Bubbling carbon dioxide through the hydroxide gives cerium(III) carbonate.

## Questions

4 Here are the symbols of some common elements arranged in order of decreasing activity. Whereabouts in this series would you place cerium?

K   Na   Ca   Mg   Al   Zn   Fe   Pb   Cu   Hg   Ag   Au

5 Write properly balanced equations for each of the following reactions:
(a) the reaction of cerium with water,
(b) the conversion of the hydroxide into the oxide $Ce_2O_3$,
(c) the reaction of the hydroxide with carbon dioxide.

6 Will the carbonate $Ce_2(CO_3)_3$ decompose easily or not? Explain your answer and write an equation for this reaction.

7 What products will be formed when the nitrate is strongly heated?

8 What do you think will happen when cerium is added to dilute nitric acid?

9 Plan an experiment to study the effect of temperature on the rate of reaction of cerium with water. Draw a labelled diagram of your apparatus and construct a table for your results.

10 Cerium is not shown on all copies of the periodic table. One student said he thought it must be another element belonging to group 2. Another said she thought it was a transition element. Try and list two or three pieces of information from the text to support each student's arguments.

# 17 Competition reactions and the extraction of metals

## 17.1 Competition reactions involving metals

**Figure 17.1** *Thermite welding of a railway track*

The photograph in figure 17.1 shows a competition. Aluminium and iron are the competitors. Oxygen is the prize. A mixture of powdered aluminium and iron oxide is put between the ends of two lengths of railway track. The mixture is then set alight. The reaction that follows produces iron. It also gives out so much heat that the iron melts and welds the track together.

The reaction can be written:

aluminium + iron oxide $\longrightarrow$ iron + aluminium oxide

You can see that aluminium is the winner of this competition. It has taken the oxygen from the iron. This section is about **competition reactions** like this one.

### What makes a good competitor?

If a metal is to be a good competitor, it must 'like' forming compounds. This is just another way of saying that it must be reactive. When we say that gold is unreactive we mean that it does not easily form compounds. It 'prefers' to remain as the element.

Sodium is a very reactive metal. It does not 'like' being an element. It would 'prefer' to react and become part of a compound.

So we can say that:

**1** reactive elements 'prefer' to form compounds,
**2** unreactive elements 'prefer' to remain as the element.

Now let us have another look at the reaction we started with.

aluminium + iron oxide ⟶ aluminium oxide + iron
element     compound        compound     element

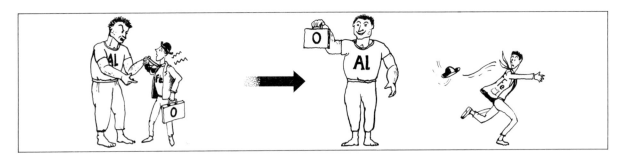

**Figure 17.2** *A competition reaction between aluminium and iron oxide*

The aluminium is taking the oxygen from the iron. Aluminium is more reactive than iron. So it 'likes' being part of a compound more than iron does. To form a compound it takes the oxygen from the iron. This leads on to a general rule about competition reactions. It is this:

> After a competition reaction, the more reactive metal is always part of a compound

## Can we predict the result of competition reactions?

To predict what will happen in a competition reaction, we must know the reactivities of the metals concerned. On the left is a list of some metals arranged in order of reactivity. The most reactive metals are at the top.

With this list we can predict the outcome of many competition reactions. Here are some examples.

potassium
sodium
calcium
magnesium
aluminium
zinc
iron
tin
lead
nickel
copper
mercury
silver
gold

decreasing reactivity

### Example 1

What will happen if a mixture of zinc powder and copper oxide are heated together?

zinc + copper oxide ⟶ zinc oxide + copper

This reaction will take place. The more reactive metal, zinc, becomes part of a compound.

## *Example 2*

What will happen if a mixture of magnesium oxide and copper powder are heated together?

copper + magnesium oxide $\longrightarrow\!\!\!\not\longrightarrow$ copper oxide + magnesium

This reaction will not happen ($\longrightarrow\!\!\!\not\longrightarrow$ shows a reaction that does not happen). If it did, the more reactive metal, magnesium, would change from a compound to an element. But we know this does not happen in a competition reaction.

### How violent are competition reactions?

We can use the list of activities to get some idea of how violent competition reactions will be. Reactions between elements close together in the activity series are rather gentle. Reactions between metals far apart in the series are violent and spectacular.

### Can competition reactions take place in solution?

**Figure 17.3** *This steel penknife has become coated with copper, because iron is higher than copper in the reactivity series*

Competition reactions between metals can take place in solution. If you dip a penknife blade into copper sulphate solution, the blade becomes coated with copper (see figure 17.3). The blade is made of steel which is mainly iron. This is what happens.

iron + copper sulphate $\longrightarrow$ iron sulphate + copper

Some of the iron in the penknife has reacted with the copper sulphate. It has taken the sulphate from the copper. The metallic copper forms a coating on the penknife blade. This reaction takes place because iron is higher in the activity series than copper. So the iron becomes part of a compound.

If you dipped a silver spoon into copper sulphate solution what would happen?

silver + copper sulphate $\longrightarrow\!\!\!\not\longrightarrow$ silver sulphate + copper

This reaction will not happen. The more reactive metal, copper, would be changed from a compound into an element. But we know this does not happen in a competition reaction.

## 17.2 Metals: rare or common?

The Earth is at least 4 500 000 000 (four thousand five hundred million) years old. Has this been long enough for

**Figure 17.4** *A gold-bearing rock. In all the millions of years that gold has been in this rock, it has not reacted with the substances around it*

all the elements to become compounds? Figure 17.4 is a photograph of a rock that contains the element gold. Even 4 500 000 000 years has not been long enough for this gold to react with the substances around it.

Most metals are more reactive than gold. At some time in the Earth's history they have reacted to form compounds. This section is about how we extract (take out) the metals from such compounds.

## What do we get metals from?

Some metals are extracted from sea water, but most are obtained from rocks. Rocks that contain metals or their compounds are called **ores**. An ore is not a pure compound. It is a mixture. Figure 17.5 shows the distribution of some of the major ore deposits in Britain.

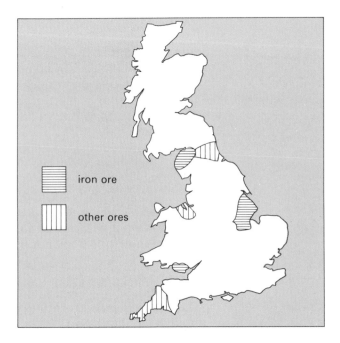

**Figure 17.5** *The principal ore resources of Britain*

| Metal | Name of ore | Chemical name | Chemical formula |
|---|---|---|---|
| potassium | carnallite | – | $KCl.MgCl_2.6H_2O$ |
| sodium | rock salt | sodium chloride | $NaCl$ |
| magnesium | carnallite | – | $KCl.MgCl_2.6H_2O$ |
| aluminium | bauxite | aluminium oxide | $Al_2O_3.2H_2O$ |
|  | cryolite | sodium aluminium fluoride | $Na_3AlF_3$ |
| zinc | zinc blende | zinc sulphide | $ZnS$ |
|  | calamine | zinc carbonate | $ZnCO_3$ |
| iron | magnetite | an oxide of iron | $Fe_3O_4$ |
|  | haematite | an oxide of iron | $Fe_2O_3$ |
| tin | tinstone or cassiterite | tin(IV) oxide | $SnO_2$ |
| lead | galena | lead sulphide | $PbS$ |
| copper | copper pyrites | – | $CuFeS_2$ |
|  | malachite | basic copper carbonate | $CuCO_3.Cu(OH)_2$ |
| mercury | cinnabar | mercury sulphide | $HgS$ |

**Table 17.1** *The ores of some metals*

Getting a metal from its ore is usually done in two stages:

1 separating the metal compound from the soil and other earthy impurities,
2 extracting (getting out) the metal from the metal compound present in the ore.

## What factors affect the price of a metal?

To decide whether a metal is likely to be expensive or cheap we must know:

1 is plenty of the metal ore available?
2 is it easy (cheap) to extract the metal from its ore?

Some examples should make this clearer.
   Silver and gold are very expensive. Ores of these two metals are rare.
   Calcium and potassium are fairly expensive. Ores containing these metals are quite common. But it is difficult and expensive to extract these metals from their ores.
   Iron is the cheapest metal. There is plenty of iron ore around and the iron is easily extracted.

## What makes some metals difficult to extract?

Highly reactive metals are difficult to extract from their ores.

Metals which are not very reactive are easy to extract from their ores.

When a metal reacts, it combines with other elements to form a compound. When you extract a metal, you do the reverse of this. The metal must be separated from other elements. Some examples will help to explain this.

### Example 3

Will it be easy to extract magnesium from magnesium oxide?

Magnesium is a reactive metal. It burns fiercely in air to form magnesium oxide. The fierceness of this reaction shows that magnesium 'likes' being joined to oxygen. So separating the magnesium from the oxygen will be difficult. Extracting magnesium from magnesium oxide will not be easy.

### Example 4

Will it be easy to extract silver from silver oxide?

Silver is a metal of low reactivity. It does not react when heated in air. From this we can see that silver 'prefers' not to combine with oxygen. So it should be easy to make silver oxide decompose into silver and oxygen. This is so. When heated silver oxide decomposes into silver metal and oxygen gas.

## 17.3  How metals are extracted

How a metal is extracted depends on the position of the metal in the activity series. Table 17.2 lists some metals in order of decreasing reactivity. It also shows how the metals are extracted from their compounds.

### The extraction of metals at the bottom of the activity series

Metals at the bottom of the activity series occur as the element. Both silver and gold have been known to humans since at least 5000 B.C. This is because they occur naturally as the metal (see figure 17.4, for example). To extract silver and gold from their ores, you just have to remove the earthy impurities.

### The extraction of metals towards the bottom of the activity series

Metals towards the bottom of the activity series are extracted by heating the metal sulphide in air.

| Metal | Method of extraction |
|---|---|
| potassium | electrolysis of potassium hydroxide |
| sodium | electrolysis of sodium chloride |
| magnesium | electrolysis of magnesium chloride |
| aluminium | electrolysis of aluminium oxide |
| zinc | heating zinc oxide with carbon |
| iron | heating iron oxide with carbon |
| tin | heating tin oxide with carbon |
| lead | heating lead oxide with carbon |
| copper | heating copper sulphide with some oxygen |
| mercury | heating mercury sulphide with some oxygen |
| silver | occurs as the element |
| gold | occurs as the element |

**Table 17.2** *The extraction of some metals*

Mercury and copper are extracted in this way. Both these metals have been known for over two thousand years. Mercury was being mined by the Romans in 400 B.C. They obtained the metal by roasting mercury sulphide in air. The following reaction takes place:

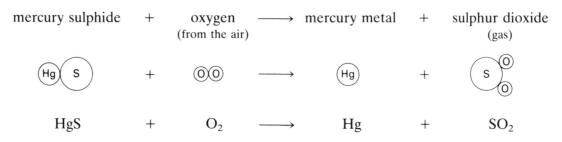

$$HgS + O_2 \longrightarrow Hg + SO_2$$

When copper sulphide is heated in air, there is a similar reaction:

$$CuS + O_2 \longrightarrow Cu + SO_2$$

### The extraction of metals by heating metal oxides with carbon

Metals near the middle of the activity series can be extracted by heating the metal oxides with carbon.

Zinc, tin and iron are all extracted in this way. The extraction is done in two stages:

**1** the ore is heated in oxygen to form the oxide,
**2** it is then heated with carbon to get the metal.

The main ore of zinc is known as zinc blende. It contains zinc sulphide. We have already seen what happens when mercury sulphide is heated in air. The sulphur leaves the mercury and combines with oxygen.
    When zinc sulphide is heated in air, both the sulphur and the zinc combine with oxygen. Instead of getting zinc we get zinc oxide.

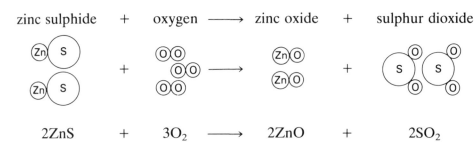

zinc sulphide    +    oxygen ⟶ zinc oxide    +    sulphur dioxide

$$2ZnS \quad + \quad 3O_2 \quad \longrightarrow \quad 2ZnO \quad + \quad 2SO_2$$

It is not really surprising that a reactive metal like zinc should combine with oxygen. The problem now is to remove the oxygen from the zinc oxide to get zinc metal. No amount of heating will make zinc oxide decompose. What is needed is a chemical that will help 'pull the oxygen away' from the zinc. Chemicals that remove oxygen like this are known as **reducing agents** (see chapter 18). The reducing agent used is carbon. This is cheaply available in the form of coke.

zinc oxide    +    carbon ⟶ zinc    +    carbon monoxide

$$ZnO \quad + \quad C \quad \longrightarrow \quad Zn \quad + \quad CO$$

This is a competition reaction (see section 17.1). The carbon 'wins' the oxygen from the zinc.
    Tin is also extracted by heating its oxide with carbon:

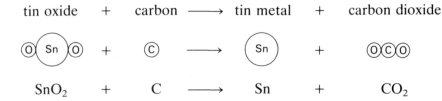

tin oxide    +    carbon ⟶ tin metal    +    carbon dioxide

$$SnO_2 \quad + \quad C \quad \longrightarrow \quad Sn \quad + \quad CO_2$$

Iron is made by heating its oxide with carbon. This extraction is carried out in a blast furnace, as shown in figure 17.6. The limestone is added to the furnace to remove the sandy impurities, which are mainly silicon dioxide.

**Stage 1**

Hot air is blown into the bottom of the furnace and the carbon (coke) burns in the air forming carbon dioxide. This helps to produce the high temperature needed for the extraction

carbon + oxygen ⟶ carbon dioxide
              (from the air)

Ⓒ + ⓄⓄ ⟶ ⓄⒸⓄ

**Stage 2**

As this carbon dioxide moves up the furnace it reacts with more carbon and changes into carbon monoxide.

carbon dioxide + carbon ⟶ carbon monoxide

ⓄⒸⓄ + Ⓒ ⟶ ⒸⓄ  ⒸⓄ

**Stage 3**

The carbon monoxide that is formed now reacts with the iron oxide, changing it to iron metal.

iron oxide + carbon monoxide ⟶ iron metal + carbon dioxide

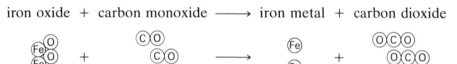

Figure 17.6 *The extraction of iron*

Figure 17.7 *A blast furnace. Iron ore, coke and limestone are taken to the furnace up the long 'shoot'. The large pipes take away the hot waste gases from the furnace*

At the high temperatures in the furnace the limestone decomposes to calcium oxide:

calcium carbonate $\longrightarrow$ calcium oxide $+$ carbon dioxide
limestone

$$CaCO_3 \longrightarrow CaO + CO_2$$

As the impure molten iron trickles down through the furnace, this calcium oxide reacts with the silicon dioxide:

calcium oxide $+$ silicon dioxide $\longrightarrow$ calcium silicate
slag

$$CaO + SiO_2 \longrightarrow CaSiO_3$$

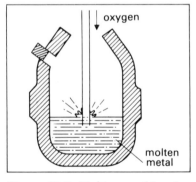

**Figure 17.8** *Making steel by the basic oxygen process*

The calcium silicate also trickles down the furnace and then floats on the layer of molten iron. This layer is called **slag**. It is tapped out through a separate hole and, when solidified, is used in road-making.

Iron from the blast furnace is really an alloy containing from 2 to 5 per cent of carbon. The carbon makes it too brittle for many uses. This high-carbon alloy can be made into usable steel in what is known as the **basic oxygen process** (see figure 17.8). It is mixed with recycled (scrap) iron and melted in a large furnace. Oxygen is now blown into the molten metal through a water-cooled 'lance'. The carbon reacts with the oxygen, forming carbon dioxide. The product that remains contains about 0.5 per cent of carbon.

## The extraction of the most reactive metals

The most reactive metals are also the most difficult to extract. You cannot extract them by heating a compound of the metal with a reducing agent. There are no reducing agents powerful enough. Because of this, the most reactive metals are extracted by electrolysis (see section 15.7). Electricity is passed through a molten compound of the metal. This makes the compound split up. The metal collects at the negative electrode (cathode). The non-metal forms at the positive electrode (anode). But there is a big problem with this method and that is the high cost of electricity. Many factories for manufacturing these metals are situated near waterfalls where cheap hydroelectric power is available (see figure 15.19).

**Figure 17.9** *Electrolysis cell used to make aluminium*

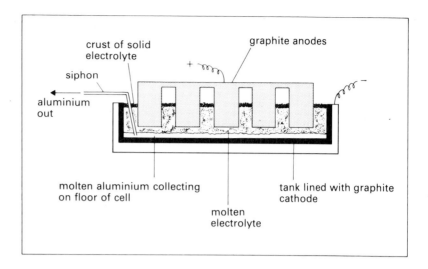

Aluminium is produced on a very large scale by electrolysis. The ore used is bauxite (hydrated aluminium oxide). Bauxite has an extremely high melting point. But it does melt at a much lower temperature when mixed with another aluminium ore called cryolite. The electrolysis cell used is shown in figure 17.9. The cell is lined with graphite which acts as the cathode. The graphite anode dips into the mixture of molten ores. When an electric current is passed, the molten bauxite is changed into aluminium metal and oxygen gas:

aluminium oxide $\longrightarrow$ aluminium metal + oxygen gas

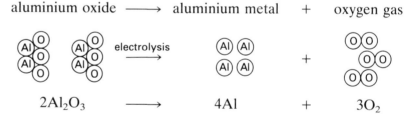

$$2Al_2O_3 \longrightarrow 4Al + 3O_2$$

The molten aluminium metal collects at the bottom of the electrolysis cell. From time to time it is tapped off.

Other metals extracted by electrolysis are magnesium, calcium, sodium and potassium. In each case the principle is the same. An electric current is passed through a molten compound. The metal collects at the cathode.

## Summary: Competition reactions and the extraction of metals

* Reactive elements easily form compounds.
* Unreactive elements form compounds with difficulty.
* The compounds of reactive elements do not easily decompose.
* The compounds of unreactive elements easily decompose.

★ In competition reactions the more reactive metal always forms a compound. The less reactive metal always finishes up as the element.
★ Metals very low in the activity series may occur naturally as the element (silver and gold).
★ Metals towards the bottom of the activity series are extracted by roasting their sulphides in air (mercury and copper).
★ Metals towards the middle of the activity series are extracted by heating their oxides with carbon (lead, tin, iron and zinc).
★ Metals towards the top of the activity series are extracted by the electrolysis of their molten compounds (aluminium, magnesium, sodium and potassium).

## Questions

1 Some of the following mixtures will react together. Others will not. For each mixture write a word equation. If there is no reaction, write 'no reaction' down as your answer.
   (a) potassium + silver oxide ⟶
   (b) zinc + magnesium oxide ⟶
   (c) aluminium + tin oxide ⟶
   (d) magnesium + nickel oxide ⟶
   (e) iron + calcium oxide ⟶
2 This question is about the extraction of iron.
   (a) Draw a diagram of a blast furnace. Clearly label the points where the raw materials are added and where the iron and slag are tapped off.
   (b) Name one form of iron ore that might be used in the blast furnace.
   (c) What is the main element present in the coke?
   (d) Why is limestone added to the furnace?
   (e) Write chemical equations to show how the iron oxide becomes iron. Say roughly where these reactions occur in the furnace.
3 If small amounts of cerium are added to the steel used to make the inside of an oven, then grease does not build up on the oven walls. So it is used in 'self-cleaning' ovens. The metal may be extracted either by the reaction of cerium fluoride with calcium or by the electrolysis of molten cerium chloride.
   (a) By considering the methods used to extract cerium, say roughly where this element should be placed in the activity series.
   (b) The diagram shows part of the equation for the reaction between cerium fluoride and calcium. Copy this diagram, and draw in the products of the reaction

   cerium fluoride + calcium ⟶

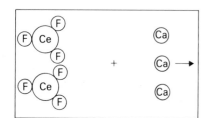

Do you think it would be possible to extract cerium by heating cerium fluoride with copper? Give reasons for your answer.

4 Sodium is extracted by electrolysing a mixture of molten sodium chloride and calcium chloride. The electrolysis cell that is used is shown below.

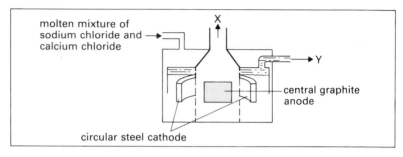

molten mixture of sodium chloride and → calcium chloride

X

Y

central graphite anode

circular steel cathode

(a) Why is molten sodium chloride used rather than sodium chloride solution?

(b) What is the purpose of adding calcium chloride to the sodium chloride? (*Hint*: A mixed electrolyte is used in the extraction of aluminium for the same reason.)

(c) What substance will be obtained at the exit X?

(d) What substance will be obtained at the exit Y?

(e) Sodium is not particularly expensive although electrolysis uses a lot of electricity. Suggest at least one reason for this.

5 Platinum occurs naturally in the form of the element. It was first discovered in 1735 in South America, where it was known to the pre-Columbian Indians. It is an expensive metal that is sometimes used in jewellery. Modern uses include plating missile nose-cones and jet engine fuel nozzles.

(a) Where would you place platinum in the activity series? Give reasons for your answer.

(b) Although platinum occurs naturally, it is very expensive. Why do you think this is?

(c) Three uses of the metal are described above. Suggest reasons why this particular metal has been chosen for these uses.

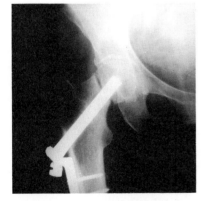

*Titanium combines strength with lightness and corrosion resistance. Here it is being used to pin a damaged thigh bone*

# Titanium: from emulsion paint to missiles

Titanium is the ninth most abundant element in the Earth's crust. Its compounds are also present on the Moon and in meteorites. The metal is as strong as steel but 45 per cent lighter. It is very reactive at high temperatures and will even burn in nitrogen! At low temperatures it gives the appearance of being a very unreactive metal. It does not corrode in water, including sea water. It is unaffected by most acids and alkalis.

The extraction of titanium presents difficulties. The most common titanium ores contain titanium dioxide, $TiO_2$. Carbon is unable to remove the oxygen from this compound and in any case carbon would react with the titanium formed at the high temperatures used to extract metals.

The oxide is very difficult to melt and so cannot be electrolysed to give the metal.

Heating titanium oxide with both carbon and chlorine gives the chloride, $TiCl_4$. Rather surprisingly this is a liquid which will not conduct electricity. It cannot therefore undergo electrolysis to give titanium metal.

Titanium is actually made by a rather unusual method: by heating titanium chloride with magnesium metal in an argon atmosphere:

$$TiCl_4 + 2Mg \longrightarrow Ti + 2MgCl_2$$

The temperature of the reaction is kept below the melting point of titanium, so the metal separates out as a solid spongy mass. The temperature is above the melting point of magnesium chloride, however, and liquid magnesium chloride therefore runs down to the bottom of the furnace.

## Questions

6 Titanium is being used more and more for aircraft and missiles, and for the reactors and pipes in chemical factories. Suggest reasons for this.

7 Titanium is the ninth most abundant element and yet it was not manufactured until 1946. Give reasons for this.

8 Even nowadays relatively few countries manufacture titanium. The estimated capacity of the five major producers is shown in the pie chart. If total World production is 100 000 tonnes, calculate the amount produced by each country.

9 Titanium is clearly a reactive metal, although it seems to be quite unreactive at normal temperature. What is the reason for this? (*Hint*: aluminium is similar.)

10 Titanium oxide has a very high melting point. What does this tell you about its chemical structure?

11 Titanium chloride liquid cannot be electrolysed. What does this tell you about its chemical structure?

12 Titanium is extracted by heating its chloride with magnesium. Name one other metal that could be used instead of magnesium.

13 Why can't titanium be extracted by electrolysing an aqueous solution of a titanium compound?

14 Titanium oxide is a white powder that is widely used as a pigment in paints. Do you think this oxide will be a stable or an unstable substance? Explain your answer.

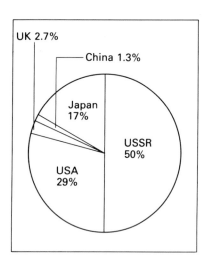

*World production of titanium metal. What is it all used for?*

# 18 Oxidation and reduction

You have already seen that chemists have a habit of putting things into groups: elements and compounds, metals and non-metals, acids and alkalis and so on.

We can also arrange different types of processes into groups. Melting, boiling and dissolving, for example, are all physical changes. Burning is a chemical change.

This grouping of processes can be taken a stage further. We can divide chemical reactions up into different types. The two most obvious types are combining and decomposing (see section 6.1). Another very important pair of processes are **oxidation** and **reduction**.

## 18.1 The addition and removal of oxygen

Oxygen reacts with most metals to form oxides (a familiar example is shown in figure 18.1). In these reactions the metal has been **oxidized**. The reaction of copper with oxygen is an example:

$$\text{copper metal} \quad + \quad \text{oxygen} \longrightarrow \text{copper oxide}$$
$$2Cu \quad + \quad O_2 \longrightarrow 2CuO$$

Here the copper has been oxidized.

When a metal loses oxygen it has been **reduced**. The decomposition of mercury oxide is an example:

$$\text{mercury oxide} \xrightarrow{\text{heat}} \text{mercury metal} + \text{oxygen}$$
$$2HgO \longrightarrow 2Hg \quad + \quad O_2$$

Here the mercury has been reduced.

Now let us look at some more complicated reactions. For example:

$$\text{copper oxide} + \text{hydrogen} \longrightarrow \text{copper metal} + \text{steam}$$
$$CuO \quad + \quad H_2 \longrightarrow Cu \quad + H_2O$$

The copper loses oxygen. It is being reduced.  The hydrogen gains oxygen. It is being oxidized.

You can see that in this reaction, a **red**uction and an **oxi**dation have occurred. Because of this, it is called a **redox reaction**.

**Figure 18.1** *This iron turnstile has rusted because iron reacts readily with oxygen in the air to form iron oxide (that is, it is easily oxidized)*

So we have seen that:

**1** when something gains oxygen it is oxidized,
**2** when something loses oxygen it is reduced,
**3** reactions in which reduction and oxidation take place are called redox reactions.

Many metals are extracted from their oxides. The oxides are being reduced. Some examples are shown below. (They are looked at in more detail in section 17.3.)

$$\text{zinc oxide} \;+\; \text{carbon} \longrightarrow \text{zinc} \;+\; \text{carbon monoxide}$$

$$\text{ZnO} \quad + \quad \text{C} \longrightarrow \text{Zn} \quad + \quad \text{CO}$$

The carbon is often called a **reducing agent**. It reduces the zinc oxide to zinc.
We could also say that the zinc oxide is an oxidizing agent. It oxidizes the carbon.
The reaction of iron with carbon monoxide is similar. Here the carbon monoxide is the reducing agent. It reduces the iron oxide to iron:

$$\text{iron oxide} \;+\; \text{carbon monoxide} \longrightarrow \text{iron metal} \;+\; \text{carbon dioxide}$$

$$\text{Fe}_2\text{O}_3 \quad + \quad 3\text{CO} \longrightarrow 2\text{Fe} \quad + \quad 3\text{CO}_2$$

## 18.2 Oxidation and reduction as electron transfer

When a metal reacts with oxygen it forms a metal oxide. Metal oxides are ionic compounds (see section 14.1). The metal is present as positive ions. This is because during the reaction the metal loses electrons to the oxygen. Take magnesium as an example:

$$\text{magnesium metal} \;+\; \text{oxygen} \longrightarrow \text{magnesium oxide}$$

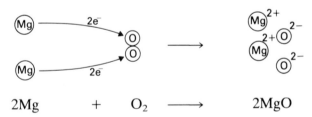

$$2\text{Mg} \quad + \quad \text{O}_2 \longrightarrow 2\text{MgO}$$

In the same way, when copper reacts with oxygen the copper loses electrons:

$$\text{copper metal} \;+\; \text{oxygen} \longrightarrow \text{copper oxide}$$

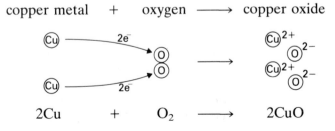

$$2\text{Cu} \quad + \quad \text{O}_2 \longrightarrow 2\text{CuO}$$

In fact, whenever any metal reacts with oxygen it loses electrons. Because of this 'being oxidized' has now come to mean 'losing electrons'.

Now let us look at the reaction of magnesium with chlorine:

magnesium metal  +  chlorine ⟶ magnesium chloride

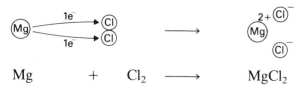

$$Mg \quad + \quad Cl_2 \quad \longrightarrow \quad MgCl_2$$

In this reaction magnesium loses electrons. So if oxidation means losing electrons, the magnesium has been oxidized in this reaction. You can see that this oxidation did not involve the element oxygen.

We saw in the previous section that reduction is the opposite of oxidation. So we can define reduction as 'gaining electrons'. In the previous example, the chlorine atoms are gaining electrons to become negative chloride ions. So the chlorine is being reduced. You can use the word **oil rig** to help you remember that:

**1 o**xidation **i**s **l**oss of electrons,
**2 r**eduction **i**s **g**ain of electrons.

Not all redox reactions reduce or oxidize a metal. For example, in the reaction of chlorine with sodium bromide the bromide is oxidized and the chlorine reduced:

sodium bromide  +  chlorine ⟶ sodium chloride  +  bromine

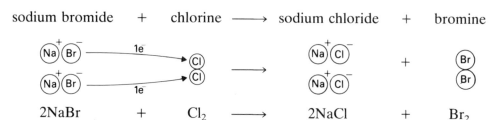

$$2NaBr \quad + \quad Cl_2 \quad \longrightarrow \quad 2NaCl \quad + \quad Br_2$$

## Oxidation numbers or oxidation states

Oxidation numbers (sometimes called oxidation states) tell us how many electrons a particular element has lost or gained. Table 18.1 (over the page) shows some commonly occurring ions.

You can see that the element itself always has an oxidation number of zero. It is only when the element reacts to form compounds that it loses or gains electrons. So only then does it have an oxidation number other than zero.

In all simple ionic compounds the oxidation number is the same as the charge on the ion.

You will often find these oxidation numbers mentioned in the names of chemicals. They are usually written in

| Ion | Oxidation number | Element | Oxidation number | Ion | Oxidation number |
|---|---|---|---|---|---|
| $Na^+$ | +1 | Na | 0 | | |
| $K^+$ | +1 | K | 0 | | |
| $Mg^{2+}$ | +2 | Mg | 0 | | |
| $Zn^{2+}$ | +2 | Zn | 0 | | |
| $Fe^{2+}$ | +2 | Fe | 0 | | |
| $Cr^{3+}$ | +3 | Cr | 0 | | |
| $Al^{3+}$ | +3 | Al | 0 | | |
| $Fe^{3+}$ | +3 | Fe | 0 | | |
| | | Cl | 0 | $Cl^-$ | −1 |
| | | Br | 0 | $Br^-$ | −1 |
| | | O | 0 | $O^{2-}$ | −2 |
| | | S | 0 | $S^{2-}$ | −2 |
| | | N | 0 | $N^{3-}$ | −3 |

**Table 18.1** *The oxidation numbers of some elements in common ions*

Roman numerals after the element they refer to. For example, the name iron(II) chloride means that in this compound iron is present in the form of 2+ ions.

## Elements that form more than one ion

Many metals have only one oxidation number in their compounds. For example, magnesium only ever forms ions with a +2 oxidation number. But some metals can form more than one type of ion. Compounds with iron having +2 and +3 numbers are well known.

The iron in iron(II) compounds has been less oxidized than that in iron(III) compounds. This is because the iron(II) atom has only lost two electrons. The iron(III) atom has lost three electrons.

So it is possible to oxidize iron(II) compounds. One such reaction is shown below:

chlorine   +   iron(II) chloride $\longrightarrow$ iron(III) chloride

$Cl_2$       +       $2FeCl_2$   $\longrightarrow$   $2FeCl_3$

Here each iron(II) atom is losing an electron (being oxidized):

$$2Fe^{2+} - 2e^- \longrightarrow 2Fe^{3+}$$

Each chlorine atom is gaining electrons (being reduced):

$$Cl_2 + 2e^- \longrightarrow 2Cl^-$$

## Summary: Oxidation and reduction

* ★ Oxidation takes place when an element combines with oxygen.
* ★ Reduction takes place when oxygen is removed from a compound.
* ★ Oxidation can also be defined as loss of electrons.
* ★ Reduction can be defined as gain of electrons.
* ★ When one substance is oxidized, another is reduced. These reactions are called redox reactions.
* ★ Oxidation numbers (or oxidation states) tell you how many electrons an atom has lost or gained.
* ★ Elements themselves have an oxidation state of 0.
* ★ Atoms that have become positive ions have positive oxidation numbers.
* ★ Atoms that have become negative ions have negative oxidation numbers.
* ★ Oxidation states are used in naming compounds. Iron(II) chloride contains $Fe^{2+}$ ions, iron(III) chloride contains $Fe^{3+}$ ions.

## Questions

1 Five words are missing from the paragraph below. Copy out the paragraph, filling the gaps with words chosen from the list.

  redox     oxidized
  reducing agent  oxidizing agent
  reduced

One way of making silicon is to heat dry powdered silicon dioxide with magnesium powder. The following reaction occurs:

silicon dioxide + magnesium ⟶ magnesium oxide + silicon

$SiO_2$ + $2Mg$ ⟶ $2MgO$ + $Si$

This is a _____ reaction in which silicon is being _____ and magnesium is being _____. The magnesium is acting as a _____ _____. We could also consider the silicon dioxide to be acting as an _____ _____.

2 Five equations are given below. Copy them out and in each case say whether the first compound in the equation is (i) reduced, (ii) oxidized, (iii) neither reduced nor oxidized.
 (a) $B_2O_3 + 3Mg \longrightarrow 2B + 3MgO$
 (b) $CO + PbO \longrightarrow CO_2 + Pb$
 (c) $2SO_2 + O_2 \longrightarrow 2SO_3$
 (d) $Na_2S + 2HCl \longrightarrow 2NaCl + H_2S$
 (e) $P_4O_{10} + 10C \longrightarrow P_4 + 10CO$

3 Copy out the formulae given below and in each case write down the oxidation state of the element that is underlined in the formula.
(a) $\underline{Mn}O_2$
(b) $\underline{Fe}_2O_3$
(c) $Ca\underline{Se}$
(d) $\underline{Si}Cl_4$
(e) $\underline{B}N$

4 Copy out the five equations given below. In each case say whether the element underlined has been (i) reduced, (ii) oxidized, (iii) neither reduced nor oxidized.
(a) $2\underline{Na} + Cl_2 \longrightarrow 2NaCl$
(b) $2K\underline{I} + Br_2 \longrightarrow 2KBr + I_2$
(c) $\underline{Al}_2O_3 + 6HF \longrightarrow 2AlF_3 + 3H_2O$
(d) $2FeBr_2 + \underline{Br}_2 \longrightarrow 2FeBr_3$
(e) $\underline{Mg} + 2HCl \longrightarrow H_2 + MgCl_2$

5 The group 7 elements are among the most reactive in the periodic table. As you would expect from their electron shell structures, halogen atoms tend to gain or share one extra electron.

Fluorine is the most reactive of the group. Metals, glass, carbon and even water can be made to burn in fluorine. It is the most powerful oxidant of all the elements.

To extract the halogens, we have to take back the electron that they have gained. This is not particularly easy.

Fluorine is extracted by the electrolysis of potassium fluoride dissolved in liquid hydrogen fluoride. Fluoride ions lose electrons at the anode:

$$2F^- - 2e^- \longrightarrow F_2$$

Bromine is extracted from sea water. The reaction shows that chlorine is a more powerful oxidant than bromine. Sea water, which contains small amounts of sodium bromide, is treated with chlorine gas. A reaction occurs giving bromine and sodium chloride:

$$2Br^- + Cl_2 \longrightarrow Br_2 + 2Cl^-$$

(a) Draw electron shell structures for (i) a fluorine atom, (ii) a fluoride ion.
(b) Why would you expect the halogen's atoms to gain or share one electron?
(c) Look up and explain the meaning of the word *oxidant*.
(d) Chlorine is said to be a more powerful oxidant than bromine. What does this mean?
(e) Why is liquid hydrogen fluoride used in the electrolysis of potassium fluoride, rather than water?
(f) In the electrolysis of potassium fluoride, are the fluoride ions being oxidized or reduced?

(g) When iron reacts with fluorine, would you expect to get iron(II) fluoride or iron(III) fluoride? Explain your answer.

## Electrochemical corrosion

In engineering great care has to be taken to avoid having certain metals in contact with each other. For example, joining iron pipe and copper pipe would be most unwise. This would be the equivalent of setting up the cell shown in the diagram.

Both metals tend to lose electrons. Iron, being more reactive than copper, has the greatest tendency to lose them. As a result iron atoms lose electrons, forming soluble iron ions:

$$Fe(s) \longrightarrow Fe^{2+} (aq) + 2e^-$$

These electrons go to the copper. Here they will probably react with hydrogen ions present in the water:

$$2H^+ (aq) + 2e^- \longrightarrow H_2(g)$$

The overall result is that the iron rapidly dissolves away: that is, it corrodes.

Sometimes engineers deliberately place a reactive metal such as magnesium or zinc in contact with an iron structure, as a way of preventing corrosion of the iron. The more reactive metal loses electrons, forming soluble ions and thus corroding. Since there are always electrons flowing to the iron from the more reactive metal, it is very difficult for the iron to lose electrons – that is, it is difficult for the iron to corrode. This technique is known as **sacrificial protection**, because the more reactive metal is sacrificed in order to prevent the less reactive metal corroding.

*Iron corrodes quickly when it is in contact with copper*

*The zinc bars will protect the ship's iron hull from corrosion*

## Questions

**6** What link is there between the reactivity of a metal and the ease with which it loses electrons?

**7** Is the iron being reduced or oxidized in the cell shown in the diagram on the previous page?

**8** Are hydrogen ions being reduced or oxidized in this cell?

**9** Explain why you might not expect such electrochemical corrosion if the copper in this cell was replaced by iron.

**10** Sometimes bridges are protected by connecting them to a large piece of magnesium buried close by. Explain, with the help of an equation like those in the text, what will happen to the magnesium.

**11** Steel dustbins are often coated in zinc metal (galvanized). If the coating becomes broken, exposing the steel underneath to the rain, would you expect the steel or the zinc to corrode? Explain your answer.

**12** Tin cans are really steel with a coating of tin. Tin is a less reactive metal than steel. If the tin coating gets broken, would you expect the tin or the steel to corrode? Draw a cell like the one in the diagram on the previous page to help you explain your answer.

# 19 Acids, bases and salts

## 19.1 What are acids and bases?

Probably the best known fact about acids is that they can be **corrosive**. This means they can 'eat away' substances including metals, stones and human flesh. This is true of the common acids: nitric, sulphuric and hydrochloric acid. They are called **strong acids**.

Some other acids are much less corrosive. Lemon juice contains citric acid, and vinegar contains ethanoic acid. Citric and ethanoic acids are only mildly corrosive. They are examples of **weak acids**.

**Figure 19.1** *Even though vinegar is an acid, it is a weak one. It can safely be used for pickling foods*

### Is it an acid?

How can you tell if a substance is an acid? Checking to see if it is corrosive does not help much. Many other substances are corrosive.

We can identify acids using substances called **indicators**. Indicators are dyes that change colour when acids are added to them. **Litmus** is one of the best known indicators. In water it is mauve. When an acid is added, it turns red. Other substances make litmus turn blue. These substances are called **alkalis**.

## Measuring acidity

Litmus can indicate if a solution is acidic, neutral or alkaline. **Universal indicator** can tell us how acidic or alkaline a solution is. It turns different colours depending upon the strength of the acid or alkali. These are shown in table 19.1.

| pH | 1 | 2 | 3 | 4 | 5 | 6 | 7 | 8 | 9 | 10 | 11 | 12 | 13 | 14 |
|---|---|---|---|---|---|---|---|---|---|---|---|---|---|---|
| **Colour of Universal indicator** | red | | pink | yellow | | green | | dark green | | light blue | | dark blue | | |
| **Type of solution** | strongly acidic | | | faintly acidic | | | neutral | | | faintly alkaline | | strongly alkaline | | |

**Table 19.1** *How the colour of Universal indicator changes with different solutions*

Electronic instruments called **pH meters** can also be used to measure acidity. A special electrode is dipped into the solution. The meter then gives a reading in pH units. Water has a pH value of 7. Acids have a pH below 7. Alkalis have a pH above 7.

## What elements do acids contain?

Acids have many reactions in common. For example, they all turn litmus red. This could mean that there is something about the structures of acids which makes them behave in the same way. Table 19.2 gives the formulae of some common acids. You will see that there is one element that is present in every one of these acids.

| Acid | Formula |
|---|---|
| hydrochloric acid | HCl |
| hydrofluoric acid | HF |
| nitric acid | $HNO_3$ |
| sulphuric acid | $H_2SO_4$ |
| ethanoic acid | $H_4C_2O_2$ |

**Table 19.2** *Some acids and their formulae*

All the acids contain the element hydrogen. But be careful not to get this statement back to front. Not all substances that contain hydrogen are acids.

## Hydrogen ions – the cause of acidity

An acid is a solution that contains a lot of hydrogen ions. The substances in table 19.2 all dissolve in water to give solutions that contain these $H^+$ ions. For example, hydrochloric acid contains $H^+$ ions and $Cl^-$ ions.

**Figure 19.2** *Hydrogen atoms in the methane molecule, and hydrogen ions in hydrochloric acid*

Many compounds contain hydrogen but do not act as acids. This happens because the hydrogen is not present in the form of $H^+$ ions. Methane, $CH_4$, is an example. The four hydrogen atoms are all present as part of a methane molecule. So methane is not acidic.

Strongly acidic solutions contain a large number of $H^+$ ions. Weakly acidic solutions contain a smaller number of $H^+$ ions.

## Bases and neutralization

When you neutralize an acid you take away its acidic properties. Substances that do this are called **bases**. We already know that $H^+$ ions make solutions acidic. So bases must be substances that remove $H^+$ ions.

Most bases belong to one of three types of compound:

**1** metal hydroxides,
**2** metal oxides,
**3** metal hydrogencarbonates or metal carbonates.

## Alkalis and bases

Most bases do not dissolve in water. Those that do dissolve are called **alkalis**. Table 19.3 and 19.4 list several bases, some of which are also alkalis.

| Name | Formula | Solubility in water |
|---|---|---|
| sodium hydroxide | NaOH | soluble |
| calcium hydroxide | $Ca(OH)_2$ | slightly soluble |
| potassium hydroxide | KOH | soluble |

**Table 19.3** *Some bases that are also alkalis*

| Name | Formula | Solubility in water |
|---|---|---|
| copper oxide | CuO | insoluble |
| calcium carbonate | $CaCO_3$ | insoluble |
| zinc hydroxide | $Zn(OH)_2$ | insoluble |

**Table 19.4** *Some bases that are not alkalis*

Notice that all the alkalis listed in table 19.3 contain the hydroxide ion, $OH^-$. Any substance that dissolves in water to give a solution containing hydroxide ions is an alkali. Strongly alkaline solutions contain many $OH^-$ ions. Weakly alkaline solutions contain a smaller number of $OH^-$ ions. All alkalis have pH values greater than 7.

**Figure 19.3** *Venn diagram for some bases and alkalis*

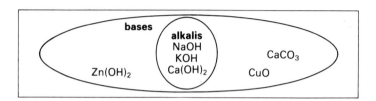

## 19.2 The reactions of acids with bases

The reactions of acids with bases are summarized below:

**1** acid + a metal hydroxide ⟶ water + a salt,
**2** acid + a metal oxide ⟶ water + a salt,
**3** acid + a metal carbonate ⟶ water + a salt + carbon dioxide.

We shall now look at some examples of these reactions in more detail.

### Reactions with metal hydroxides

Most metal hydroxides are bases. They can neutralize acids. Those that dissolve in water, called alkalis, are convenient bases to use.

This equation shows what happens when sodium hydroxide neutralizes hydrochloric acid:

| sodium hydroxide | + | hydrochloric acid ⟶ | water | + | sodium chloride |
|---|---|---|---|---|---|
| a base | | an acid | | | a salt |

$$NaOH(aq) + HCl(aq) \longrightarrow H_2O(aq) + NaCl(aq)$$

You can see that the $OH^-$ ion from the alkali combines with the $H^+$ ion from the acid. As a result water is formed. The other product is sodium chloride. This belongs to a group of compounds called **salts**.

Many metal hydroxides do not dissolve in water. They are bases but not alkalis. These insoluble hydroxides must be used in the form of a solid. Let us look at what happens when solid zinc hydroxide reacts with sulphuric acid:

zinc hydroxide    +    sulphuric acid $\longrightarrow$    water    +    zinc sulphate
a base                        an acid                                                    a salt

$$Zn(OH)_2(s) \quad + \quad H_2SO_4(aq) \quad \longrightarrow \quad 2H_2O(l) \quad + \quad ZnSO_4(aq)$$

You can see that both hydrogen ions from the acid have combined with the hydroxide ions from the base to form water. The other substance formed during the reaction is a salt called zinc sulphate.

You should also notice the state symbols (s), (aq) and (l) in the equation for this reaction. These give you information about the physical state of these compounds. Their meaning is explained in section 8.1.

## Reactions with metal oxides

Metal oxides are bases, that is, they remove $H^+$ ions from solution. They do this by converting the $H^+$ ions into water.

copper oxide    +    sulphuric acid $\longrightarrow$    water    +    copper sulphate
a base                        an acid                                                    a salt

$$CuO(s) \quad + \quad H_2SO_4(aq) \quad \longrightarrow \quad H_2O(l) \quad + \quad CuSO_4(aq)$$

In the same way magnesium oxide will neutralize hydrochloric acid.

magnesium oxide    +    hydrochloric acid $\longrightarrow$    water    +    magnesium chloride
a base                                an acid                                                        a salt

$$MgO(s) \quad + \quad 2HCl(aq) \quad \longrightarrow \quad H_2O(l) \quad + \quad MgCl_2(aq)$$

## Reactions with metal carbonates

Carbonates will neutralize acids. When they do, the reaction mixture effervesces (fizzes) and carbon dioxide gas is given off. This happens when calcium carbonate reacts with hydrochloric acid.

calcium carbonate  +  hydrochloric acid  ⟶  water  +  calcium chloride  +  carbon dioxide
                 a base                            an acid                                                a salt

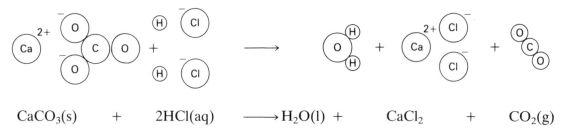

$$CaCO_3(s) \quad + \quad 2HCl(aq) \quad \longrightarrow H_2O(l) + \quad CaCl_2 \quad + \quad CO_2(g)$$

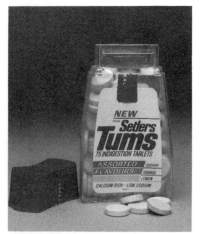

**Figure 19.4 (above)** *Indigestion mixtures work by neutralizing the acids in your stomach*

**Figure 19.5 (right)** *The salt copper sulphate*

Like all reactions between a base and an acid, water and a salt have been formed. But when acids react with carbonates there is an extra product. This is carbon dioxide gas.

## 19.3 Salts

### What are salts?

Whenever an acid is neutralized a salt is produced.

Salts contain most of the elements present in the acid from which they were made. What they do not contain is $H^+$ ions. These are replaced by one or more metal ions. An example is shown in figure 19.5.

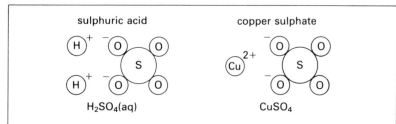

All salts made from sulphuric acid contain the $SO_4^{2-}$ ion. The names of such salts all finish with the word **sulphate**.

Salts made from nitric acid all contain the $NO_3^-$ ion. The names of these salts all finish in **nitrate**. An example is sodium nitrate, shown in figure 19.6.

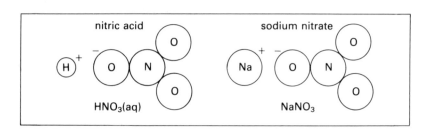

**Figure 19.6** *The salt sodium nitrate*

Hydrochloric acid gives salts called **chlorides**. These all contain the $Cl^-$ ion. Potassium chloride is shown in figure 19.7.

**Figure 19.7** *The salt potassium chloride*

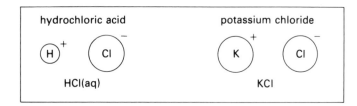

You can see that all salts are made up of ions. Like all ionic compounds, salts are solids with high melting points. Many salts are soluble in water.

## The solubility of some salts

Table 19.5 shows which salts are soluble in water.

| Soluble | Insoluble |
| --- | --- |
| all sodium, potassium and ammonium salts | – |
| all nitrates | – |
| most chlorides, bromides and iodides | silver and lead chlorides, bromides and iodides |
| most sulphates | lead sulphate, barium sulphate (calcium sulphate is very slightly soluble) |
| sodium, potassium and ammonium carbonates | most other carbonates |

**Table 19.5** *Patterns of solubility*

## Ways of making salts

There are four ways of making salts from acids. The general equations of these methods are:

1 metal + acid $\longrightarrow$ hydrogen + a salt
2 metal carbonate + acid $\longrightarrow$
$\qquad\qquad\qquad\qquad$ water + a salt + carbon dioxide
3 metal oxide + acid $\longrightarrow$ water + a salt
4 metal hydroxide + acid $\longrightarrow$ water + a salt

Making any salt is done in two stages:

1 the calculated amounts of the chemicals are allowed to react together,
2 the salt is separated from the solution.

The final solution should contain only the salt and water. So we do not want unreacted acid left or too much base in the solution. This means that we need some method of telling when all the acid has just been neutralized. When we reach this point, only the salt and water will be present. Table 19.6 shows how we can tell when all the acid has just reacted. Once this point has been reached, the method of carrying out the reaction is the same in each case, except for reactions involving alkalis.

| Substance reacting with the acid | What happens when the acid has been neutralized |
| --- | --- |
| metals | bubbles of hydrogen no longer form |
| metal carbonates | bubbles of carbon dioxide no longer form |
| insoluble metal oxides | metal oxide stops dissolving |
| insoluble metal hydroxides | metal hydroxide stops dissolving |
| soluble metal hydroxides (alkalis) | no visible change, an indicator must be used |

**Table 19.6** *How to tell when the acid has been neutralized*

## Soluble salts from insoluble bases

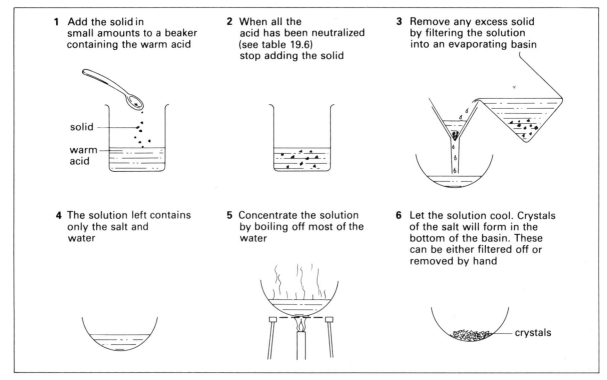

**1** Add the solid in small amounts to a beaker containing the warm acid

solid

warm acid

**2** When all the acid has been neutralized (see table 19.6) stop adding the solid

**3** Remove any excess solid by filtering the solution into an evaporating basin

**4** The solution left contains only the salt and water

**5** Concentrate the solution by boiling off most of the water

**6** Let the solution cool. Crystals of the salt will form in the bottom of the basin. These can be either filtered off or removed by hand

crystals

**Figure 19.8** *Making a soluble salt from an insoluble base*

## Soluble salts from alkalis

When you neutralize an acid using an alkali, there is no easy way of seeing when all the acid has reacted. But this problem can be overcome by using an indicator. A suitable method is shown in figure 19.9.

**Figure 19.9** *Making a soluble salt from an alkali*

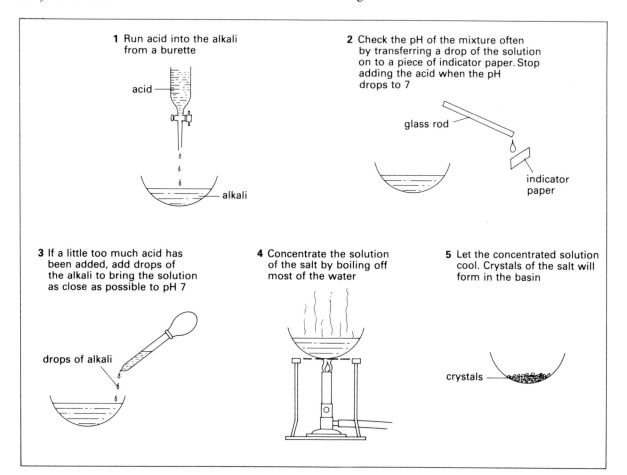

**1** Run acid into the alkali from a burette

acid

alkali

**2** Check the pH of the mixture often by transferring a drop of the solution on to a piece of indicator paper. Stop adding the acid when the pH drops to 7

glass rod

indicator paper

**3** If a little too much acid has been added, add drops of the alkali to bring the solution as close as possible to pH 7

drops of alkali

**4** Concentrate the solution of the salt by boiling off most of the water

**5** Let the concentrated solution cool. Crystals of the salt will form in the basin

crystals

## Making insoluble salts

The methods we have looked at so far assumed that the salt is soluble. This is not always the case. Sometimes we want to form salts that are insoluble. Insoluble salts are prepared by a different method. This method does not necessarily use an acid or a base. Let us look at the preparation of lead sulphate.

First we must start with two soluble compounds. One must contain lead ions and the other must contain sulphate ions. Look at table 19.5 of the solubilities of salts. From this you can see that we could use lead nitrate and sodium sulphate. When solutions containing these compounds are mixed together, insoluble lead sulphate is precipitated. (This means that it forms little bits of the solid that tend to settle to the bottom.)

sodium sulphate   +   lead nitrate   $\longrightarrow$   lead sulphate   +   sodium nitrate

$$Na_2SO_4(aq) \quad + \quad Pb(NO_3)_2(aq) \quad \longrightarrow \quad PbSO_4(s) \quad + \quad 2NaNO_3(aq)$$

The method used is shown in figure 19.10.

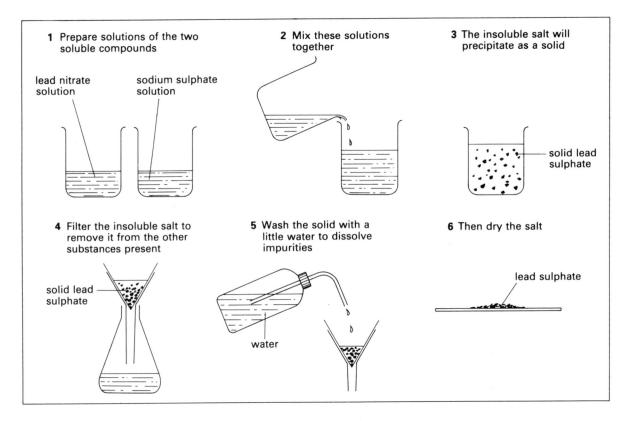

**1** Prepare solutions of the two soluble compounds

lead nitrate solution      sodium sulphate solution

**2** Mix these solutions together

**3** The insoluble salt will precipitate as a solid

solid lead sulphate

**4** Filter the insoluble salt to remove it from the other substances present

solid lead sulphate

**5** Wash the solid with a little water to dissolve impurities

water

**6** Then dry the salt

lead sulphate

**Figure 19.10** *Making an insoluble salt (lead sulphate)*

# 19.4 Volumetric analysis: titrations

In section 19.3 we saw how to prepare a salt by neutralizing a solution of an alkali with an acid. In this section we look in more detail at reactions between solutions, in particular solutions of known concentrations.

A **molar solution** contains 1 mole of solute per $dm^3$ (see section 7.3). When using ionic compounds like acids, bases and salts, we need to be clear what this means. A molar solution of sodium chloride, for instance, contains 1 mole $(23 + 35.5 = 58.5\,g)$ of NaCl per $dm^3$. When an ionic compound such as sodium chloride dissolves in water the ions separate from each other. So a 1M solution of NaCl contains 1 mole of $Na^+$ ions and 1 mole of $Cl^-$ ions in every $dm^3$ – that is, it is 1M with respect to each of its ions (see figure 19.11).

**Figure 19.11** *The molarity of ionic solutions*

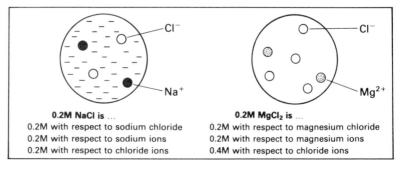

**0.2M NaCl is** …
0.2M with respect to sodium chloride
0.2M with respect to sodium ions
0.2M with respect to chloride ions

**0.2M MgCl₂ is** …
0.2M with respect to magnesium chloride
0.2M with respect to magnesium ions
0.4M with respect to chloride ions

Similarly, a 1M solution of magnesium chloride, $MgCl_2$, contains 1 mole of $MgCl_2$ per $dm^3$. It contains 1 mole of $Mg^{2+}$ ions per $dm^3$ too, but two moles of $Cl^-$ ions per $dm^3$ (since $MgCl_2$ contains two $Cl^-$ ions for every $Mg^{2+}$ ion). So a 1M solution of magnesium chloride will be a 1M solution of $Mg^{2+}$ ions, but a 2M solution of $Cl^-$ ions.

## The procedure for a titration

**Titration** is a procedure used to find out what volumes of different solutions will react together. The most common titrations involve an acid and an alkali. We shall look at an example.

**Stage 1:** The alkali is measured into a conical flask using a pipette.

**Stage 2:** A few drops of an indicator are added. This is to show us when the solution has become neutral.

**Stage 3:** The acid is slowly added from a burette until the indicator just changes colour. The amount of acid used is then carefully noted.

## What factors affect the volume of acid used?

There are three factors to consider:

1 the volume of alkali used,
2 the concentration of the acid and alkali solutions,
3 the chemical equation for the reaction.

Let us look at some more examples.

### Example 1

In a titration it was found that 25 cm³ of 0.1M ethanoic acid was needed to just neutralize 25 cm³ of 0.1M sodium hydroxide. How much acid would be needed to neutralize 50 cm³ of 0.1M sodium hydroxide?

**Figure 19.12** *Measuring liquid safely, using a pipette*

**Figure 19.13** *Carrying out a titration safely*

The answer is of course 50 cm$^3$. If we use twice as much alkali then we need twice as much acid to neutralize it.

Now let us consider the effect of concentrations.

### Example 2

In a titration 25 cm$^3$ of 0.1M ethanoic acid just neutralized 25 cm of 0.1M sodium hydroxide. What volume of 0.2M ethanoic acid would have been needed in this titration?

We can see that the 0.2M acid is twice as concentrated as the previously used acid. So we should only need half as much of it, that is, 25 ÷ 2 = 12.5 cm$^3$.

Finally, let's look at how the chemical equation affects the volumes of solutions needed.

### Example 3

What volume of 0.1M hydrochloric acid will just neutralize 10 cm$^3$ of 0.1M sodium hydroxide?

First, we write the equation for the neutralization reaction:

$$NaOH(aq)  +  HCl(aq)  \longrightarrow  H_2O(l)  +  NaCl(aq)$$
$$\text{1 mole} \qquad\quad \text{1 mole}$$

We can see from the equation that we need equal numbers of moles of the alkali and acid.
  We know that we have 10 cm$^3$ of 0.1M alkali.
  So we need to know what volume of 0.1M hydrochloric acid solution will contain the same number of moles. The answer is obviously 10 cm$^3$ of the acid.

Not all problems are as straightforward as this.

### Example 4

What volume of 1M sulphuric acid will just neutralize 20 cm$^3$ of 1M potassium hydroxide solution?

The equation for the neutralization reaction is:

$$2KOH(aq)  +  H_2SO_4(aq)  \longrightarrow  2H_2O(l)  +  K_2SO_4(aq)$$
$$\text{2 moles} \qquad\quad \text{1 mole}$$

Here we can see from the equation that we need only half as many moles of acid as we do of alkali.
  We know we have 20 cm$^3$ of 1M alkali.
  So we need to know what volume of the acid contains half as many moles as this does. The answer is that we need 10 cm$^3$ of 1M sulphuric acid.

## An equation for titration calculations

Here is a list of the important values in any titration:

volume of alkali used (in cm³)    $V_1$
concentration of alkali (in M)    $M_1$
moles of alkali shown in the
chemical equation    $N_1$
volume of acid used (in cm³)    $V_2$
concentration of acid (in M)    $M_2$
moles of acid shown in the
chemical equation    $N_2$

These values can be linked by a mathematical equation:

$$\frac{V_1 \times M_1}{N_1} = \frac{V_2 \times M_2}{N_2}$$

This can be used to solve any titration problem. Here are a few examples.

### *Example 5*

What volume of 0.2M hydrochloric acid will just neutralize 10 cm³ of 1M sodium carbonate solution?

The equation is:

$$Na_2CO_3(aq) + 2HCl(aq) \longrightarrow H_2O(l) + CO_2(g) + 2NaCl(aq)$$

Here    $V_1 = 10$        $M_1 = 1$    $N_1 = 1$
$V_2 =$ unknown    $M_2 = 0.2$    $N_2 = 2$

Putting these values into our mathematical equation we get:

$$\frac{10 \times 1}{1} = \frac{V_2 \times 0.2}{2}$$

Thus    $V_2 = \dfrac{10 \times 1 \times 2}{1 \times 0.2}$

$$= \frac{20}{0.2} = 100$$

So we need 100 cm³ of 0.2M hydrochloric acid.

The equation can also be used to find the concentration of a solution where the amounts of solutions are known.

### *Example 6*

25 cm³ of 0.1M sulphuric acid just neutralized 10 cm³ of sodium hydroxide solution. What is the concentration of the sodium hydroxide?

$$2NaOH(aq) + H_2SO_4(aq) \longrightarrow H_2O(l) + Na_2SO_4(aq)$$

$V_1 = 10 \quad M_1 = \text{unknown} \quad N_1 = 2$
$V_2 = 25 \quad M_2 = 0.1 \qquad\quad N_2 = 1$

Putting these in our mathematical equation:

$$\frac{10 \times M_1}{2} = \frac{25 \times 0.1}{1}$$

Thus $\quad M_1 = \dfrac{25 \times 0.1 \times 2}{1 \times 10}$

$$= \frac{5}{10} = 0.5$$

So the sodium hydroxide solution is 0.5M.

## Summary: Acids, bases and salts

* Acids give solutions with a pH below 7.
* Alkalis give solutions with a pH above 7.
* Acid solutions always contain hydrogen ions.
* A base is a substance that neutralizes acids. It does so by reacting with the $H^+$ ions.
* The most common bases are oxides, hydroxides and carbonates.
* Bases that dissolve in water are called alkalis.
* An acid + a base $\longrightarrow$ water + a salt.
* When carbonates neutralize acids they form carbon dioxide as well as water and a salt.
* Salts can also be made by adding a reactive metal to an acid:
   reactive metal + acid $\longrightarrow$ hydrogen + a salt.
* Insoluble salts can be prepared by precipitation reactions.
* All metal nitrates are soluble in water.
* All metal sulphates are soluble in water, *except* those of lead, barium, mercury and calcium.
* All metal chlorides are soluble in water, *except* those of silver, mercury and lead.
* All sodium, potassium and ammonium salts are soluble in water.
* Titration is a procedure used to find out what volumes of different solutions will react together.

## Questions

**1** Select the pH values that each of the substances in the questions might have.

> pH 1
> pH 4
> pH 7
> pH 10
> pH 14

(a) water
(b) a strong acid
(c) a weak alkali
(d) a weak acid
(e) a strong alkali

**2** The names of various chemicals are given below. Alongside each name is the pH of an aqueous solution of the chemical.

> sodium hydroxide     pH 13
> sodium nitrate       pH 7
> hydrogen bromide     pH 2
> potassium iodide     pH 7
> calcium hydroxide    pH 11
> silver nitrate       pH 6–7
> barium hydroxide     pH 12
> hydrogen iodide      pH 2
> magnesium sulphate   pH 7

(a) Which of the above substances are alkalis?
(b) Which of the above substances are acids?
(c) Which of the above substances are salts?
(d) What ion do all acids contain?
(e) What ion do all alkalis contain?

**3** Calcium carbonate does not turn Universal indicator solution blue and yet it will neutralize acids. How do you explain this?

**4** Copy out and complete the word equations shown below. Under each word equation give the full balanced chemical equation.

(a) potassium hydroxide + hydrochloric acid $\longrightarrow$
(b) calcium hydroxide + sulphuric acid $\longrightarrow$
(c) calcium oxide + sulphuric acid $\longrightarrow$
(d) silver oxide + nitric acid $\longrightarrow$
(e) calcium carbonate + hydrochloric acid $\longrightarrow$
(f) nickel carbonate + hydrochloric acid $\longrightarrow$
(g) sodium hydrogencarbonate + nitric acid $\longrightarrow$
(h) magnesium metal + hydrochloric acid $\longrightarrow$
(i) iron metal + hydrochloric acid $\longrightarrow$

**5** Write out word and chemical equations for reactions that you could use to make the following three salts:

(a) potassium sulphate,
(b) mercury chloride,
(c) zinc nitrate.

Explain, with words and diagrams, the practical details of the method you would use to make each of these three salts.

6 (a) What volume of 1M sodium hydroxide will just neutralize 25 cm$^3$ of 1M sulphuric acid?

$$2NaOH + H_2SO_4 \longrightarrow H_2O + Na_2SO_4$$

(b) What volume of 0.2M hydrochloric acid will just neutralize 10 cm$^3$ of 0.4M sodium hydroxide?

$$NaOH + HCl \longrightarrow H_2O + NaCl$$

(c) What volume of 0.1M nitric acid will just neutralize 20 cm$^3$ of 0.2M calcium hydroxide?

$$Ca(OH)_2 + 2HNO_3 \longrightarrow 2H_2O + Ca(NO_3)_2$$

(d) What volume of 0.2M potassium chloride will react exactly with 100 cm$^3$ of 0.1M silver nitrate?

$$KCl + AgNO_3 \longrightarrow AgCl + KNO_3$$

(e) What volume of 0.4M aluminium chloride will react exactly with 25 cm$^3$ of 0.8M silver nitrate?

$$AlCl_3 + 3AgNO_3 \longrightarrow 3AgCl + Al(NO_3)_3$$

7 It is known that 1 mole of ethanoic acid reacts with 1 mole of sodium hydroxide. In an investigation of a sample of vinegar, which contains ethanoic acid, 25 cm$^3$ of the vinegar neutralized 21.6 cm$^3$ of 1M sodium hydroxide. What is the concentration of the ethanoic acid in this vinegar? (Give your answer in mols per dm$^3$.)

8 A 5 cm$^3$ sample of household ammonia solution reacted exactly with 28.7 cm$^3$ of 1M hydrochloric acid. The equation for the reaction is:

$$NH_3(aq) + HCl(aq) \longrightarrow NH_4Cl(aq)$$

What is the molar concentration of the ammonia solution?

9 A 25 cm$^3$ sample of lemonade was boiled for a few minutes. This drove off the carbon dioxide but left the citric acid. After cooling the boiled lemonade reacted exactly with 18.3 cm$^3$ of 0.1M sodium hydroxide. The equation for the reaction is

$$C_6H_8O_7(aq) + 3NaOH(aq) \longrightarrow Na_3C_6H_5O_7(aq) + 3H_2O(l)$$

What was the molar concentration of citric acid in the lemonade?

10 Limewater is saturated calcium hydroxide solution, $Ca(OH)_2(aq)$. 25 cm$^3$ of limewater was neutralized with 10.2 cm$^3$ of 0.1M hydrochloric acid. What was the molar concentration of the calcium hydroxide solution?

## The extraordinary element

Hydrogen is an element that defies classification. Under normal conditions it is a gas but it is thought that it exists as a metallic solid on the planet Jupiter.

Looking at the structures of hydrogen compounds does little to make things clearer. Many organic compounds contain hydrogen covalently bonded to carbon. This is what one would expect for a molecule consisting of two non-metals.

Compounds also exist in which the hydrogen is present as $H^-$ ions, such as lithium hydride, $Li^+H^-$. Again this is not too surprising as we might expect to get an ionic compound from a metal and a non-metal. We should perhaps add that this compound is far from stable. It reacts vigorously with water forming lithium hydroxide and hydrogen gas, in which the hydrogen atoms are covalently bonded.

Finally there are acids. When pure, or when dissolved in an organic solvent such as petrol, they are covalently bonded. They have no effect upon indicator papers and show few of the reactions that we expect of acids.

In water all this changes. The molecules of acid ionize, giving $H^+$ ions (which are, of course, just protons).

$$HCl(g) + aq \longrightarrow H^+(aq) + Cl^-(aq)$$

The water clearly plays an important role: the $H^+$ ion only forms if it can attach itself to a water molecule. Because of this the ionization is often written as:

$$HCl + H_2O \longrightarrow H_3O^+ + Cl^-$$

It is the $H_3O^+$ (or $H^+(aq)$) ions which are the real cause of acidity of aqueous acids. Indeed, many books define acids as **proton donors**. Bases are the opposite of acids and so they must be **proton acceptors**.

## Questions

**11** Give the name and structure of a compound in which hydrogen is covalently bonded to carbon.

**12** Explain why an $H^+$ ion is a proton.

**13** Why is an aqueous solution of citric acid a better conductor than a propanone solution of it?

**14** Tins of health salts usually contain a powder containing citric acid and the base sodium hydrogencarbonate.
(a) Why don't these chemicals react in the tin?
(b) What gas will be given off when this mixture is put into water?

**15** 'When hydrogen chloride is dissolved in water, the water could be considered to be acting as a base.' Explain this.

**16** Try to explain why the reaction of lithium hydride with water is an acid–base reaction.

# 20 Ammonia and fertilizers

## 20.1 The need for fertilizers

The number of people alive today is larger than all the people that have previously lived on Earth. Look back to figure 1.4 (on page 3). This will give you some idea of how the world's population has increased over the last 1000 years.

Feeding a population that is growing at this rate is quite a problem. This section is about some of the ways in which chemists have helped to tackle this problem.

### How can we get more food?

Chemists have helped to increase the world's food supply in three main ways:

1 by providing fertilizers – this enables more food to be grown,
2 by protecting crops from pests and diseases,
3 by developing new kinds of food.

**Figure 20.1** *Spraying crops to protect them from pests and diseases*

### What are fertilizers?

Growing plants remove **plant nutrients** from the soil. Plant nutrients are the chemicals the plant uses to grow. Fertilizers are substances that contain such plant nutrients.

## What makes a good fertilizer?

There are three important considerations. A fertilizer must:

1 be cheap,
2 have suitable solubility in water,
3 contain a high percentage of the elements needed by the plants.

Very soluble fertilizers are 'fast acting'. The fertilizer dissolves in rain water and is soon absorbed through the plant roots. These highly soluble fertilizers have some disadvantages. They soon get washed out of the soil into streams and rivers, which is wasteful and expensive. It also pollutes the rivers. Less soluble fertilizers are slower acting. But they provide a cheap and steady supply of nutrients over a longer period of time.

## What elements do plants need?

Table 20.1 shows some of the elements needed by plants.

| Elements needed in large amounts | | Trace elements | |
|---|---|---|---|
| carbon | potassium | zinc | manganese |
| hydrogen | phosphorus | boron | molybdenum |
| oxygen | nitrogen | copper | |

**Table 20.1** *Some of the elements that all plants need*

Some elements are needed in very small traces (amounts). These are called the **trace elements**. Most soils contain more than enough trace elements.

Although plants need large amounts of carbon, hydrogen and oxygen, these do not have to be included in fertilizers. They can be obtained from the carbon dioxide in the air and the water in the soil.

Compounds of potassium and phosphorus are also needed by plants. These are taken from the soil through the plant roots. Unless these elements are replaced, the soil will eventually run out of them. So they are present in many fertilizers.

Nitrogen is the other main element needed for plant growth. Farmland needs regular treatment with fertilizer containing nitrogen. This may seem rather strange, since the plants are surrounded by air which is 80 per cent nitrogen. Unfortunately nitrogen gas is too unreactive for most plants to make use of. But all plants can make use of compounds of nitrogen, as these are more reactive. So the plants take these compounds from the soil. For the soil to remain fertile, the nitrogen compounds must be returned to it.

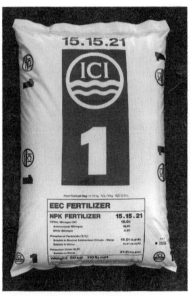

**Figure 20.2** *Nitrogen (N), phosphorus (P) and potassium (K) are the three main elements in fertilizers*

## 20.2 The nitrogen cycle

Some areas of uncultivated land are covered with wild
flowers year after year. Such soil does not seem to need
fertilizers. The reason for this is simple. Wild plants die
where they grow. When they decay (rot) the nitrogen
compounds within them are returned to the soil (see
figures 20.3 and 20.4).

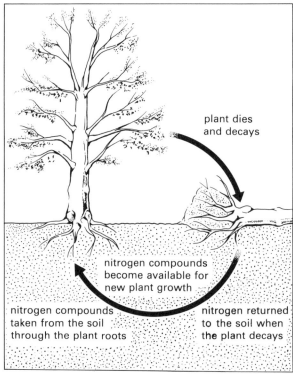

**Figure 20.3** *Uncultivated woodland does not need
fertilizer, because wild plants die where they grow.
They put back nitrogen compounds into the soil*

**Figure 20.4** *Plant growth and decay*

The nitrogen content of this type of uncultivated land may
actually increase. One reason for this is thunderstorms.
Lightning passing through the air makes nitrogen and oxygen
combine together. The nitrogen oxides produced dissolve in
the rain forming a natural 'liquid fertilizer'. The nitrogen
content of the soil may be also be increased by plants called
**legumes**. Clover, peas and beans are all legumes. These
legumes contain special bacteria in their roots. The bacteria
cause swellings called nodules on the roots. Inside the
nodules, reactions take place in which nitrogen from the air
is used to make nitrogen compounds. In effect the nodules
provide the plant with a 'personal fertilizer factory'.
Eventually these plants die and decay. The soil is then richer
in nitrogen compounds than it was before these plants grew
there.

So now you can see why uncultivated land does not need fertilizers. Nitrogen may be taken from the soil by plants but soon these nitrogen compounds are returned back again. This series of changes form part of what is known as the **nitrogen cycle**. Figure 20.5 shows a simple version of the nitrogen cycle.

**Figure 20.5** *The nitrogen cycle*

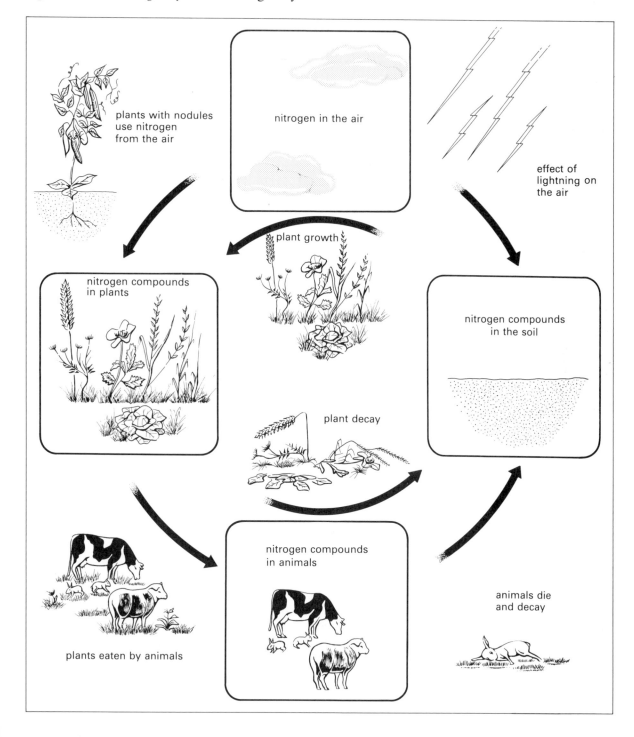

Farming disturbs this cycle. Crops are harvested and taken far away from the fields in which they grew. Sheep and cattle also remove nitrogen from the soil. Few of these animals die and decay in the same fields in which they feed. So fertilizers are necessary because nitrogen compounds and other plant nutrients are permanently removed from the soil.

## 20.3 Synthetic fertilizers

### Ammonia as a fertilizer

Ammonia ($NH_3$) is usually a gas but it can be changed into a liquid by compressing (squeezing) it. Sometimes liquid ammonia is 'injected' into the soil as a fast-acting fertilizer. Most of the ammonia dissolves in water that is present in the soil. Some will change back into a gas and escape into the air.

Ammonia is very soluble in water, so it tends to get washed out of the soil by the rain. It is also more difficult to apply to the soil than solid fertilizers. Even so, ammonia is widely used as a fertilizer in the USA. It is less widely used in Britain, where most farmers seem to prefer solid fertilizers.

### Making solid fertilizers

When ammonia dissolves in water it forms an alkali. All alkalis react with acids to form solids known as salts (see section 19.3). When ammonia reacts with acids, compounds called **ammonium salts** are formed. These can be used as solid fertilizers. The most commonly used salt is ammonium nitrate. This is made by the reaction of ammonia with nitric acid:

ammonia    +    nitric acid    $\longrightarrow$    ammonium nitrate

$$NH_3 \quad + \quad HNO_3 \quad \longrightarrow \quad NH_4NO_3$$

Ammonium sulphate is another commonly used fertilizer.

### The manufacture of ammonia

In South America there are large amounts of minerals that contain nitrogen compounds. From the middle of the 1800s onwards these were imported into Europe. But by 1900, scientists realized that supplies of these minerals would run out.

Within twenty years a German scientist, Fritz Haber, had developed a process for making nitrogen compounds using nitrogen from the air. In his process nitrogen was made to react with hydrogen, producing ammonia:

hydrogen  +  nitrogen  $\xrightarrow{\text{heat}}$  ammonia

$$3H_2 \quad + \quad N_2 \quad \xrightarrow{\text{heat}} \quad 2NH_3$$

But remember that nitrogen is a very unreactive element. It took a great deal of research to find the right conditions under which the nitrogen would react with hydrogen. A special catalyst was found, which made the reaction occur at a reasonable speed. (A **catalyst** is a substance which can speed up a reaction but which does not get used up in the process.) Quite high temperatures are also used to help speed up the reaction. It is very important to choose just the right temperature, high enough but not so high as to decompose the ammonia.

**Figure 20.6** *A modern ammonia factory*

Table 20.2 shows some of the properties of ammonia.

| |
|---|
| colourless, choking gas |
| less dense (lighter) than air |
| extremely soluble in water giving alkaline solutions |

**Table 20.2** *Properties of ammonia*

## 20.4 Ammonia in the laboratory

### Making ammonia in the laboratory

Ammonia cannot easily be made from nitrogen and hydrogen in the laboratory. Instead we must start with an ammonium compound, such as ammonium sulphate. When this is heated with an alkali (you could use calcium hydroxide) ammonia gas is given off:

calcium hydroxide $+$ ammonium sulphate $\longrightarrow$ calcium sulphate $+$ water $+$ ammonia

$$Ca(OH)_2 \quad + \quad (NH_4)_2SO_4 \quad \longrightarrow \quad CaSO_4 \quad + \; 2H_2O \; + \quad 2NH_3$$

The ammonia gas is dried by passing it over lumps of calcium oxide. As ammonia is not as dense as air, it is collected in an inverted (upside-down) test tube (see figure 20.7).

**Figure 20.7** *Apparatus for making ammonia*

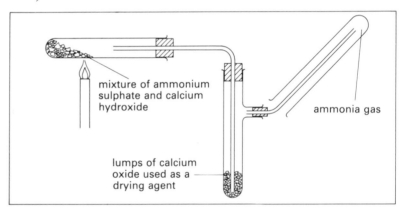

mixture of ammonium sulphate and calcium hydroxide

ammonia gas

lumps of calcium oxide used as a drying agent

**Figure 20.8** *Ammonia, like most alkalis, can remove grease*

### Ammonia dissolves in water

Ammonia is very soluble in water. It forms an alkaline solution with a pH value of about 11.

Most of the ammonia will dissolve in the water to form ammonia solution. A small amount will react with the water to form ammonium ions, $NH_4^+$, and hydroxide ions, $OH^-$. This reaction is shown in the equation below. The double arrow symbol in the equation is used for reactions in which not all of the reactants become products (see section 28.2).

$$NH_3 + H_2O \rightleftharpoons NH_4^+ + OH^-$$

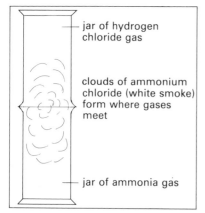

**Figure 20.9** *Preparation of solid ammonium chloride*

## Ammonia is a base

Ammonia reacts with acids to form salts. The reaction between ammonia and nitric acid is mentioned in section 20.3 above. A similar reaction occurs between ammonia and sulphuric acid to give ammonium sulphate:

$$2NH_3 + H_2SO_4 \longrightarrow (NH_4)_2SO_4$$

When ammonia solution reacts with hydrochloric acid, ammonium chloride is formed:

$$NH_3(aq) + HCl(aq) \longrightarrow NH_4Cl(aq)$$

In this equation the symbol 'aq' (aqueous) means that the chemicals are dissolved in water.

Solid ammonium chloride can be prepared directly. To do this ammonia gas is mixed with hydrogen chloride gas (see figure 20.9):

$$NH_3(g) + HCl(g) \longrightarrow NH_4Cl(s)$$

## Ammonia is a reducing agent

This means that ammonia can take oxygen away from other compounds. The reaction of ammonia with copper oxide is a good example:

copper oxide + ammonia $\longrightarrow$ copper + water + nitrogen
$$3CuO \quad + \quad 2NH_3 \quad \longrightarrow \quad 3Cu \quad + 3H_2O + \quad N_2$$

The apparatus shown in figure 20.10 can be used for this experiment.

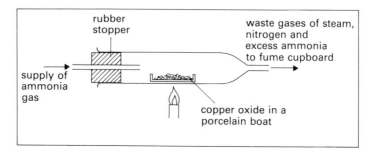

**Figure 20.10** *Apparatus to show that ammonia is a reducing agent*

# 20.5 Nitric acid

## The manufacture of nitric acid

Over two million tonnes of nitric acid are used each year just to make ammonium nitrate fertilizer. Large amounts of the acid are also used in the manufacture of explosives, such as nitroglycerine and TNT.

Most nitric acid is made from ammonia. There are three main stages in the process.

**Figure 20.11** *Large amounts of nitric acid are used in explosives such as TNT and nitroglycerine*

**1** Ammonia from the Haber process and oxygen from the air react together to form nitrogen oxide gas (NO):

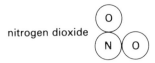

**2** The nitrogen oxide gas reacts with more oxygen to form nitrogen dioxide gas ($NO_2$):

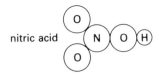

**3** Further reaction between nitrogen dioxide, oxygen and water gives nitric acid:

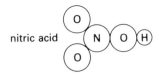

Figure 20.12 (opposite) summarizes the process. The first reaction takes place as the gases pass over a platinum–rhodium catalyst at 900 °C. The equation for the reaction is:

$$4NH_3(g) + 5O_2(g) \longrightarrow 4NO(g) + 6H_2O(g)$$

The gas is then cooled and mixed with more oxygen. In the oxidizing tower it reacts to form nitrogen dioxide:

$$4NO(g) + 2O_2(g) \longrightarrow 4NO_2(g)$$

This gas, along with excess air, then passes into an absorption tower. This is where the nitric acid is formed:

$$4NO_2(g) + O_2(g) + 2H_2O(l) \longrightarrow 4HNO_3(l)$$

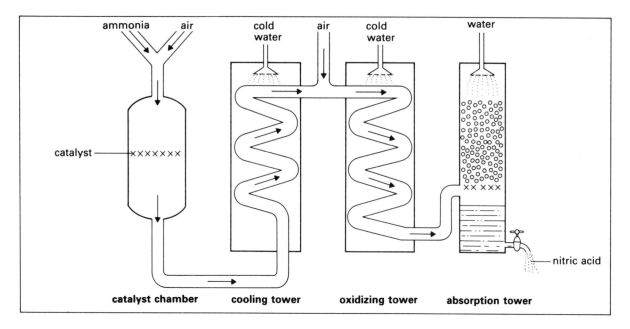

| catalyst chamber | cooling tower | oxidizing tower | absorption tower |

**Figure 20.12** *The manufacture of nitric acid*

## Some properties of nitric acid

We saw above how nitric acid could be made from ammonia. In many ways nitric acid is a typical acid. It reacts with bases to form salts called **nitrates**. For example:

sodium hydroxide + nitric acid ⟶ water + sodium nitrate
  a base          an acid              a salt

$$NaOH + HNO_3 \longrightarrow H_2O + NaNO_3$$

When nitric acid reacts with carbonates, the products are water, a salt and carbon dioxide:

potassium carbonate + nitric acid ⟶ water + potassium nitrate + carbon dioxide

$$K_2CO_3 + 2HNO_3 \longrightarrow H_2O + 2KNO_3 + CO_2$$

## An unusual property of nitric acid

Nitric acid almost never produces hydrogen gas when it reacts with metals. But it does react with metals, even those quite low in the activity series. This is because nitric acid, particularly when it is concentrated, is an oxidizing agent. You can see this if we look at the reaction of nitric acid with copper. We can think of it as taking place in two stages.

In the first stage, the nitric acid acts as an oxidizing agent. Oxygen is transferred from the nitrate ion ($NO_3^-$) to the copper. Copper oxide and nitrogen dioxide ($NO_2$) are two of the products formed.

Now copper oxide is a base. So in the second stage of the reaction, this oxide neutralizes the nitric acid. Copper nitrate and water are formed.

Overall we get the equation:

copper + nitric acid $\longrightarrow$ copper nitrate + water + nitrogen dioxide

$$Cu + 4HNO_3 \longrightarrow Cu(NO_3)_2 + 2H_2O + 2NO_2$$

The nitrate ion in the acid can lose more than one of its oxygen atoms. If this happens nitrogen oxide (NO) gas or even nitrogen ($N_2$) may be formed in place of the nitrogen dioxide. Which gas is produced depends on the conditions of the reaction.

### Salts of nitric acid: the nitrates

The salts of nitric acid are called nitrates. All metal nitrates dissolve in water. This can be useful to remember.

Nitrates are not particularly stable. When they are heated, they decompose (break down). Table 16.6 on page 198 gives more detail about their thermal decomposition.

## Summary: Ammonia and fertilizers

* Plants need nitrogen in order to grow. They cannot use nitrogen from the air because it is too inert.
* Plants need nitrogen compounds such as ammonium compounds or nitrates. They obtain these from the soil.
* In nature the nitrogen cycle keeps the amount of nitrogen in the soil at roughly the same level all the time.
* Modern farming removes nitrogen from the soil.
* Fertilizers replace soil nutrients removed by farming.
* In the Haber process nitrogen reacts with hydrogen to form ammonia, $NH_3$.
* Ammonia gas can be used as a fertilizer.
* Ammonia is very soluble in water. It forms an alkaline solution.
* Ammonia reacts with acids to form ammonium salts. These contain the $NH_4^+$ ion.
* Ammonium salts are used as solid fertilizers.
* Ammonia gas can reduce certain metal oxides to the metal.
* Ammonia can be converted to nitric acid by its reaction with oxygen and water.
* Nitric acid forms salts called nitrates. These contain the $NO_3^-$ ion.
* Nitrates are used in fertilizers and in explosives.
* Nitric acid is an oxidizing agent. So it can react with metals that are low in the activity series, such as copper.

# Questions

1 The next paragraph has five blank spaces. Copy out the paragraph and choose words from the list to fill the spaces.

ammonium    legumes    hydrogen    cycle    soluble

In uncultivated land the nitrogen compounds removed from the soil are eventually returned to it. The nitrogen content may even increase if plants known as ⎯⎯⎯⎯⎯⎯ are present. Farming disturbs the nitrogen ⎯⎯⎯⎯⎯⎯ by permanently removing nitrogen compounds from the soil. As a result, fertilizers are often used. Ammonia is sometimes used as a fertilizer. It is made by allowing nitrogen and ⎯⎯⎯⎯⎯⎯ to react together. Ammonia is an alkali. It reacts with acids to form ⎯⎯⎯⎯⎯⎯ salts. If these salts are very ⎯⎯⎯⎯⎯⎯ then the fertilizer acts quickly. Such fertilizers can, however, cause pollution problems.

2 Questions (a) to (e) are about the compounds described in the table below.

| Compound | Physical state | Solubility in water | Percentage nitrogen | Other comments |
|---|---|---|---|---|
| liquid ammonia | liquid | high | 82 | |
| ammonium nitrate | solid | high | 35 | |
| urea | solid | high | 47 | |
| calcium cyanamide | solid | insoluble | 35 | slowly reacts to form ammonia |
| ammonium phosphate | solid | soluble | 21 | |

(a) State one advantage of using liquid ammonia as a fertilizer.

(b) State in what way liquid ammonia is an inconvenient substance to use as a fertilizer.

(c) Both ammonium nitrate and calcium cyanamide contain 35 per cent nitrogen. Explain why calcium cyanamide is preferred for some uses but ammonium nitrate for others.

(d) Ammonium phosphate contains only 21 per cent nitrogen and yet it is still used in fertilizers. What other advantage does it have?

(e) Many commercial fertilizers are called NPK fertilizers. What do these letters tell you about the elements contained in this kind of fertilizer?

**3** Questions (a) to (e) are about a reaction between ammonia and copper oxide. The reaction was carried out in the apparatus shown in the diagram.

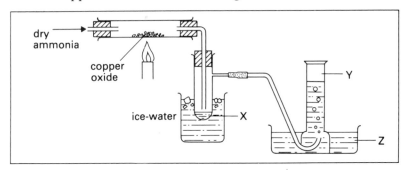

(a)  Write a word equation for the reaction.

(b)  Write a full chemical equation for the reaction.

(c)  What would collect in tube X of the apparatus shown in the diagram?

(d)  How could you identify this substance?

(e)  What would collect in the gas jar Y of the apparatus shown?

(f)  In what way would the appearance of the copper oxide change?

(g)  Some ammonia passes through the apparatus without reacting with the copper oxide. Why does this ammonia not collect in the gas jar at Y?

(h)  If a few drops of Universal indicator were added to the trough Z at the end of the experiment, what colour would it go?

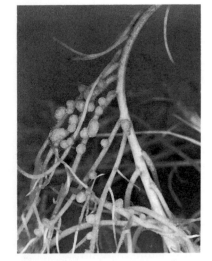

*Some plants, such as peas, have nodules like these on their roots. The nodules help the plant to use nitrogen in the air. The plant can then make nitrogen compounds for its own use. Genetic engineering could enable other crops to do this. What would be the impact on a company that owned a large fertilizer factory?*

## Ammonia and fertilizer production

Every year 25 million tonnes of nitrogen are fixed by atmospheric lightning and another 100 million tonnes by biological processes. Even so, the artificial conversion of nitrogen gas into its compounds continues to increase dramatically. In Britain alone about 6000 tonnes of ammonia are manufactured every day and 80 per cent of this is used for fertilizer production.

Almost all ammonia is manufactured by the Haber process in which nitrogen and hydrogen react together in the presence of an iron catalyst:

$$N_2(g) + 3H_2(g) \longrightarrow 2NH_3(g)$$

The nitrogen is obtained by the liquefaction and distillation of air. Hydrogen is produced by the reaction of natural gas (mainly methane, $CH_4$) with steam over a nickel catalyst at about 750 °C:

$$CH_4(g) + 2H_2O(g) \longrightarrow 4H_2(g) + CO_2(g)$$

Some carbon monoxide is also formed in this reaction. This has to be removed, otherwise it would 'poison' the iron catalyst used in the main ammonia-forming reaction. It is removed by allowing it to react with more steam to form carbon dioxide and more hydrogen.

The nitrogen (from the air) and the hydrogen (from the natural gas and water) are now mixed together and compressed to about 200 times normal atmospheric pressure. The actual synthesis of ammonia is carried out by passing the mixed gases over pellets of a ceramic material impregnated with iron.

The conditions chosen for the reaction are a compromise. At higher temperatures the gases combine together faster. The ammonia produced, however, decomposes more rapidly back into hydrogen and nitrogen. The effects of temperature and pressure upon the extent of conversion of the reactants to ammonia can be seen in the graphs.

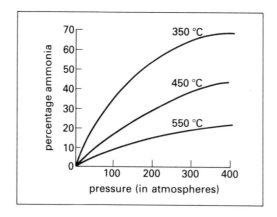

## Questions

4 Describe one biological process which fixes nitrogen.
5 Write a balanced equation for the reaction of carbon monoxide with steam.
6 What is a catalyst?
7 What is a catalyst poison?
8 Suggest a method of separating carbon dioxide from hydrogen.
9 Describe a reasonably economic way of making hydrogen if natural gas were not available.
10 Why is the catalyst used in pellet form?
11 The reaction between nitrogen and hydrogen is usually carried out at about 400 °C and 200 atmospheres pressure. Explain the reason for choosing these conditions.
12 In November 1914 there was a major sea battle off the coast of Chile between the British and German navies. It finished in victory for the British just off the Falkland Islands, giving Britain control over nitrate exports from Chile. Name one use for nitrates other than as fertilizers.

# 21 Sulphur and sulphur compounds

## 21.1 How we get sulphur

All the objects shown in figure 21.1 have something in common. Sulphur or a sulphur compound was used in their manufacture.

**Figure 21.1** *Sulphur or sulphur compounds were used to make these products*

### Where does sulphur come from?

Many minerals and some metal ores contain sulphur compounds. Sulphur compounds are also found in crude oil and natural gas, and in the gases emitted by volcanoes. The element sulphur itself is often found near volcanoes. It occurs in huge quantities in Texas, USA.

### How do we extract the element sulphur?

Much of the world's sulphur needs comes from the underground deposits in Texas. This sulphur is about 150 metres below ground. It is mainly covered by fine sand. So it cannot be mined in the normal way. One method of obtaining this sulphur was invented by an American named Frasch. His idea was to melt the sulphur below the ground, and then to force the liquid sulphur to the surface by compressed air. Figure 21.2 shows the apparatus used. Three pipes fit inside

**Figure 21.2** *The Frasch sulphur extraction pump*

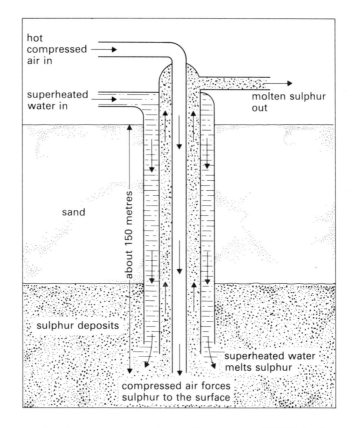

each other and superheated water at 180 °C is forced down the outermost pipe. This melts the sulphur (melting point 115 °C). Then hot compressed air is forced down the innermost pipe. This pushes the liquid sulphur up the third pipe. At the surface the molten sulphur is sprayed into vats where it cools and solidifies.

## 21.2 Heating sulphur

When sulphur is heated, it changes in several unusual ways.

Below 115 °C sulphur is a yellow solid. It is made of sulphur molecules arranged in a regular pattern. Each sulphur molecule contains eight atoms joined in a ring.

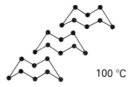

100 °C

At its melting point of 115 °C the regular arrangement of the sulphur molecules breaks down. The sulphur atoms are still arranged in rings, but now these ring molecules are no longer packed neatly next to each other.

120 °C

At about 140 °C the rings split open. So we get chains eight atoms long. This does not affect the appearance or physical properties of the molten sulphur very much. It is now amber-coloured and fairly 'runny'.

140 °C

Above 155 °C the sulphur becomes more and more viscous (treacly). This happens because the chains of eight atoms begin joining together. By the time the temperature has reached 185 °C, as many as half a million chains may have joined together. These very long chains cannot easily flow past each other. They get tangled up. As a result the liquid becomes viscous.

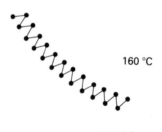

160 °C

Above about 250 °C these long chains begin to break down again, mainly into chains eight atoms long. Then the liquid becomes less viscous (more runny).

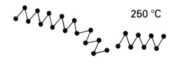

250 °C

The boiling point of sulphur is 444 °C. By the time this temperature is reached the long chains have all broken down to short chains. If you carry on heating them, they will leave the test tube as sulphur vapour. They may also catch fire!

440 °C

## 21.3 Allotropes of sulphur

We saw in the last section what happens when sulphur is heated to its boiling point. When you cool boiling sulphur, though, one of two things can happen. It all depends on how you cool it.

### Monoclinic sulphur

If you cool the sulphur slowly then the changes shown in the diagrams in section 21.2 happen in reverse. The chains eventually change back into rings of eight atoms. At the melting point these ring molecules arrange themselves into a regular pattern. The crystals that form are long and needle-like. They are called **monoclinic sulphur** or **β-sulphur**. These crystals can be made in the laboratory by the method shown in figure 21.3.

**Figure 21.3** *Making monoclinic sulphur from molten sulphur*

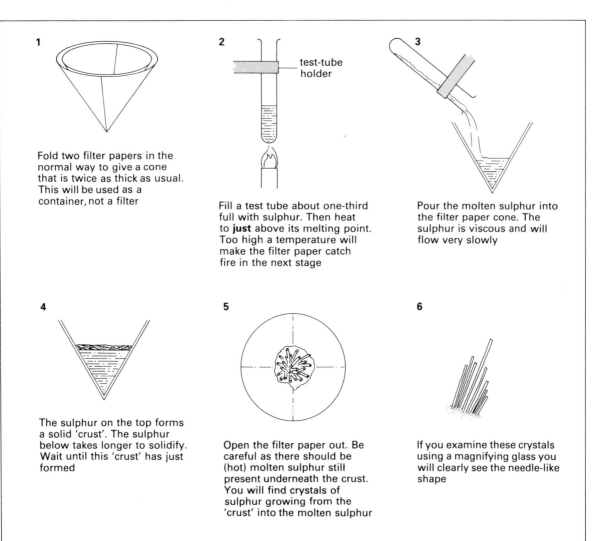

1 Fold two filter papers in the normal way to give a cone that is twice as thick as usual. This will be used as a container, not a filter

2 Fill a test tube about one-third full with sulphur. Then heat to **just** above its melting point. Too high a temperature will make the filter paper catch fire in the next stage

test-tube holder

3 Pour the molten sulphur into the filter paper cone. The sulphur is viscous and will flow very slowly

4 The sulphur on the top forms a solid 'crust'. The sulphur below takes longer to solidify. Wait until this 'crust' has just formed

5 Open the filter paper out. Be careful as there should be (hot) molten sulphur still present underneath the crust. You will find crystals of sulphur growing from the 'crust' into the molten sulphur

6 If you examine these crystals using a magnifying glass you will clearly see the needle-like shape

## Plastic sulphur

If boiling sulphur is poured into cold water a completely different form of sulphur is obtained. The cooling takes place so rapidly that only some of the changes shown in section 21.2 have time to take place. In the boiling sulphur are short chains of sulphur atoms. When you cool the sulphur very quickly, these short chains join together to give the longer chains. But by the time this has happened the molecules have lost most of their energy to the cold water. Because they do not have much energy left, the long chains cannot change any further. So the sulphur atoms are trapped in the form of a long chain. The substance obtained is brown and rubbery. It is called **plastic sulphur**. Plastic sulphur can be made using the method shown in figure 21.4.

**Figure 21.4** *Making plastic sulphur. The experiment must be done in a fume cupboard*

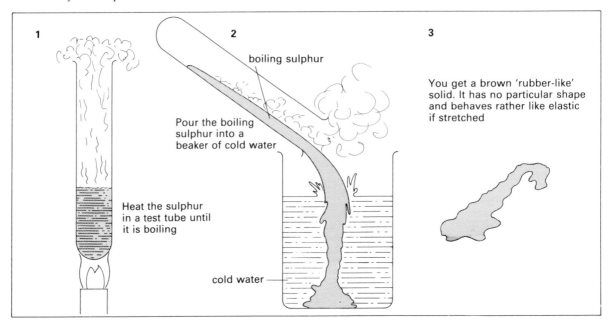

1. Heat the sulphur in a test tube until it is boiling

2. boiling sulphur

   Pour the boiling sulphur into a beaker of cold water

   cold water

3. You get a brown 'rubber-like' solid. It has no particular shape and behaves rather like elastic if stretched

Monoclinic sulphur and plastic sulphur both contain only sulphur atoms. But the arrangement of atoms within these two forms of sulphur is very different. Elements that can exist in different forms are said to have different **allotropes** (see section 14.3). Monoclinic sulphur and plastic sulphur are two allotropes of sulphur.

## Rhombic sulphur

Sulphur will not dissolve in water. But it will dissolve in a liquid such as methylbenzene. A saturated solution of sulphur can be made by heating excess sulphur with methylbenzene. Some sulphur does not dissolve. This can be removed by filtering the hot solution. To obtain crystals of sulphur, we must let the hot solution cool. Crystals that form

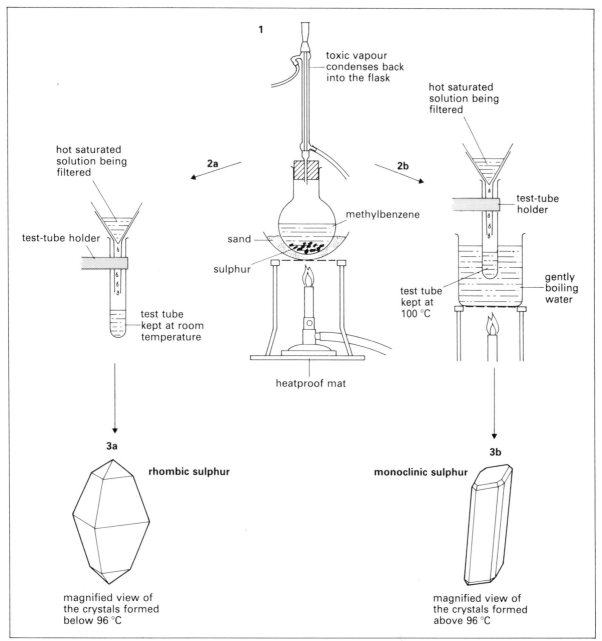

**Figure 21.5** *Making rhombic sulphur and monoclinic sulphur*

from a solution with a temperature above 96 °C (the **transition temperature** are needle-like. This is monoclinic sulphur, mentioned already. Crystals that form below 96 °C are rhombic-shaped (shaped like a rhombus) (see figure 21.5). This is another allotrope of sulphur. It is called **rhombic sulphur** or **α-sulphur**.

Monoclinic sulphur and rhombic sulphur both contain the same type of molecule. These molecules are made of eight sulphur atoms joined in a ring. What is different is the way in which these rings pack next to each other. In the rhombic-shaped crystals the rings are packed more closely together.

## 21.4 Sulphur dioxide and metal sulphites

If sulphur is heated in a bunsen flame it melts and then catches fire. It burns with a blue flame. As it does, it gives off an invisible, choking gas. It is reacting with oxygen in the air to form sulphur dioxide:

$$\text{sulphur} + \text{oxygen} \longrightarrow \text{sulphur dioxide}$$
$$S + O_2 \longrightarrow SO_2$$

### Making sulphur dioxide

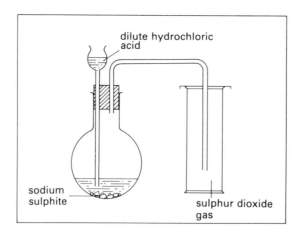

**Figure 21.6** *Apparatus for making sulphur dioxide. The apparatus must be in a fume cupboard*

Most sulphur dioxide that is made for industry is made by burning sulphur. In the laboratory there is a more convenient method. A dilute acid is added to sodium sulphite, using the apparatus shown in figure 21.6. The following reaction occurs:

hydrochloric acid + sodium sulphite $\longrightarrow$ sodium chloride + water + sulphur dioxide

$$2HCl + Na_2SO_3 \longrightarrow 2NaCl + H_2O + SO_2$$

Table 21.1 lists some of the properties of sulphur dioxide.

| colourless, choking gas |
|---|
| much denser (heavier) than air |
| highly soluble in water giving an acidic solution that can act as a bleach, sometimes called sulphurous acid |

**Table 21.1** *Properties of sulphur dioxide*

## Sulphur dioxide and corrosion

When coal and also some other fuels are burned, a lot of sulphur dioxide is produced. This dissolves in rain water, making an acidic solution. When this 'acidic rain' falls, it corrodes both stone and metal (see figure 21.7).

**Figure 21.7** *A badly corroded statue on a building*

## Metal sulphites

When sulphur dioxide is bubbled into an aqueous alkali, a **metal sulphite** is produced. These are compounds that contain the sulphite ion $SO_3^{2-}$. This reaction can be described by a general equation:

aqueous alkali + sulphur dioxide $\longrightarrow$ water + metal sulphite

The diagram equation below shows the formation of sodium sulphite:

sodium hydroxide + sulphur dioxide $\longrightarrow$ sodium sulphite + water
alkali

$$2NaOH \quad + \quad SO_2 \quad \longrightarrow \quad Na_2SO_3 \quad + \quad H_2O$$

## Some uses of sulphur dioxide

A reducing agent is a chemical that can take oxygen away from other substances (see section 18.1). Sulphur dioxide solution is a reducing agent. It gains oxygen to form sulphuric acid:

$$SO_2 + H_2O + O \longrightarrow H_2SO_4$$

Many coloured substances lose their colour when they are reduced. Because of this, aqueous sulphur dioxide can be used as a bleach. It is used to bleach paper (see figure 21.8).

**Figure 21.8 (right)** *Wood pulp for paper-making is bleached using metal sulphites. The huge rolls will be used for printing newspapers*

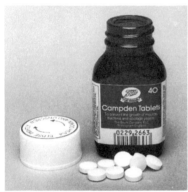

**Figure 21.9** *These Campden tablets are added to home-made wine. They release sulphur dioxide, which stops fermentation and helps to preserve the wine*

Sulphur dioxide kills bacteria and fungi. It is therefore used in the preservation of food (see figure 21.9).

## 21.5 Sulphuric acid

### The manufacture of sulphuric acid

Sulphuric acid is manufactured from sulphur in the steps shown in figure 21.10.

**Figure 21.10** *The manufacture of sulphuric acid*

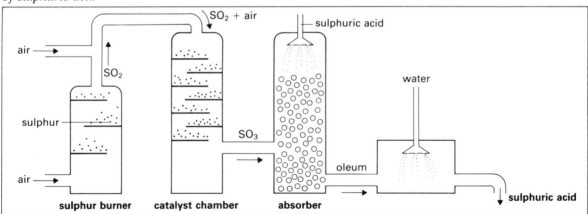

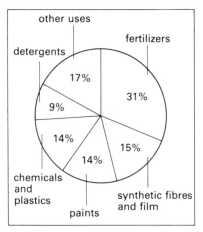

**Figure 21.11** *Some uses of sulphuric acid*

| dense, colourless, oily liquid |
|---|
| produces a lot of heat when mixed with water |
| solution in water is called dilute sulphuric acid and behaves as a typical acid |
| concentrated acid is a good dehydrating agent |
| concentrated acid is an oxidizing agent |

**Table 21.2** *Properties of sulphuric acid*

Sulphur and oxygen react in the sulphur burner to form sulphur dioxide:

sulphur + oxygen $\longrightarrow$ sulphur dioxide

$$S + O_2 \longrightarrow SO_2$$

The sulphur dioxide that is formed is mixed with more air. This is then heated to 220 °C and passed into the catalyst chamber. Here the sulphur dioxide reacts with more oxygen to form sulphur trioxide:

sulphur dioxide + oxygen $\longrightarrow$ sulphur trioxide

$$2SO_2 + O_2 \longrightarrow 2SO_3$$

A catalyst of vanadium(v) oxide is used to make this reaction occur at a reasonable speed.

The sulphur trioxide would react with water to make sulphuric acid. But if this is done, a lot of steamy, acid fumes are produced. It is better to dissolve the sulphur trioxide first in concentrated sulphuric acid. This gives an oily liquid called **oleum**. The oleum is then mixed with enough water to react with the dissolved sulphur trioxide:

sulphur trioxide + water $\longrightarrow$ sulphuric acid

$$SO_3 + H_2O \longrightarrow H_2SO_4$$

Sulphuric acid is an extremely important chemical. Some of its uses are shown in the pie chart in figure 21.11.

Table 21.2 lists some of the properties of sulphuric acid.

## Diluting sulphuric acid

### Never add water to concentrated sulphuric acid!

When concentrated sulphuric acid is mixed with water a lot of heat is given out. Adding water to the acid may cause a spray of hot corrosive acid to be formed (see figure 21.12).

Adding the acid to water is safer, but even then gloves and safety spectacles are essential.

**Figure 21.12** *How to dilute sulphuric acid*

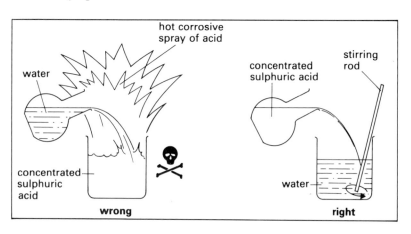

**Figure 21.13** *Concentrated sulphuric acid can remove water from sugar to leave just carbon*

## Sulphuric acid as a dehydrating agent

Concentrated sulphuric acid is a good dehydrating agent. This means that it can remove water from other substances. Many gases are dried by bubbling them through the concentrated acid.

Sulphuric acid is such a good dehydrating agent that with some compounds it will even remove the elements that make water. For example, it reacts with sugar, which is a carbohydrate. The molecules of carbohydrates consist of carbon, hydrogen and oxygen atoms. Sulphuric acid removes the hydrogen and oxygen from these compounds and black carbon is left (see figure 21.13).

## Sulphuric acid – a typical acid

Dilute sulphuric acid behaves as a typical acid. It reacts with many metals to produce hydrogen gas. For example:

magnesium metal + sulphuric acid $\longrightarrow$ hydrogen + magnesium sulphate

$$Mg + H_2SO_4 \longrightarrow H_2 + MgSO_4$$

It also reacts with bases to form salts called **sulphates**. For example, it reacts with sodium hydroxide to form sodium sulphate:

sodium hydroxide + sulphuric acid $\longrightarrow$ water + sodium sulphate

$$2NaOH + H_2SO_4 \longrightarrow 2H_2O + NaSO_4$$

## Sulphuric acid – an oxidizing agent

Hot concentrated sulphuric acid is an oxidizing agent. It will give oxygen to other substances. When it does, water and sulphur dioxide are produced. For example, the acid reacts with metals low down in the activity series, like copper.

We can think of the reaction as taking place in two stages. First, the metal is converted to its oxide:

copper + sulphuric acid $\longrightarrow$ copper oxide + water + sulphur dioxide

$$Cu + H_2SO_4 \longrightarrow CuO + H_2O + SO_2$$

Next, the oxide reacts with more acid to form the salt copper sulphate:

copper oxide + sulphuric acid $\longrightarrow$ water + copper sulphate

$$CuO + H_2SO_4 \longrightarrow H_2O + CuSO_4$$

## Summary: Sulphur and sulphur compounds

★ Most sulphur is obtained by the Frasch process.
★ When you heat sulphur, it undergoes a series of reversible changes. In these changes the arrangement of the sulphur atoms is altered.
★ There are three allotropes of sulphur at room temperature: monoclinic sulphur, plastic sulphur and rhombic sulphur.
★ When sulphur is burned, sulphur dioxide ($SO_2$) is formed.
★ Sulphur dioxide can also be made by the reaction of metal sulphites with acid.
★ Metal sulphites are made by the reaction of sulphur dioxide with an aqueous alkali.
★ Sulphur dioxide in the air causes corrosion of stonework and metals.
★ Sulphur dioxide can be used as a bleach.
★ When you add a catalyst to sulphur dioxide, it reacts with oxygen to form sulphur trioxide, $SO_3$.
★ Sulphur trioxide reacts with water to form sulphuric acid, $H_2SO_4$.
★ Sulphuric acid is a dehydrating agent.
★ Hot concentrated sulphuric acid is a good oxidizing agent.

## Questions

1 Copy out the sentences below and choose numbers from the list to fill in the blank spaces.

    8
    96
    140
    444
    millions of

(a) The transition temperature between rhombic and monoclinic sulphur is _____ °C.
(b) Sulphur molecules begin splitting open from rings into chains at about _____ °C.
(c) In a sulphur molecule at 185 °C there will be _____ atoms joined together.
(d) In a molecule of rhombic sulphur there are _____ atoms joined together.
(e) Sulphur boils at a temperature of _____ °C.

2 The following notes were found in an old chemistry notebook.

'I added the sulphur to a flammable liquid called carbon disulphide. Not all of the sulphur dissolved. I removed the undissolved sulphur and this left a clear yellow solution. I left the beaker containing the solution in the fume cupboard. By the following day the carbon disulphide had all disappeared. The bottom of the beaker had lots of yellow sulphur crystals over it.'

(a) State one precaution that you would have to take in this experiment.

(b) Draw a labelled diagram of the apparatus that you would use to separate the undissolved sulphur.

(c) The carbon disulphide 'disappeared' overnight. What word would be better than 'disappeared'?

(d) Would the crystals formed at the bottom of the beaker be rhombic or monoclinic? Explain your answer.

(e) Draw the shape of this type of crystal.

3 This question is about the experiment described below.

A small amount of sulphur was heated on a deflagrating spoon. It melted and then caught fire. The burning sulphur was lowered into a gas jar. The bottom of the gas jar contained water and a few drops of litmus solution. This solution turned red. When the sulphur had stopped burning, sodium hydroxide solution was added to the jar. Several drops were needed to make the indicator change to its original blue colour.

(a) What substance was formed when the sulphur burned?

(b) Write a word equation and a full equation for the burning reaction.

(c) What substance was formed when sodium hydroxide was added to the aqueous solution at the bottom of the gas jar? (The litmus was not involved in this reaction.)

(d) Write an equation for the reaction between the sodium hydroxide and the aqueous solution.

(e) State one industrial use of sulphur dioxide solution.

4 This question is about sulphuric acid and some of its reactions.

(a) Describe the contact process for the manufacture of sulphuric acid. Make sure that your answer includes chemical equations for each of the reactions taking place.

(b) Magnesium reacts with dilute sulphuric acid to form hydrogen gas and magnesium sulphate solution. Explain, with the help of diagrams, how you could obtain (i) a sample of hydrogen gas, (ii) crystals of magnesium sulphate.

(c) Say what you would see when concentrated sulphuric acid is added to sugar. Explain what is happening.

(d) Sulphuric acid can be made to react with metals that are low down in the activity series. Give an example of such a reaction. Explain what is happening, giving as much detail as possible.

(e) It used to be said that a country's prosperity could be measured by its sulphuric acid production. Give reasons why this should be so.

## Sulphur production – a changing scene

The production of sulphur and sulphur compounds nicely illustrates how the chemical industry can change. In 1900 almost all sulphur came from Italian sulphur mines. Today these produce only 10 000 tonnes a year, which is barely 0.03 per cent of world sulphur requirements.

In Britain the only sulphur compound that occurs in large amounts is anhydrite, which is a form of calcium sulphate. Heating the crushed rock with coke to 1450 °C was once a major source of sulphur dioxide. This is now uneconomic.

Not long ago the impressively efficient Frasch process dominated world sulphur production. Now it is surpassed by other methods of making sulphur or its compounds. Some of these methods have been forced on the industry by outside factors.

Large amounts of sulphur are obtained by the desulphurization of natural gas. Although the British gas contains relatively little sulphur, some French and Canadian sources contain more than 15 per cent of hydrogen sulphide. This must be removed to leave a usable fuel.

Sulphur dioxide is formed as a by-product in the roasting of metal ores. This is a vital step in the extraction of metals from ores such as zinc blende and iron pyrites. This is now a major source of sulphur dioxide.

Considerable amounts of sulphur dioxide are released into the air in the smoke from coal-fired power stations. There is pressure on the electricity boards to reduce the amounts emitted. If all this sulphur were removed from the smoke it could provide most of the needs of British industry.

*Volcanoes were the first producers of acid rain and of sulphur. Coal-burning power stations contribute to today's acid rain. Perhaps they will be tomorrow's source of sulphur*

Absorbing sulphur dioxide cheaply and efficiently is extremely difficult, however. A way must be found to recycle the absorbing material if huge quantities of waste are to be avoided.

In one method, the gases are passed into a suspension of magnesium hydroxide. The sulphur dioxide reacts with magnesium hydroxide, giving magnesium sulphite. Upon heating the sulphite decomposes to give back the (now concentrated) sulphur dioxide, together with carbon dioxide and magnesium oxide. The sulphur dioxide is treated with hydrogen sulphide to give sulphur. The magnesium oxide is reconverted to the hydroxide for further use.

## Questions

5 Write an equation for the reaction of calcium sulphate with carbon. Suggest why this is now uneconomic as a method of producing sulphur.

6 Briefly describe the Frasch process. Why is it now a less important source of sulphur compounds than it was?

7 Write equations for the roasting of zinc blende, $ZnS$, and of iron pyrites, $FeS_2$.

8 Why is a solution of sulphur dioxide in water acidic?

9 Using words, diagrams of normal laboratory apparatus and chemical equations, explain how you could absorb sulphur dioxide from the air using magnesium hydroxide and finish up with concentrated sulphur dioxide. You must also finish with the same amount of magnesium hydroxide that you started with.

10 Suggest reasons why power stations should reduce emissions of sulphur dioxide.

# 22 Chlorine and hydrogen chloride

Swimming pools always seem to have a special smell. This is because chlorine is present to kill any bacteria in the water. Only very small amounts are needed. Larger quantities are harmful to people. This property of chlorine was used in the Great War in 1915. Over 15 000 men were killed or injured by chlorine gas released over their trenches (see figure 22.1).

**Figure 22.1** *Shells containing poisonous chlorine gas were used in the First World War*

**Figure 22.2** *Rock salt (naturally occurring sodium chloride) being cut from a massive underground deposit*

## 22.1  Making chlorine

### The manufacture of chlorine

Most chlorine is made from sodium chloride (common salt). There are large amounts of salt in underground deposits in Cheshire. Some salt is obtained by **solution mining**. Deep holes are drilled in the ground and water is pumped down them. The salt dissolves to form **brine** (salt solution). This is then pumped to the surface. Electrolysis of this salt solution produces chlorine at the anode of the electrolysis cell (see section 15.7). Salt is also obtained by underground mining (see figure 22.2).

## How chlorine is made in the laboratory

Potassium manganate(VII) (sometimes called potassium permanganate) is an oxidizing agent (see section 18.1). It gives oxygen to other substances. When it is mixed with hydrochloric acid a reaction occurs. The hydrogen from the acid reacts with oxygen from the potassium manganate(VII). The chlorine from the acid is set free as chlorine gas:

hydrochloric acid     +     oxygen     $\longrightarrow$     water     +     chlorine gas

from potassium
manganate(VII)

2HCl          +          [O]     $\longrightarrow$     $H_2O$     +          $Cl_2$

Chlorine can also be made from manganese(IV) oxide and hydrochloric acid:

manganese(IV) oxide   +   hydrochloric acid $\longrightarrow$ chlorine   +   manganese(II) chloride   +   water

$MnO_2$          +          4HCl     $\longrightarrow$     $Cl_2$     +          $MnCl_2$          + $2H_2O$

The apparatus is the same for each method of making the gas. It is shown in figure 22.3. The main difference is that when using manganese(IV) oxide, the mixture has to be warmed.

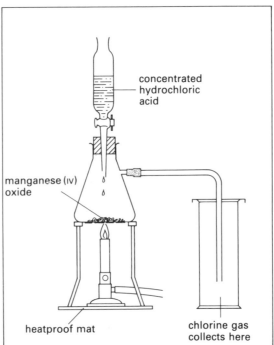

concentrated
hydrochloric
acid

manganese (IV)
oxide

chlorine gas
collects here

heatproof mat

**Figure 22.3** *Apparatus for making chlorine in the laboratory. The apparatus must be in a fume cupboard*

## 22.2 **Properties and reactions of chlorine**

Table 22.1 shows some of the properties of chlorine.

| |
|---|
| greenish-yellow, choking gas |
| much denser than air |
| fairly soluble in water, giving an acidic solution that can act as a bleach |
| reacts with a lot of metals and non-metals to form chlorides |

**Table 22.1** *Properties of chlorine*

### Solutions of chlorine – chlorine water

When chlorine is bubbled into water a solution called
**chlorine water** is obtained. The chlorine dissolves in the
water and reacts with it. What happens is this:

chlorine gas    +    water  $\longrightarrow$  hydrochloric acid    +    chloric(I) acid

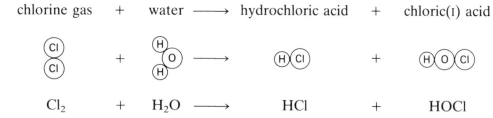

$$Cl_2 \quad + \quad H_2O \longrightarrow \quad HCl \quad + \quad HOCl$$

You can see that two acids are formed in this reaction. One
of these acids, the chloric(I) acid (sometimes called
hypochlorous acid), is unstable (see figure 22.4). It slowly
loses oxygen, as shown below:

chloric(I) acid  $\longrightarrow$  hydrochloric acid    +    oxygen gas

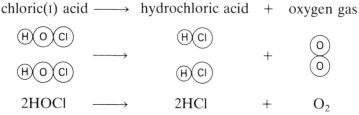

$$2HOCl \quad \longrightarrow \quad 2HCl \quad + \quad O_2$$

Because chloric(I) acid easily loses oxygen it is an oxidizing
agent (see section 18.1). This explains the interesting way in

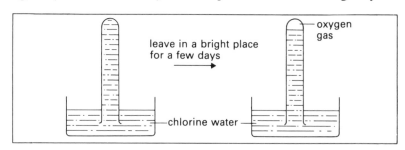

**Figure 22.4** *How chloric(I)
acid decomposes*

which chlorine water reacts with litmus. At first the litmus
goes red. This is not surprising. But shortly afterwards the
litmus loses its colour altogether. It has been bleached. This
happens because the chloric(I) acid oxidizes the litmus and
changes it into a colourless substance. Many coloured
substances lose their colour when oxidized. Because of this,
chloric(I) acid and its salts are used to make bleach.

## Chlorides

Chlorine is a very reactive non-metal. It reacts with all
metals, including gold. The compounds formed are called
**metal chlorides**. For example:

magnesium metal    +    chlorine gas $\longrightarrow$ magnesium chloride

$$Mg \quad + \quad Cl_2 \quad \longrightarrow \quad MgCl_2$$

and

aluminium metal    +    chlorine gas $\longrightarrow$ aluminium chloride

$$2Al \quad + \quad 3Cl_2 \quad \longrightarrow \quad 2AlCl_3$$

Figure 22.5 shows how this reaction can be used to make
aluminium chloride.

**Figure 22.5** *Apparatus for making aluminium chloride*

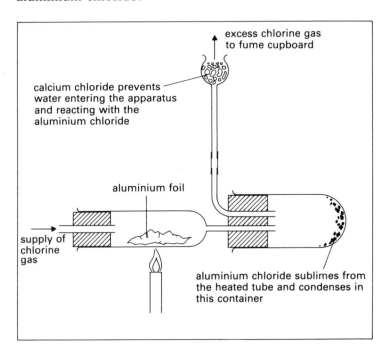

Chlorine also reacts with many non-metals. White
phosphorus catches alight when put in a gas jar of chlorine.
White clouds of phosphorus(V) chloride are formed:

phosphorus    +    chlorine $\longrightarrow$ phosphorus(V) chloride

$$2P \quad + \quad 5Cl_2 \quad \longrightarrow \quad 2PCl_5$$

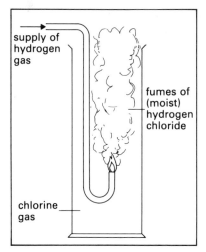

**Figure 22.6 (above)** *Burning hydrogen in chlorine. The experiment must be done in a fume cupboard*

**Figure 22.7** *Making hydrochloric acid*

Hydrogen 'burns' in chlorine to form hydrogen chloride gas (see figure 22.6):

hydrogen   +   chlorine  ⟶  hydrogen chloride

$$H_2 \quad + \quad Cl_2 \quad \longrightarrow \quad 2HCl$$

## 22.3 Hydrogen chloride gas and hydrochloric acid

### The manufacture of hydrogen chloride

Hydrogen chloride gas is made industrially by burning hydrogen in chlorine. Most of the hydrogen chloride is then dissolved in water to give a solution of hydrochloric acid (see figure 22.7).

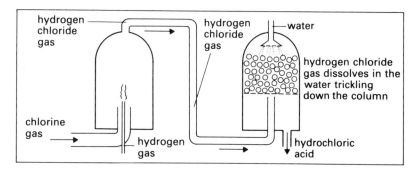

### Making hydrogen chloride in the laboratory

In the laboratory hydrogen chloride is usually made by the reaction of sodium chloride with concentrated sulphuric acid:

$$NaCl + H_2SO_4 \longrightarrow HCl + NaHSO_4$$

Traces of water that may be present in the gas can be removed by bubbling it through concentrated sulphuric acid. The hydrogen chloride is then collected as shown in figure 22.8.

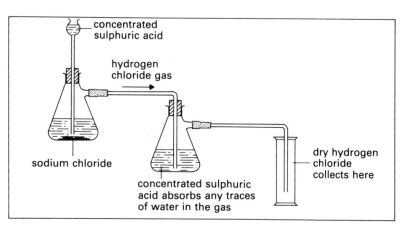

**Figure 22.8** *Making dry hydrogen chloride in the laboratory*

Table 22.2 lists some of the properties of hydrogen chloride.

| colourless, choking gas |
|---|
| denser than air |
| forms misty fumes in moist air |
| extremely soluble in water, forming very acidic solutions |

**Table 22.2** *Properties of hydrogen chloride*

## Dissolving hydrogen chloride gas in water

If a gas jar of hydrogen chloride gas is placed upside down in a trough of water, the jar immediately fills up with water, as shown in figure 22.9. This happens because hydrogen chloride gas is so soluble. It dissolves in the water leaving a vacuum in part of the gas jar. The water then rises to fill this vacuum.

**Figure 22.9** *Dissolving hydrogen chloride gas in water*

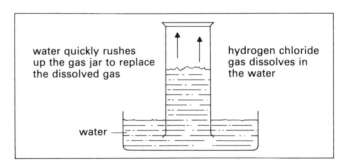

water quickly rushes up the gas jar to replace the dissolved gas

hydrogen chloride gas dissolves in the water

water

## Hydrochloric acid: a typical acid

Hydrochloric acid reacts with the more reactive metals to give off hydrogen gas. For example:

magnesium + hydrochloric acid ⟶ hydrogen + magnesium chloride

Mg + 2HCl ⟶ $H_2$ + $MgCl_2$

The acid reacts with alkalis to form chlorides. For example:

hydrochloric acid + sodium hydroxide ⟶ water + sodium chloride

HCl + NaOH ⟶ $H_2O$ + NaCl

It reacts with carbonates to give off carbon dioxide gas. For example:

calcium carbonate + hydrochloric acid ⟶ water + calcium chloride + carbon dioxide

$CaCO_3$ + 2HCl ⟶ $H_2O$ + $CaCl_2$ + $CO_2$

## 22.4 Uses of some chlorine compounds

**Figure 22.10**

(a) TCP (trichlorophenol) is a good antiseptic. It can be used to kill germs in skin wounds. You can also gargle with it to kill the germs that cause sore throats

(b) PVC (polyvinyl chloride) is waterproof and can be used to make protective clothing

(c) Trichloroethene is a liquid used to 'dry clean' clothes

(d) Sodium chlorate(I) is a good bleach

(e) The tablets used to sterilize babies' feeding bottles contain chlorine compounds

## Summary: Chlorine and hydrogen chloride

★ Chlorine is manufactured by the electrolysis of sodium chloride solution.
★ Chlorine can be prepared in the laboratory by the reaction of hydrochloric acid with manganese(IV) oxide.
★ Chlorine water contains chloric(I) acid. It acts as a bleach.
★ Chlorine reacts with metals and with non-metals to form chlorides.
★ Hydrogen chloride is manufactured by burning hydrogen in chlorine.

★ Hydrogen chloride is prepared in the laboratory by the reaction of sodium chloride with concentrated sulphuric acid.
★ Hydrogen chloride is very soluble in water. The solution is called hydrochloric acid.

## Questions

1 One way of making chlorine is to drop concentrated hydrochloric acid on to solid potassium manganate(VII). A pupil used the apparatus shown below to try and prepare some dry chlorine.

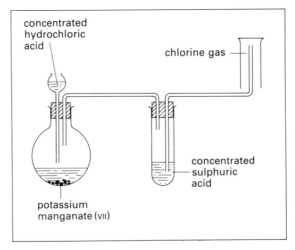

(a) These are three serious mistakes in the apparatus shown. What are they? Draw a diagram of a suitable apparatus without these faults.
(b) What is the purpose of the concentrated sulphuric acid?
(c) If no potassium manganate(VII) were available, what other chemical could be used instead? If this chemical was used, what extra apparatus would be needed?
2 Hydrogen 'burns' in chlorine to form hydrogen chloride gas. Write an equation for this reaction and include the state symbols. Draw an apparatus suitable for carrying out this experiment.
3 When chlorine gas is passed over heated silicon a reaction occurs. The silicon chloride is formed as a gas but condenses to a liquid when cooled to room temperature. Write a word equation for the reaction and draw a suitable apparatus for carrying out the reaction and collecting a sample of silicon chloride.
4 Write a chemical equation to show what happens when chlorine gas is passed into water. Explain what would happen if litmus solution was added to the chlorine water.

5 Describe, with a labelled diagram, how you would prepare some hydrogen chloride gas.

6 What would you expect to happen when hydrochloric acid is added to the following substances?
   (a) Universal indicator solution,
   (b) calcium carbonate (marble chips),
   (c) zinc metal,
   (d) gold.

7 Imagine that you have two unlabelled gas jars. One contains chlorine and the other contains hydrogen chloride. Describe at least three methods by which you could distinguish the gases.

8 Make a list of all the things in your home that contain chorine.

9 Figure 22.4 shows an experiment in which chloric(I) acid decomposes to form oxygen. Design an experiment to find out how the reaction is affected by
   (a) the brightness of the light,
   (b) the temperature,
   (c) the concentration of the chloric(I) acid.
   Draw a diagram of the apparatus you would use and construct a table showing clearly what measurements you would expect to make.

## What chemists can do with salt

The raw materials used by the chemical industry range from simple substances like air to highly complex mixtures like oil. Undoubtedly one of the simplest and yet most important raw materials is salt (or sodium chloride to give it its chemical name).

Salt itself is probably best known as a food additive. It enhances the flavour of savoury foods. A small amount is essential to our diet. But there is some evidence that eating too much salt may be bad for people with high blood pressure or heart disease.

The United Kingdom chemical industry electrolyses 2.5 million tonnes of salt per year. The electrolysis products are chlorine, hydrogen and sodium hydroxide. All three substances are important and to make a profit the chemical companies must be able to sell them all.

Three main types of electrolysis cell are used for the process. The mercury cell shown in figure 15.18 is the cheapest to run, but it carries some environmental risks. The other cells produce either less pure or less concentrated sodium hydroxide.

Most people know that chlorine is added to water to kill bacteria (see section 10.2). In fact this is a relatively minor use, taking only about 4 per cent of chlorine production.

*The salt resources of Britain*

The biggest use of chlorine (30 per cent of output) is in the manufacture of the plastic PVC. This is used for items ranging from records to floor tiles.

A further 27 per cent of chlorine output is used in making chemicals such as polyurethanes, polyesters, pesticides, herbicides and dyes.

Some 23 per cent of chlorine production is used in the manufacture of chlorinated hydrocarbons. These are made from petroleum products – hydrocarbons (compounds of carbon and hydrogen) obtained from oil. Some chlorinated hydrocarbons are used for dry cleaning, degreasing and paint stripping. Others are used as intermediates in the manufacture of fluorocarbons which form the non-stick coatings on kitchen pans. You may have noticed that the solvent in typists' correction fluid (Tipp-Ex) is 1,1,1-trichloro-ethane, which is a chlorinated hydrocarbon too.

The production of hydrochloric acid and related chemicals, although it is among the best known uses of chlorine, accounts for only about 12 per cent of output.

### Questions

**10** Draw a pie chart showing the uses of chlorine.

**11** In what everyday products will you find (a) polyurethanes, (b) polyesters?

**12** (a) What are herbicides?
(b) What are pesticides?

**13** 'Chlorinated hydrocarbons are intermediates in fluorocarbon manufacture.' What does this mean?

**14** The mercury cell is said to involve environmental risks. What is meant by this?

**15** The major chlorine production factories in the United Kingdom are all near natural salt deposits. In contrast, the silicon-based micro-chip industry is not centred around silicon deposits. Explain this difference.

**16** It is suggested that the electrolysis cells producing dilute sodium hydroxide are at a disadvantage compared to the mercury cell. Suggest reasons for this.

**17** It is not possible to alter the ratio of sodium hydroxide to chlorine that the electrolysis cells produce from sodium chloride. Why might this cause commercial difficulties?

# 23 Carbon and some simple carbon compounds

Carbon is found in the air, the sea, the rocks and in all living things. It is present in every part of your body and in the food you eat. Nearly all the world's energy is obtained by burning carbon and its compounds. Plastics, clothing, medicines, detergents, cosmetics and even this book all consist mainly of carbon. So life without carbon would be unthinkable!

Because carbon forms so many compounds we have spread its chemistry over two chapters. This chapter is about the element carbon and simple compounds containing carbon.

## 23.1 Different forms of carbon

**Charcoal** is an impure form of carbon. It is made by heating wood without air being present. **Coal** is another impure form of carbon. It contains between 65 per cent and 95 per cent carbon. **Coke** is made by heating coal in the absence of air. It can contain even more carbon.

There are two pure forms of carbon. These are **graphite** and **diamond**. Graphite is made by heating coke in an electric furnace for many hours. The carbon atoms in the coke gradually take up new positions and black slippery crystals of graphite are formed.

If graphite is heated to 3000 °C and put under tremendously high pressure, the atoms once again take up different positions. This time diamond crystals are formed.

Graphite and diamond are allotropes (different forms – see section 14.3) of carbon. Both contain only carbon atoms. The difference between them is the way in which these atoms are arranged (see figure 23.1).

### Uses of graphite

Graphite is greasy and slippery to the touch. It is slippery because of its layer structure. These layers can slide very easily over each other. This makes graphite a useful

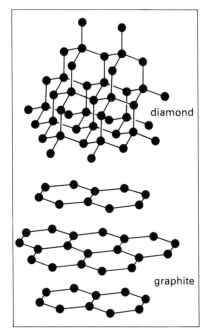

**Figure 23.1** *The structures of diamond and graphite*

lubricant. It can be used either on its own or added to oils and greases.

Graphite is unique in being the only non-metal that will conduct electricity. In industry it is often used for making electrodes. Graphite electrodes are used in aluminium production, for example.

## Uses of diamond

As you know, diamond is transparent and sparkles in the light. Because diamonds are both attractive and expensive, they are used a lot in jewellery.

**Figure 23.2** *These oil-well drilling bits have many diamonds in them. This is because diamond is hard and can easily cut through rock*

Diamond is very hard – it is the hardest of all substances. This hardness makes it useful for cutting, grinding and polishing other hard materials (see figure 23.2). Diamond-tipped drills are used when drilling for oil and natural gas. They are also used by your dentist.

## Uses of impure carbon

Graphite is expensive and diamond (even synthetic diamond) is very expensive. But coke, an impure form of carbon, is very cheap. This form of carbon has two main uses:

1 as a fuel,
2 as a reducing agent in the extraction of metals (see section 17.3).

## 23.2 The oxides of carbon

When carbon is burned in air, it can form either carbon monoxide or carbon dioxide.

Carbon dioxide is essential to life on Earth. Plants use this gas for photosynthesis (see section 24.4).

But carbon monoxide is a deadly poisonous gas. If you breathe it in, it stops your red blood cells from carrying oxygen around your body and suffocates you.

## The industrial preparation of carbon dioxide

There are several different ways this gas can be made. It is obtained in the liquefaction of air (see section 9.1) and is formed as a by-product in the manufacture of alcohol (see section 24.6).

It is produced in lime kilns (see section 23.4) and may also be made by burning carbon in plenty of air:

$$\text{carbon} \quad + \quad \text{oxygen} \longrightarrow \text{carbon dioxide}$$
$$C \quad + \quad O_2 \longrightarrow CO_2$$

Naturally, we would use coke, and not graphite or diamond, in this reaction!

## Making carbon dioxide in the laboratory

Air contains only 0.03 per cent of carbon dioxide. So it is not practical to try and obtain this gas from the air in the laboratory. Instead it is made by the reaction of a metal carbonate with an acid:

$$\text{metal carbonate} + \text{acid} \longrightarrow \text{water} + \text{a salt} + \text{carbon dioxide}$$

All metal carbonates react in this way, but the substance most often used is marble chips. This is a cheap and reasonably pure form of calcium carbonate. The reaction is as follows:

$$\text{calcium carbonate} + \text{hydrochloric acid} \longrightarrow \text{water} + \text{calcium chloride} + \text{carbon dioxide}$$
$$CaCO_3 \quad + \quad 2HCl \longrightarrow H_2O + CaCl_2 + CO_2$$

The apparatus used for carrying out this reaction is shown in figure 23.3.

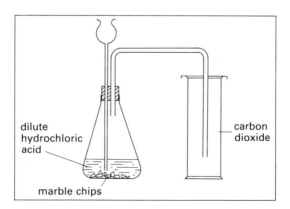

**Figure 23.3** *Apparatus for making carbon dioxide in the laboratory*

## Properties of carbon dioxide

Table 23.1 shows some of the properties of carbon dioxide gas.

**Figure 23.4** *The carbon dioxide in some natural mineral waters comes from the limestone (calcium carbonate) rocks through which they have passed*

| |
|---|
| colourless gas |
| denser than air |
| soluble in water, giving a weakly acidic solution |
| turns limewater milky – this is a test for carbon dioxide |
| taken in by plants in sunlight |
| exhaled by animals |
| a gas in which few things will burn |

**Table 23.1** *Properties of carbon dioxide gas*

## Testing for carbon dioxide gas

A gas that extinguishes a lighted spill is quite likely to be carbon dioxide. It could, however, be some other unreactive gas such as nitrogen.

The best way to identify carbon dioxide is to bubble it through limewater. A reaction occurs in which tiny pieces of solid calcium carbonate are formed, making the mixture look chalky or milky:

carbon dioxide $\quad+\quad$ limewater $\quad\longrightarrow\quad$ calcium carbonate $\quad+\quad$ water

$$CO_2 \quad + \quad Ca(OH)_2 \quad \longrightarrow \quad CaCO_3 \quad + \quad H_2O$$

## Solutions of carbon dioxide

Carbon dioxide dissolves in water and reacts with it. The reaction that takes place produces $H^+$ ions. These make the solution mildly acidic.

carbon dioxide $\quad+\quad$ water $\quad\rightleftharpoons\quad$ hydrogen ions $\quad+\quad$ hydrogencarbonate ions

$$CO_2 \quad + \quad H_2O \quad \rightleftharpoons \quad H^+ \quad + \quad HCO_3^-$$

The $\rightleftharpoons$ symbol means that only some of the carbon dioxide is used up in this reaction.

Carbon dioxide is present in 'fizzy' drinks such as beer, cola and cider (see figure 23.4). This makes the drinks slightly acid, giving them a 'sharp' taste.

## Carbon dioxide as a fire extinguisher

Carbon dioxide is a much less reactive gas than oxygen. It is contained in many fire extinguishers. A stream of the gas is directed just above the fire. Being denser than air, the gas sinks and covers the flames. So it extinguishes the fire by preventing oxygen from reaching the flames.

You cannot extinguish burning magnesium using carbon dioxide. It will still burn, although less vigorously. This happens because magnesium is reactive enough to take oxygen away from the carbon dioxide:

magnesium + carbon dioxide $\longrightarrow$ magnesium oxide + carbon

$$2Mg + CO_2 \longrightarrow 2MgO + C$$

## Carbon monoxide

Heating carbon in a limited supply of air gives carbon monoxide:

carbon + oxygen $\longrightarrow$ carbon monoxide

$$2C + O_2 \longrightarrow 2CO$$

Badly adjusted boilers and oil heaters may produce this gas. It is also present in the exhaust fumes of badly tuned cars and lorries.

Table 23.2 shows some of the properties of carbon monoxide.

**Figure 23.5** *The air in mines is continuously checked for carbon monoxide and other dangerous gases. This picture shows a mineworker checking a filter in a tube near an air-sampling point*

| |
|---|
| colourless, odourless gas |
| similar density to air |
| highly poisonous |
| burns in air to form carbon dioxide |
| good reducing agent (it can take oxygen away from many other substances) |

**Table 23.2** *Properties of carbon monoxide*

Carbon monoxide can often be seen burning to form carbon dioxide at the top of fires burning smokeless fuels. It burns with a blue flame.

Carbon monoxide can remove oxygen from other substances. This property is made use of in the blast furnace. Here it reduces iron oxide to iron (see figure 17.6):

iron oxide + carbon monoxide $\longrightarrow$ iron + carbon dioxide

## 23.3 Carbon in rocks

Carbon is present in many rocks in the form of metal carbonates. Table 23.3 lists some of these.

| Name of rock | Carbonate present |
|---|---|
| chalk | calcium carbonate |
| limestone | calcium carbonate |
| marble | calcium carbonate |
| magnesite | magnesium carbonate |
| calamine | zinc carbonate |
| malachite | copper carbonate |

**Table 23.3** *Carbonates and the rocks they come from*

By far the commonest of these is calcium carbonate (see figure 23.6). You can see that chalk, limestone and marble are all forms of this compound. Although these substances differ in hardness, their chemical reactions are alike.

**Figure 23.6** *Calcium carbonate is found in huge quantities. These chalk cliffs are made up largely of this chemical*

### Heating metal carbonates

Most metal carbonates decompose when heated (see tables 16.5 and 16.6 on pages 197 and 198):

$$\text{metal carbonate} \xrightarrow{\text{heat}} \text{metal oxide} + \text{carbon dioxide}$$

The carbonates of reactive metals are harder to decompose than those of less reactive metals.

Sodium carbonate does not decompose even at high temperatures. Copper carbonate decomposes on gentle heating. The reactivity of calcium is somewhere between

**Figure 23.7** *Baking powder contains sodium hydrogencarbonate and a weak acid. These react together in the moist cake mixture. The carbon dioxide given off makes the cake rise*

that of sodium and copper. So calcium carbonate decomposes, but only on strong heating.

## Metal hydrogencarbonates

Metal carbonates usually react with water and carbon dioxide to form metal hydrogencarbonates. These hydrogencarbonates decompose more easily than the carbonates do. Because of this, sodium hydrogencarbonate is sometimes used to make cakes rise. It is more often used by cooks in the form of baking powder, however (see figure 23.7).

Because hydrogencarbonates tend to decompose, we get stalactites and stalagmites in some caves.

Limestone (calcium carbonate) reacts with rain water containing dissolved carbon dioxide:

calcium carbonate $+$ water $+$ carbon dioxide $\longrightarrow$ calcium hydrogencarbonate

$$CaCO_3 \quad + \quad H_2O \quad + \quad CO_2 \quad \longrightarrow \quad Ca(HCO_3)_2$$

The hydrogencarbonate dissolves in the rain water, which carries it away. Some of this water drips from cave roofs. The water evaporates and the hydrogencarbonate decomposes to form solid limestone again. In this way pillars are formed from the floor (stalagmites) and ceiling (stalactites) – see figure 23.8.

**Figure 23.8** *Stalactites and stalagmites are formed by the decomposition of calcium hydrogencarbonate*

An easy way to remember which is which is: 'little *mites* (children) grow upwards' and 'you have to hold *tight* to the ceiling'.

## Carbonates as bases

**Figure 23.9** *The base calcium carbonate is widely used to neutralize acids present in the soil – here it is being applied to an orchard*

Both carbonates and hydrogencarbonates are bases. They neutralize acids to form water, carbon dioxide and a salt (see figure 23.9). These compounds are often included in indigestion tablets to neutralize excess acid in the stomach. For example:

sodium hydrogencarbonate + hydrochloric acid ⟶ water + carbon dioxide + sodium chloride

$$NaHCO_3 \quad + \quad HCl \quad \longrightarrow H_2O + \quad CO_2 \quad + \quad NaCl$$

## 23.4 The uses of limestone

Limestone is one of the world's most common minerals. Because it is cheap and readily available it is an important raw material in the chemical industry.

Figure 23.10 shows how many materials depend upon limestone for their manufacture.

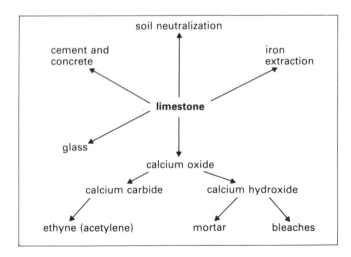

**Figure 23.10** *The uses of limestone*

## Quicklime (calcium oxide)

Calcium oxide can be made by heating calcium carbonate (limestone) in huge lime kilns (see figure 23.11). This makes the carbonate decompose to calcium oxide (quicklime) and carbon dioxide. Much of this calcium oxide is converted into calcium hydroxide.

**Figure 23.11** *Lime kilns*

## Slaked lime (calcium hydroxide)

Calcium hydroxide (slaked lime) is made by adding a small amount of water to calcium oxide.

It is the cheapest base available and so it is widely used where acids have to be neutralized before they can be disposed of.

Calcium hydroxide is only slightly soluble in water. The solution that is formed is called **limewater**.

Two major uses of calcium hydroxide are the manufacture of mortar and cement.

## Mortar

Mortar is a mixture of slaked lime and sand. It is mixed with water and used as a thick paste for bricklaying. The water dries out first. Later the calcium hydroxide reacts with carbon dioxide from the air to form calcium carbonate. This second process makes the mortar much harder.

calcium hydroxide  +  carbon dioxide $\longrightarrow$ calcium carbonate  +  water

$$Ca(OH)_2 \quad + \quad CO_2 \quad \longrightarrow \quad CaCO_3 \quad + \quad H_2O$$

The main disadvantage of mortar is that it is corroded by rain water. This process is very much like the dissolving of limestone rocks in rain water (see section 23.3 above).

## Cement

Cement is made by heating calcium carbonate with clay. This forms a mixture of calcium and aluminium silicates. The cement is mixed with sand and water to form a paste which can then be used for bricklaying. Unlike mortar, cement is not corroded by rain water.

## Concrete

Concrete is a mixture of cement with sand and gravel or stones, mixed with water. It is very widely used in building. Often the concrete is strengthened (reinforced) by allowing it to set around steel rods.

## Glass

Limestone is also used in the manufacture of sodium carbonate. This chemical is needed to make another building material, glass.

Sand and sodium carbonate are mixed and heated to 1500 °C. They react to form sodium silicate, which is glass:

sodium carbonate + silicon dioxide $\longrightarrow$ sodium silicate + carbon dioxide
sand                                                                            glass

$$Na_2CO_2 \quad + \quad SiO_2 \quad \longrightarrow \quad Na_2SiO_3 \quad + \quad CO_2$$

The molten mixture is cooled quite quickly. The liquid does not form crystals, however. Instead it becomes less and less runny, and finally hardens into glass.

**Figure 23.12** *Glass, concrete and cement are all made from limestone*

## Summary: Carbon and simple carbon compounds

★ Carbon has two allotropes: graphite and diamond.
★ Graphite is used as a lubricant and an electric conductor.
★ Diamond is used for cutting and grinding and for jewellery.
★ Coke is an impure form of carbon. It is used as a fuel and as a reducing agent in the extraction of metals.
★ Carbon dioxide, $CO_2$, is obtained from the air and from fermentation.
★ In the laboratory carbon dioxide is made by the reaction of a metal carbonate with acid.
★ Carbon dioxide turns limewater milky.
★ Aqueous carbon dioxide is weakly acidic.
★ Carbon dioxide is taken in by plants during photosynthesis.
★ Carbon dioxide is used in some fire extinguishers.
★ Carbon monoxide, $CO$, is a very poisonous gas.
★ Metal carbonates occur in several rocks, especially chalk and limestone.
★ When heated, most metal carbonates decompose to carbon dioxide and the metal oxide.

* Limestone reacts with aqueous carbon dioxide to give calcium hydrogencarbonate solution.
* Carbonates and hydrogencarbonates react with acids to give water, a salt and carbon dioxide.
* Heating limestone produces quicklime (calcium oxide).
* Adding water to quicklime gives slaked lime (calcium hydroxide).
* Limestone is essential in the manufacture of mortar, cement, concrete and glass.

## Questions

1 This question is about different forms of carbon.
(a) Graphite and diamond are allotropes of carbon. What are allotropes?
(b) Give two physical properties of graphite and two physical properties of diamond.
(c) Diamond is denser than graphite. Give a reason.
(d) Graphite is sometimes used to make the 'brushes' that carry electricity to the moving part of an electric motor. What two properties of graphite make it suitable?
(e) Name one use of diamond other than in jewellery. What makes diamond especially suitable for this use?
(f) Write a word equation for the reaction that would occur if diamonds were burned in oxygen. Write a similar word equation for the reaction of graphite with oxygen.
2 Describe, with a labelled diagram, how you would make several gas jars of carbon dioxide gas in the laboratory.
3 A solution of carbon dioxide in water is mildly acidic.
(a) Why is the solution acidic?
(b) What approximate pH value would carbon dioxide solution have?
4 What do you see happen when carbon dioxide is bubbled into limewater? Write a chemical equation for the reaction that takes place. Include physical state symbols.
5 What substance would you expect to be formed when carbon dioxide is bubbled into sodium hydroxide solution? Write a word equation to describe what is happening.
6 Carbon dioxide can be used in fire extinguishers. Explain how it helps to put out fires. Are there any burning materials that it would not extinguish? Could carbon monoxide also be used as a fire extinguisher? Give reasons for your answers.
7 Strontianite is an ore found in North Germany and western Scotland. It contains strontium, which comes from the same group of the periodic table as calcium. Strontium compounds are similar to calcium compounds. Here are the results of a pupil's investigations of strontianite.

'I heated 0.296 g of the solid. It gave off 48 cm$^3$ of gas. When this gas was bubbled through limewater the mixture went milky. I also tried treating 0.296 g of the solid with an excess of hydrochloric acid. Only 43 cm$^3$ of gas were given off. I added a further 0.296 g of solid to the same acid and this time I got a further 48 cm$^3$ of gas. I took the solid that had been formed by heating the strontianite and added it to water. The mixture got quite warm, and a solution was produced that turned Universal indicator blue.'

(a)  What gas is given off when the ore is heated?

(b)  What is the chemical name for the ore strontianite?

(c)  Give a word equation for the reaction when strontianite is heated. Also give a chemical equation.

(d)  Give a word equation for the reaction that takes place when strontianite is added to hydrochloric acid. Also give a chemical equation.

(e)  Draw a diagram of an apparatus suitable for carrying out the reaction described between strontianite and hydrochloric acid.

(f)  Suggest reasons why the pupil got 43 cm$^3$ of gas from the first reaction with hydrochloric acid but 48 cm$^3$ from the second.

## Carbon dioxide: friend or foe?

Carbon dioxide makes up about 0.03 per cent of the air. It is needed for photosynthesis and so is essential to plant life. All animal life depends directly or indirectly on plants, so carbon dioxide is essential to humans too.

Even so, carbon dioxide can be considered a pollutant. The air in a badly ventilated room can contain up to 1 per cent of carbon dioxide. This causes drowsiness and headaches. Concentrations from 8 to 12 per cent of carbon dioxide can cause death by asphyxiation. When carbon dioxide dissolves in the blood it increases the blood acidity. This change in acidity reduces the ability of the red cells of the blood to transport oxygen around the body.

In manned spacecraft, arrangements have to be made to extract carbon dioxide from the air. The Apollo missions used an absorption system employing lithium hydroxide.

Many scientists are concerned that the carbon dioxide level in the air is increasing and that this will bring a problem which has been called the 'greenhouse effect'. Much of the energy received from the Sun is re-emitted from the Earth in the form of infra-red rays. Carbon dioxide absorbs these rays, and so an increase in carbon dioxide in the air would reduce the Earth's heat loss and result in a temperature rise. This could dramatically affect climate and agriculture. It could melt the polar icecaps, raising the sea level some 80–100 metres. This would put half of Britain under water!

*Carbon dioxide must be removed from the air to prevent these astronauts suffering from drowsiness and eventual suffocation*

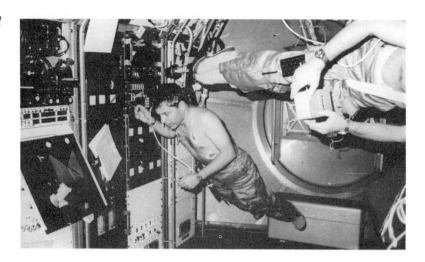

*Methane is produced in the digestive systems of cattle, sheep and goats, and also contributes to the 'greenhouse effect'. An increased methane concentration in the atmosphere might be the result of an increased number of farm animals*

There are various reasons why the level of carbon dioxide might increase. One is the chopping down of vast areas of tropical forests in recent years. Another is the burning of fossil fuels. Yet another is the manufacture of quicklime from limestone:

$$CaCO_3 \xrightarrow{\text{heat}} CaO + CO_2$$

It is estimated that if all the limestone were eventually treated in this way there would be six times as much carbon dioxide in the air as there is now. At least some of the uses to which quicklime is put eventually result in carbon dioxide being taken out of the air again, however. For example, mortar contains calcium hydroxide which reacts with carbon dioxide from the air.

Fortunately carbon dioxide levels do not seem to be rising as quickly as scientists at one time expected. This is probably because more of it is dissolving in the sea, which is estimated to contain sixty times as much carbon dioxide as the atmosphere.

## Questions

**8** Why does carbon dioxide form acidic solutions?

**9** Why can 10 per cent of carbon dioxide in the air cause death even though oxygen is still present?

**10** Write an equation for the reaction of carbon dioxide with lithium hydroxide solution. What type of reaction is this?

**11** Name another chemical that could be used in spacecraft in place of lithium hydroxide.

**12** Why should forests affect carbon dioxide levels in the air?

**13** Write equations showing how using limestone for mortar should not affect the amount of carbon dioxide in the air.

**14** Both limestone and quicklime can be used to neutralize acids in the soil. Write ionic equations showing how they remove $H^+$ ions.

# 24 Organic chemistry

Compounds from organisms which are alive, or were once alive, usually contain carbon. Because of this the chemistry of carbon compounds is often called **organic chemistry**.

## 24.1 The alkanes

The **alkanes** are a **family** of organic compounds. We call them a family because they are alike in many ways. They contain only carbon and hydrogen and are therefore known as **hydrocarbons**. Their atoms are all held together by single covalent bonds (see section 13.4). Table 24.1 lists some of the properties of alkanes.

| |
|---|
| do not dissolve in water |
| burn in oxygen or air, producing large amounts of heat |
| those with less than five carbon atoms per molecule are gases at room temperature |

**Table 24.1** *Properties of alkanes*

From table 24.2 (opposite), you can see that these compounds form a series. Each differs from the next by one carbon and two hydrogen atoms.

### What we get alkanes from

Most alkanes are obtained from the two fossil fuels, oil and natural gas. North Sea natural gas is about 95 per cent methane, 3 per cent ethane and 2 per cent propane and butane together.

Oil is a much more complicated substance. It contains many different hydrocarbons, some of which have more than one hundred carbon atoms in a molecule. At an oil refinery like the one shown in figure 24.1, this complicated mixture is split up into several simpler ones.

| Compound | Formula | Structure |
|---|---|---|
| methane | $CH_4$ | H<br>\|<br>H—C—H<br>\|<br>H |
| ethane | $C_2H_6$ | H  H<br>\|  \|<br>H—C—C—H<br>\|  \|<br>H  H |
| propane | $C_3H_8$ | H  H  H<br>\|  \|  \|<br>H—C—C—C—H<br>\|  \|  \|<br>H  H  H |
| butane | $C_4H_{10}$ | H  H  H  H<br>\|  \|  \|  \|<br>H—C—C—C—C—H<br>\|  \|  \|  \|<br>H  H  H  H |

**Table 24.2** *Some alkanes*

**Figure 24.1** *Chemical plant for the fractional distillation of oil at a refinery*

## Fractional distillation of oil

**Fractional distillation** (see section 4.1) on a very large scale is used to separate oil into simpler 'fractions'.

First, heated crude oil is passed into a large column that is hot at the bottom and cool at the top (see figure 24.2).

**Figure 24.2** *The fractional distillation of oil*

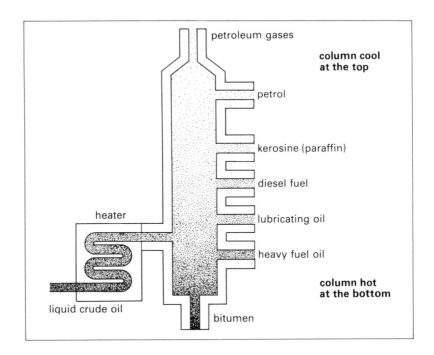

At the bottom all but the largest hydrocarbons boil. The vapours produced move up the column and get cooled on the way. When they have cooled down enough, they condense back to a liquid. Hydrocarbons with a high boiling point condense when they are still quite low down, in the hotter part of the column. Hydrocarbons with a lower boiling point only condense further up in the cooler part of the column. In this way, the crude oil is separated into 'fractions' with different boiling points.

As table 24.3 shows, the boiling points of hydrocarbons depend mainly upon the size of the molecules. For example, a molecule with six carbon atoms will have a higher boiling point than a molecule with only four.

| Substance | Number of carbon atoms | Boiling range (°C) |
|---|---|---|
| gas | 1–4 | below 40 |
| petrol | 4–12 | 40–175 |
| kerosine | 9–16 | 150–240 |
| diesel oil | 15–25 | 220–275 |
| lubricating oil | 20–70 | 250–350 |
| bitumen | more than 70 | above 350 |

**Table 24.3** *The boiling points of some hydrocarbons*

## Isomerism

Some alkanes can have the same formula as each other but different arrangements of atoms. For example, there are two substances with the formula $C_4H_{10}$ (see figure 24.3).

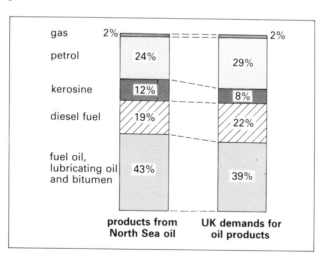

**Figure 24.3** *Two substances with the formula $C_4H_{10}$*

The diagrams in figure 24.3 are often called **structural formulae** because they tell you the structure of such compounds.

Substances that have the same atoms arranged differently in their molecules are called **isomers**.

## Uses of alkanes

Most of the alkanes that are produced are used as fuels. This is looked at in more detail in the next chapter. Alkanes are also used in the manufacture of **alkenes** (see section 24.2, below).

Fractional distillation gives us substances like petrol and kerosine. Unfortunately, it does not give us these in the proportions that we need. Figure 24.4 shows this. You will see, for example, that North Sea oil does not give as much petrol as we use.

**Figure 24.4** *Why North Sea oil cannot meet all the UK demand for some oil products*

| | products from North Sea oil | UK demands for oil products |
|---|---|---|
| gas | 2% | 2% |
| petrol | 24% | 29% |
| kerosine | 12% | 8% |
| diesel fuel | 19% | 22% |
| fuel oil, lubricating oil and bitumen | 43% | 39% |

To get more of the smaller molecules a process called catalytic cracking is used (this is described below). It is also possible to join smaller molecules into larger ones by a process called reforming.

## 24.2 The alkenes

The **alkenes** form another family of organic compounds. Like the alkanes they all contain only two elements: hydrogen and carbon. So they are also hydrocarbons. But they are very different from the alkanes in their reactions. This is because the alkenes all contain a double covalent bond (see section 13.4).

Table 24.4 gives the names, formulae and structures of some alkenes.

| Compounds | Formula | Structure |
|-----------|---------|-----------|
| ethene | $C_2H_4$ | 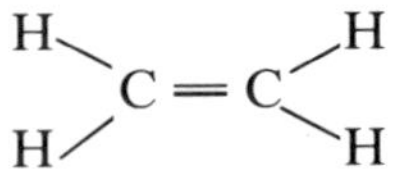 |
| propene | $C_3H_6$ | 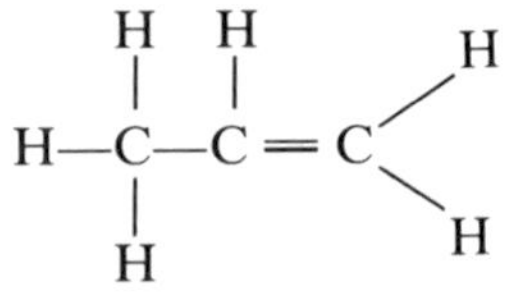 |
| butene | $C_4H_8$ | 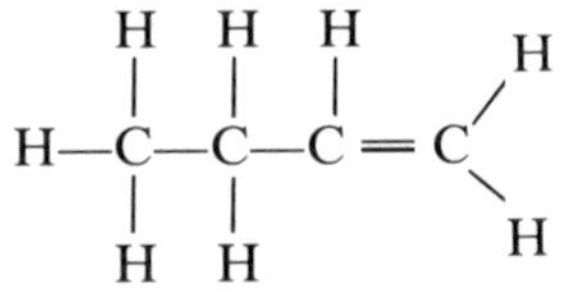 |

Table 24.4 *Some alkenes*

### Where do we get alkenes from?

Most of the alkenes that we use are obtained by breaking up alkane molecules. This is usually done by passing the alkanes over a heated catalyst. The process is called **catalytic cracking**. Figure 24.5 shows how ethene can be formed by the cracking of hexane molecules. Whenever any alkane is cracked (broken down), at least one alkene is formed.

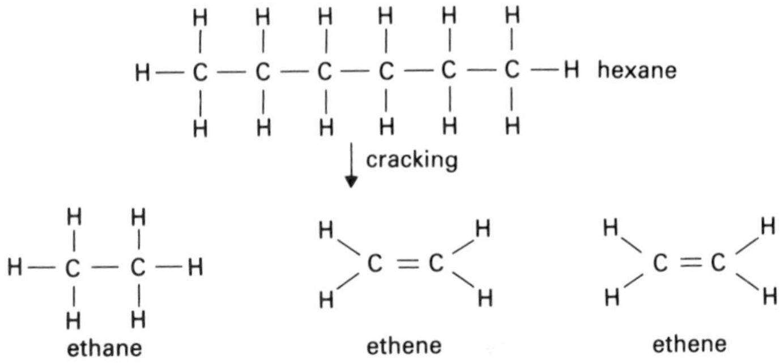

**Figure 24.5** *The formation of ethene*

Table 24.5 shows some of the properties of alkenes.

| |
|---|
| do not dissolve in water |
| burn in oxygen or air, giving out a lot of heat |
| decolorize bromine water |
| will polymerize |

**Table 24.5** *Properties of alkenes*

## Why are alkenes different from alkanes?

Have another look at table 24.2 (on page 297) which shows the structures of alkanes. You can see that each carbon atom in an alkane molecule is bonded (joined) to four other atoms. Four is the maximum number of bonds that a carbon atom can form. Because of this it is impossible for more atoms to join on to an alkane molecule. The molecules are **saturated** (cannot take any more).

Now look at the structures of the alkenes, shown in table 24.4. These all contain a pair of carbon atoms joined by a double bond. These carbon atoms are each joined to only three atoms. So more atoms can join on to an alkene molecule. Alkenes are called **unsaturated**. They are much more reactive than alkanes are.

## Bromine water – a test for unsaturated compounds

It is easy to tell whether a hydrocarbon is an alkane or an alkene. You just add a few drops of the hydrocarbon to some bromine water. If it is an alkene a reaction takes place in which bromine joins on to the alkene. This results in the bromine water losing its colour. This reaction does not take place with alkanes.

$$
\begin{array}{ccc}
\text{H} \quad \text{H} & & \text{H} \\
\ \ \diagdown \ \diagup & & | \\
\ \ \ \text{C} & \text{Br} & \text{H---C---Br} \\
\ \ \ \| & + \ \ | & \longrightarrow \ \ \ \ \ | \\
\ \ \ \text{C} & \text{Br} & \text{H---C---Br} \\
\ \ \diagup \ \diagdown & & | \\
\text{H} \quad \text{H} & & \text{H}
\end{array}
$$

As the equation shows, the bromine molecule 'adds on' across the double bond.

## Uses of alkenes

Many plastics are made from alkenes. These are described in section 24.3.

Ethanol can be made from ethene by reaction with water. The water molecule adds on across the double bond. A catalyst is necessary.

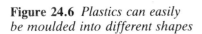

## 24.3 Synthetic polymers and plastics

**Polymers** are very large molecules that have been made by joining many smaller molecules together. The word polymer means 'many units'. Polymers that you will already know include polythene. This should be called polyethene, because it is made of many ethene molecules. Polystyrene (many styrene molecules) is another well-known polymer.

Polymers that can be easily moulded into different shapes are called **plastics** (see figure 24.6).

**Figure 24.6** *Plastics can easily be moulded into different shapes*

### 'Making ends meet' – polymerization

**Polymerization** is a reaction in which many small molecules become joined together. The small molecule that we start off with is called the **monomer** (one unit). The large molecule formed is a polymer (many units).

Alkenes can become polymers. First, two molecules become joined together, then three, then four and so on.

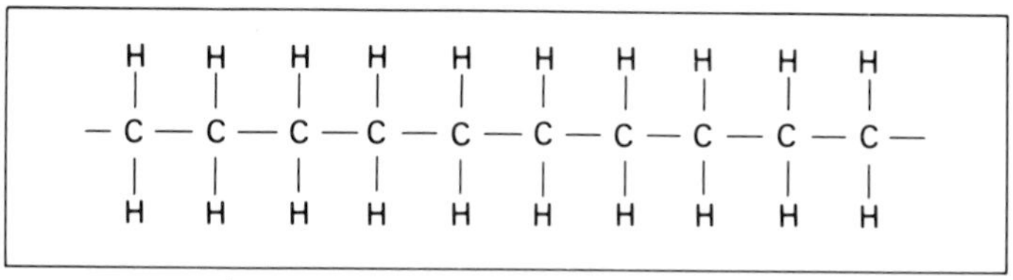

**Figure 24.7** *Part of the polymer formed from ethene*

Eventually a chain is formed with as many as 50 000 alkene molecules. Figure 24.7 shows part of the polymer molecule formed when ethene polymerizes.

Another way of representing this molecule is shown in figure 24.8. The unit inside the bracket is like an ethene molecule except that it no longer has a double bond. The '*n*' outside the bracket means some very large number.

**Figure 24.8** *Representing a polymer*

## More addition polymers

The polymers formed from alkenes are called **addition polymers**. This is because during the polymerization the molecules add on to each other. Figure 24.9 shows you some addition polymers and the monomers they are made from.

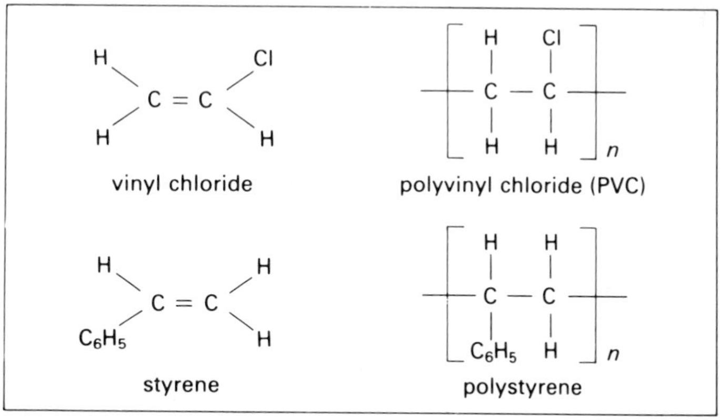

**Figure 24.9** *Two addition polymers and the monomers they are made from*

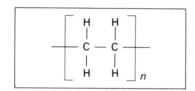

vinyl chloride      polyvinyl chloride (PVC)

styrene      polystyrene

PVC is a tougher plastic than polyethene. It is used for making hoses, electrical insulation and car upholstery.

Polystyrene is widely used as ceiling tiles and packing material. These are made from 'expanded' polystyrene. The white plastic is filled with thousands of bubbles held in a thin coating of polystyrene. This is what makes it so light.

It is also possible to make polystyrene without bubbles. This polystyrene is used for making model aircraft, ball-point pens, rulers and protractors.

## 24.4 Carbon in living things

All green plants take in carbon dioxide and give out oxygen:

carbon dioxide $\longrightarrow$ carbon + oxygen

   taken from           kept for     given out
    the air            use within     to the
                the plant      air

Plants use carbon from the air and water from the soil to synthesize (build up) compounds within the plant. Carbohydrates are the simplest of these compounds. They contain carbon, hydrogen and oxygen in the approximate ratio 1:2:1. The formulae of some carbohydrates are shown in table 24.6. They may look rather complicated but you can see that they could – in theory, at least – all be made from carbon and water. This is why they were named carbohydrates. In fact, the starting materials are usually carbon dioxide and water.

| glucose | $C_6H_{12}O_6$ |
| --- | --- |
| sucrose | $C_{12}H_{22}O_{11}$ |
| starch | approximately $C_{1000}H_{2000}O_{1000}$ |

**Table 24.6** *Some carbohydrates*

For example, in the synthesis of glucose:

$$6CO_2 \;+\; 6H_2O \;\xrightarrow{\text{sunlight}}\; C_6H_{12}O_6 \;+\; 6O_2$$

taken from    taken from      glucose     given out to
 the air       the soil                   the air

A lot of energy is needed to synthesize glucose. Plants obtain this energy by absorbing sunlight. By changing carbon dioxide and water into glucose the plants store energy from the Sun. This process is called **photosynthesis**.

**Figure 24.10** *Plants get the energy they need from sunlight*

**Figure 24.11** *These caterpillars consume the carbohydrates that the plant has produced during photosynthesis*

Animals, including humans, eat plants. Inside the animal the plants' carbohydrates react with oxygen breathed in by the animal. Carbon dioxide and water are produced. For example:

$$\text{glucose} + \text{oxygen} \longrightarrow \text{carbon dioxide} + \text{water}$$

$$C_6H_{12}O_6 + 6O_2 \longrightarrow 6CO_2 + 6H_2O$$

When this reaction takes place a great deal of energy is given out. Animals are able to use this energy for warmth and movement. The use of foods such as glucose in this way is known as **tissue respiration**.

You may have noticed that plants and animals are carrying out exactly opposite reactions. Plants take in carbon dioxide and also absorb energy from the Sun. Within the plant, large molecules are built up. Animals break down these big molecules back into carbon dioxide and water. At the same time energy is given out in the form of warmth and movement.

Neither plants nor animals use up the carbon. They are all simply part of the **carbon cycle** shown in figure 24.12.

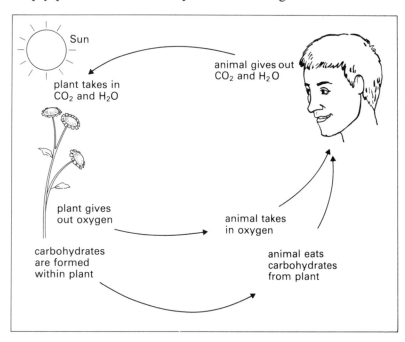

**Figure 24.12** *The carbon cycle*

## 24.5 Breaking down some organic molecules

In section 24.4 we saw how plants can synthesize glucose from carbon dioxide and water. They can also polymerize glucose into a polymer called **starch**.

### How to test for starch

Starch is present in many foods. Starch and foods containing it go deep blue when iodine solution is added to them. This can be used as a test for starch.

### Breaking down starch to glucose

Starch is a polymer. It can be broken down to its monomer by heating starch solution in a boiling water bath for an hour or so (see figure 24.13). Hydrochloric acid is added to make the reaction go faster. The acid acts as a catalyst. You can tell when all the starch has been broken down by removing some solution and adding it to iodine. When no blue colour develops you know that all the starch has gone.

**Figure 24.13** *Breaking down starch to glucose*

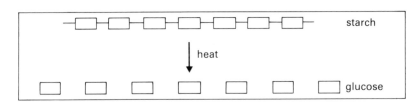

### Breaking down starch to maltose

The saliva in your mouth contains **enzymes**. Enzymes are catalysts made by plants or animals. The enzyme in saliva breaks down starch. But the starch does not break down to glucose. Instead it breaks into molecules containing two glucose molecules joined together (see figure 24.14). This is a **maltose** molecule.

**Figure 24.14** *Breaking down starch to maltose*

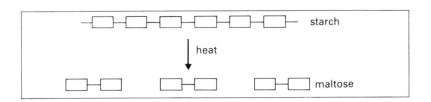

This reaction is easily carried out in the laboratory. You just mix starch solution and saliva in a test tube at room temperature. Chromatography (see section 4.3) can be used to show that the product is maltose.

### Breaking down glucose to ethanol

Animals break down carbohydrates by making them react with oxygen to form carbon dioxide and water.

Yeasts can break glucose down if there is very little oxygen present. When this happens **ethanol** (alcohol) is formed:

$$\text{glucose} \xrightarrow{\text{yeast}} \text{ethanol} + \text{carbon dioxide}$$

$$C_6H_{12}O_6 \longrightarrow 2C_2H_5OH + 2CO_2$$

This reaction has been known for thousands of years. It is used to make drinks such as beer and wine.

## 24.6 Alcohols and organic acids

### Alcohols

The **alcohols** are another family of organic compounds. Table 24.7 lists the structures and formulae of some alcohols.

| Compound | Formula | Structure |
|----------|---------|-----------|
| methanol | $CH_3OH$ | H<br>\|<br>H—C—OH<br>\|<br>H |
| ethanol | $C_2H_5OH$ | H  H<br>\|  \|<br>H—C—C—OH<br>\|  \|<br>H  H |
| propanol | $C_3H_7OH$ | H  H  H<br>\|  \|  \|<br>H—C—C—C—OH<br>\|  \|  \|<br>H  H  H |

**Table 24.7** *Some alcohols*

Usually, when people talk about 'alcohol' they mean ethanol. This is the alcohol that is present in beer, wines and spirits.

Some alcohols, such as methanol, are extremely poisonous. Methanol is added to methylated spirits to make it undrinkable.

### Making ethanol

Ethanol is usually made by **fermentation**. In this process yeast is added to glucose or some other sugar (see figure 24.15). The yeast contains enzymes which break down the sugars to ethanol and carbon dioxide:

$$\text{sugars} \xrightarrow{\text{yeast}} \text{ethanol} + \text{carbon dioxide}$$

**Figure 24.15** *Yeast breaks down glucose to make ethanol. Here fermentation is being carried out in a brewery*

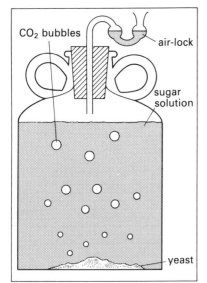

**Figure 24.16** *Making ethanol*

This reaction is best carried out at about 25 °C. At lower temperatures the reaction becomes very slow. Heating the reaction mixture much above 45 °C is likely to kill the yeast.

It is also important that air and vinegar-flies are kept out of the mixture. These would make the alcohol turn into a acid, vinegar. To prevent this, wine makers usually use an air-lock, as shown in figure 24.16. This lets the carbon dioxide out but does not let other things in.

### Organic acids

Vinegar is a dilute solution of an **organic acid**. The first two members of this organic family are shown in table 24.8.

**Methanoic acid** is found in ant 'bites' and stinging nettles. **Ethanoic acid** is present in vinegar.

| Compound | Formula | Structure |
|----------|---------|-----------|
| methanoic acid | HCOOH | 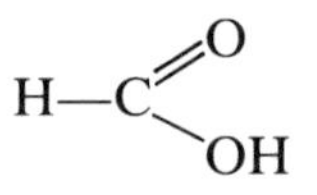 $H-C\overset{O}{\underset{OH}{\diagdown}}$ |
| ethanoic acid | CH$_3$COOH | 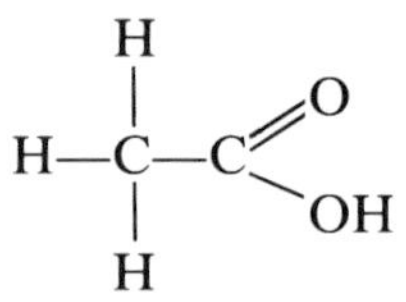 $H-\underset{H}{\overset{H}{C}}-C\overset{O}{\underset{OH}{\diagdown}}$ |

**Table 24.8** *Two organic acids*

### Some properties of ethanoic acid

Ethanoic acid, sometimes called by its old name acetic acid, behaves as a typical weak acid. Its solutions have a pH of about 3 (see section 19.1). It reacts with bases to give salts. For example:

sodium hydroxide + ethanoic acid ⟶ water + sodium ethanoate

$$NaOH + CH_3COOH \longrightarrow H_2O + CH_3COONa$$

Organic acids, including ethanoic acid, react with alcohols to form **esters**. These are discussed in section 24.7.

## 24.7 Esters, soaps and detergents

### Esters

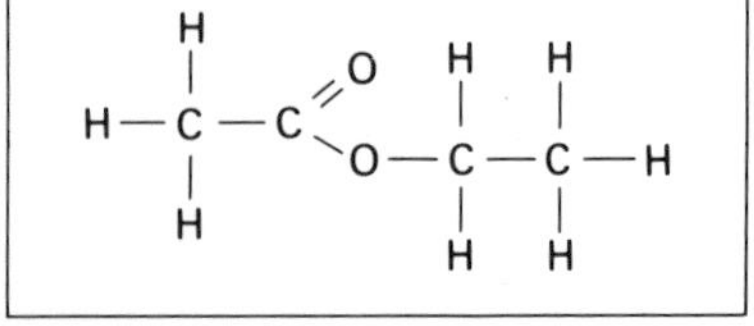

Organic acids react with alcohols to form **esters**. For example, ethanoic acid reacts with ethanol to form ethyl ethanoate (figure 24.17). A few drops of concentrated sulphuric acid are also needed as a catalyst.

**Figure 24.17** *Structure of ethyl ethanoate*

ethanoic acid  +  ethanol  $\rightleftharpoons$  ethyl ethanoate  +  water

$$CH_3COOH + C_2H_5OH \rightleftharpoons CH_3COOC_2H_5 + H_2O$$

The $\rightleftharpoons$ symbol means that not all of the reactants change into products. Some acid and alcohol is always left.

Most esters have strong, often pleasant, smells. They are present in flowers and fruit. They are also used to make perfumes and to flavour some foods.

## Soaps

Fats are complicated esters. When they are boiled with alkalis, soaps are formed. This process has been known for many centuries and is called **saponification**. The equation below shows how glyceryl stearate (a fat) reacts with sodium hydroxide (an alkali):

glyceryl stearate  +  sodium hydroxide  $\longrightarrow$  sodium stearate  +  glycerol

## How soap works

Soap is often thought of as having two ends to the molecule. It is ionic at the $-COO^- Na^+$ end. This part 'likes' to dissolve in water.

It is covalent at the hydrocarbon $C_{17}H_{35}-$ end. This part 'likes' to mix with grease and dirt.

As a result the grease is attracted to the soap and the soap dissolves in the water, bringing the grease with it.

## Soapless detergents

Many washing powders are not made from animal fats and vegetable oils. Instead they are made from petroleum hydrocarbons. Like soaps, these detergents have an ionic and a covalent end. So they work in the same way. Many are unaffected by hard water (see section 10.3). In this way they are better than soaps.

## Summary: Organic chemistry

★ The alkanes are a family of hydrocarbons. Their molecules contain only single bonds.
★ Most alkanes are obtained by fractional distillation of crude oil.
★ Alkanes are used mainly as fuels.
★ Isomers are compounds which contain the same atoms but in different arrangements.
★ The alkenes are hydrocarbons with molecules which contain a double bond between two carbon atoms.
★ Alkenes can be obtained by cracking (breaking up) larger hydrocarbons.
★ Alkenes are unsaturated, so other atoms can be added to their molecules.
★ Unsaturated compounds decolorize bromine water.
★ A polymer is a large molecule formed by joining together many small molecules.
★ Alkenes can be polymerized to form products such as polythene, polystyrene and PVC.
★ Photosynthesis is a process in which plants synthesize carbohydrates from carbon dioxide and water.
★ Starch gives a blue colour with iodine.
★ Starch is broken down in aqueous acid to form glucose.
★ In the presence of saliva, starch breaks down to maltose.
★ Fermentation is a process in which yeast converts glucose into ethanol (alcohol) and carbon dioxide.
★ Vinegar contains ethanoic acid.
★ Ethanoic acid reacts with alcohols to form compounds called esters.
★ Esters are often used in perfumes and flavourings.
★ Soap is made by the reaction of animal fats or vegetable oils with aqueous alkali.
★ Synthetic detergents are made from petroleum hydrocarbons.

## Questions

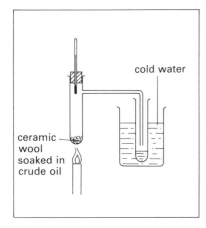

ceramic wool soaked in crude oil

cold water

1 Crude oil is a mixture of hydrocarbons, each with a different boiling point. It can be separated into different portions by using the apparatus shown. (Do not try this experiment. Heating oil like this can be dangerous.)

The crude oil was separated into four portions. Details are given in the table below.

| Portion | Boiling range (°C) |
|---------|--------------------|
| 1 | room temperature to 70 |
| 2 | 70–120 |
| 3 | 120–170 |
| 4 | 170–230 |

(a)  Name the two main elements in crude oil.
(b)  What is the name given to the process used to separate the crude oil into portions?
(c)  Draw a diagram of the equipment used in industry to separate crude oil into 'fractions'. Label the diagram fully. Include the names of the substances obtained and the temperature ranges over which they boil.
(d)  Which of the four portions obtained in the above equipment is most like petrol (gasoline)?
(e)  Octane is one of the major ingredients of petrol. It is an alkane with eight carbon atoms. Draw a diagram of an octane molecule. (Use table 24.2 to help you.)

2  An experiment was carried out to produce ethene from decane.

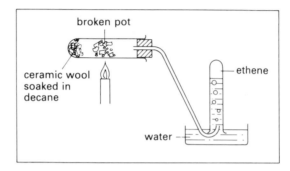

Decane is a colourless liquid hydrocarbon. It is insoluble in water and has no effect upon bromine water. Ethene is a colourless gas which decolorizes bromine water.
(a)  How can you tell from the above information that decane molecules are bigger than ethene molecules?
(b)  What is the name given to the process in which large hydrocarbons are broken down into smaller ones?
(c)  What does the word *saturated* mean? Which of the two hydrocarbons is saturated? Give reasons for your answer.
(d)  What purpose does the broken pot serve in the apparatus shown above?
(e)  At the end of the experiment some decane was found floating on the cold water in the trough. Explain how it got there. (It was not accidentally spilled!)

3  Propene is a hydrocarbon with the formula $C_3H_6$. It is the starting material for the manufacture of polypropene, sometimes called polypropylene. This is formed by a chain reaction which will only take place if a suitable catalyst is present.
(a)  What sort of reaction is the conversion of propene to polypropene?
(b)  What is a catalyst?
(c)  Why is the formation of polypropene described as a chain reaction?

(d) Explain, with the help of diagrams, how polypropene is formed from propene.

(e) Would you expect propane to undergo a similar reaction? Give reasons for your answer.

4 The experiments below were carried out on separate portions of starch solution.

Experiment A   A few drops of iodine solution were added to the starch. A blue–black colour was formed.

Experiment B   Saliva was added to the starch solution. After a few minutes, iodine was added. A pale yellow solution was formed.

Experiment C   Dilute hydrochloric acid was added to the starch solution and this was boiled. A pale yellow solution was formed when iodine was added to this mixture.

Experiment D   Yeast was added to the starch solution and the mixture was placed somewhere warm but not hot. Within a few days the solution had started to become frothy and it was shown to contain ethanol.

(a) What is an enzyme? Which two of the above experiments involve enzymes?

(b) Using only information given above, what can you say about the starch that was present in experiments B and C?

(c) Use word equations to describe what is happening to the starch in experiments B and C.

(d) What gas caused the frothiness in experiment D?

(e) Draw a labelled diagram of an apparatus suitable for separating the ethanol (boiling point 80 °C) and water that are present at the end of experiment D.

5 Soaps are made by the action of strong alkali upon fats or vegetable oils. Soapless detergents are generally made by reacting petroleum hydrocarbons with concentrated sulphuric acid. Both types of detergent contain a 'head' and a 'tail' as shown below.

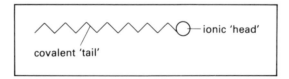

(a) Name two strong alkalis that would be suitable for making soap.

(b) What is the 'tail' made of in a soapy detergent?

(c) Name any vegetable oil that is used in the manufacture of soaps.

(d) What is the main advantage of soapless detergents over soaps?

(e) The diagram below shows some grease molecules attached to a cloth that is in some water. Copy this diagram but draw in a detergent molecule as well. Show clearly which part of the detergent will mix with the water, and which mixes with the grease.

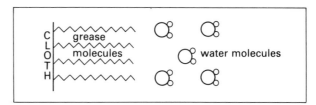

**6** Ethene is obtained mainly from the catalytic cracking of petroleum. A typical cracking product might contain: methane (15 per cent), ethene (40 per cent) and propene (20 per cent).
(a) What is catalytic cracking?
(b) Why are certain petroleum fractions cracked in this way?
(c) Which of the above cracking products are unsaturated?
(d) How could you test for such unsaturation experimentally?
(e) In the USA ethene is often made from ethane. What other product is formed?
(f) Which of the above cracking products will polymerize?
(g) It is possible to carry out the following reaction sequence. Find out what additional chemicals are needed to do so.

ethene ⟶ ethanol ⟶ ethanoic acid ⟶ ethyl ethanoate

## Polystyrene – an adaptable plastic

Polystyrene is a plastic that comes in many disguises. It can be stiff and transparent like the clear outer casing of cheap ball-point pens. It may have colouring added like the grey plastic model aircraft. It can also be very thin and flexible as in plastic yogurt pots.

Although all these forms seem fairly different, there is a further form that is so different you might not think it was the same substance at all. 'Expanded' polystyrene is white, quite spongy and very, very light. Blocks of expanded polystyrene are used in the packaging of fragile things like television sets.

All forms of polystyrene are made from the same starting material. It is a liquid called styrene (or sometimes phenylethene). Another substance, called an initiator, is also

*Here polystyrene is being used in roadbuilding in Norway*

needed. This is usually an organic peroxide. Such peroxides easily decompose into two halves. Each half then needs something else to combine with, and soon finds an unsaturated styrene molecule. This initiator starts off a chain reaction. The styrene that has added on the peroxide now adds itself on to another styrene molecule. Soon the third, fourth and fifth molecules become joined together. Before long a polymer chain has formed, thousands of styrene molecules long:

$$n \begin{array}{c} H \\ \diagdown \\ C \\ \diagup \\ H \end{array} = \begin{array}{c} C_6H_5 \\ \diagup \\ C \\ \diagdown \\ H \end{array} \longrightarrow \left[ \begin{array}{cc} H & C_6H_5 \\ | & | \\ -C- & C- \\ | & | \\ H & H \end{array} \right]_n$$

To get clear, stiff polystyrene we just add an initiator to the styrene, as described above. Making expanded polystyrene is slightly different. In one method a volatile liquid such as pentane is added to the mixture of styrene and initiator. The styrene polymerizes, and as it does so the temperature rises because of the heat given out.

We would expect the pentane to boil off, but this is prevented by carrying out the reaction in a strong sealed container, rather like a big pressure cooker. When the styrene molecules have all formed long chains the container is opened. The result is rather like shaking a can of coke and opening it. The pentane that is now dissolved in the hot polystyrene is free to escape – and this is just what it does. Millions of bubbles form within the mixture, producing millions of popcorn-like beads of polystyrene. These can be moulded into shapes, such as ceiling tiles, by warming the beads in a suitable mould.

### Questions

7  In what ways is expanded polystyrene different from normal polystyrene?
8  Explain the meaning of the word *unsaturated*.
9  Why is the organic peroxide called an *initiator*?
10  Styrene is a liquid. Polystyrene is a solid. Explain why.
11  What is meant by 'a volatile liquid'?
12  Suggest an alternative liquid to pentane for the manufacture of polystyrene.
13  Expanded polystyrene is fairly cheap and is an excellent heat insulator. Why isn't it used for loft insulation?
14  During polymerization the polystyrene chain may stop growing before all of the styrene has been used up. Suggest one possible reason why the reaction should end in this way.

# 25 Energy changes and fuels

We live in a world where there is an ever-increasing demand for energy. This chapter is about the chemical reactions that provide this energy. We shall also be looking at how we can measure the energy changes that take place during chemical reactions.

## 25.1 What are fuels?

Fuels are chemicals that provide us with energy. At the moment most of our energy comes from the fossil fuels: coal, oil and natural gas or substances made from them.

All these fuels contain carbon and hydrogen. When they burn they react with oxygen to form carbon dioxide and water:

$$\text{fossil fuel} + \text{oxygen} \longrightarrow \text{carbon dioxide} + \text{water}$$

We are using up our reserves of fossil fuels at a tremendous rate. Within the next hundred years or so, we shall have to develop new energy sources. Some new sources are already being used – an example is shown in figure 25.2.

**Figure 25.1** *As the costs of oil, gas and electricity rise, people are again using coal to heat their homes*

**Figure 25.2** *Solar panels provide energy for this satellite*

### Why do we need fuels?

**Figure 25.3** *The ways we use energy from fuels*

Fuels provide us with energy. Some of the ways we use this energy are shown in figure 25.3.

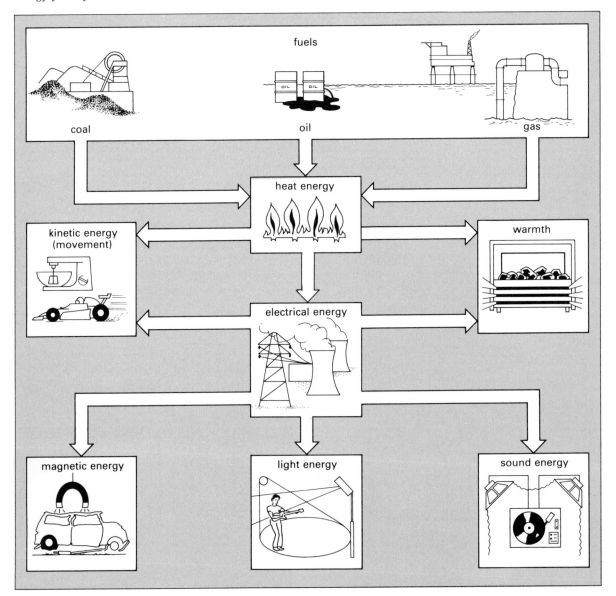

### What makes a good fuel?

The perfect fuel will:

1 be cheap,
2 burn easily and produce large amounts of heat,
3 burn without producing smoke, ashes or other substances that cause pollution.

Needless to say, the perfect fuel does not exist!

## 25.2 How do we get fuels?

(a)                                                        (b)

**Figure 25.4** *Two ways of getting oil: (a) drilling for oil in the desert, (b) a production platform in the North Sea*

(a)                                                        (b)

**Figure 25.5** *Two ways of getting coal: (a) cutting coal underground, (b) opencast coal mining*

## 25.3 Energy values of fuels

### What do energy values tell us?

When a good fuel is burned it will give out a large amount of heat energy. To decide which fuel is best, you need to know just how much energy it does give out.

## Measuring the temperature rise produced by burning a fuel

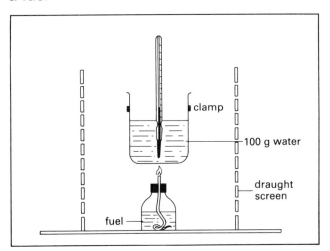

Figure 25.6 shows an apparatus in which a liquid fuel is being burned. The heat from the flame is used to raise the temperature of a liquid contained in the small can.

If the fuel gives out a lot of energy, there will be a large increase in temperature.

### Heat energy and temperature changes

Simply measuring the temperature rise in the above experiment will not tell us how much heat energy has been given out. There are other important factors. One of these is the amount of liquid in the can. The more liquid that we heat, the lower the temperature rise will be.

The type of liquid in the can is also important. For example, raising the temperature of cooking oil by 20 °C will take much less energy than raising the temperature of the same amount of water by 20 °C. This is because cooking oil has a lower **specific heat capacity** than water.

### What units do we measure heat energy in?

All forms of energy, including heat energy, are measured in units called **joules**, symbol J. The energy changes that occur in chemical reactions are often equal to many thousands of joules. Because of this, chemists use the kilojoule, kJ:

1 kilojoule = 1000 joules

To give you some idea of the size of a kilojoule, it takes about 300 kJ to boil a kettle of water.

### Calculating the heat given out

To calculate the heat given out in a reaction you must know three things:

1 the temperature rise that took place (in °C),
2 the mass of liquid that was heated (in grams),
3 the specific heat capacity of the liquid that was heated (in J/g/°C).

The liquid that is heated is nearly always water. This has a specific heat capacity of 4.2 J/g/°C. Putting this another way, it takes 4.2 joules to heat 1 gram of water through 1 °C.

To calculate the heat given out you simply multiply these three pieces of information together:

heat given out = mass of water heated × specific heat capacity × temperature change

A few examples will make this clearer.

### Example 1

How much heat is needed to raise the temperature of 2 g of water by 1 °C?

Heat given out = 2 × 4.2 × 1
= 8.4 J

### Example 2

How much heat is needed to raise the temperature of 200 g of water by 10 °C?

Heat given out = 200 × 4.2 × 10
= 8 400 J or 8.4 kJ

### The energy values of some fuels

The following results were obtained using the apparatus shown in figure 25.6. From them we can calculate the energy value of the fuel.

| | |
|---|---|
| Mass of lamp at start | 45.2 g |
| Mass of lamp at end | 45.0 g |
| Mass of fuel burned | 0.2 g |
| Temperature of water at start | 20 °C |
| Temperature of water at end | 30 °C |
| Temperature change | 10 °C |
| Mass of water in can | 100 g |

We can calculate the heat given out by 0.2 g of fuel.

heat given out = mass of water × specific heat capacity × temperature change
= 100 × 4.2 × 10
= 4 200 J or 4.2 kJ

From this answer we can work out how much heat would be obtained by burning 1 g of fuel. As $0.2 = \frac{1}{5}$ g, we get five

times as much energy from burning 1 g,

$$5 \times 4.2 = 21.0\,kJ \text{ per g of fuel}$$

### Energy values for some common fuels

Figure 25.7 shows the energy values for a range of commonly used fuels. Values for some foods are also given in figure 25.8. These foods provide the body with energy – that is, they are fuels for the body.

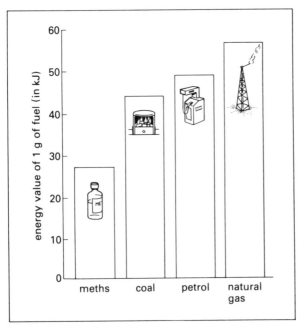

**Figure 25.7** *The energy values of some fuels*

**Figure 25.8** *The energy values of some foods*

**Figure 25.9 (above)** *The burning of candle wax is an exothermic reaction*

**Figure 25.10 (right)** *Dissolving sulphuric acid in water is an exothermic process*

## 25.4 Exothermic and endothermic reactions

All reactions in which a fuel burns are **exothermic**. This means that they give out heat energy to their surroundings. ('Ex' means 'out', as in 'exit'.)

Many chemical reactions not involving fuels are also exothermic. If magnesium reacts with acid, the reaction

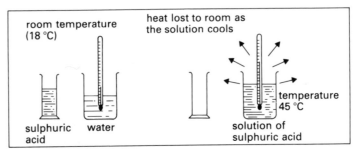

mixture gets warm. Before long this heat is lost to the surroundings.

Adding sulphuric acid to water is a highly exothermic process. The temperature rise can be very great.

Some chemical reactions are **endothermic**. The temperature of these reaction mixtures drops below room temperature during the reaction. Eventually the mixture warms back up to room temperature by taking in energy from the surroundings.

Dissolving ammonium nitrate in water is an example of an endothermic process.

**Figure 25.11** *Dissolving ammonium nitrate in water is an endothermic process*

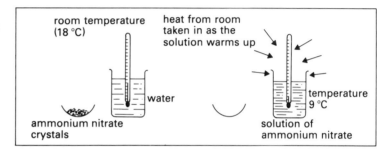

## Energy-level diagrams

Energy-level diagrams are a convenient way of showing whether chemicals are losing or gaining energy during a reaction.

In an exothermic reaction the chemicals give out heat to the surroundings. This means that the products of these reactions have less energy than the reactants (see figure 25.12). In fact the chemicals have lost energy.

When an endothermic reaction takes place heat is taken in from the surroundings. The products of these reactions have more energy than the reactants (see figure 25.13).

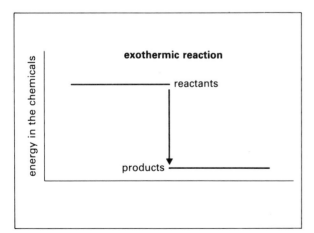

**Figure 25.12** *Energy-level diagram for an exothermic reaction*

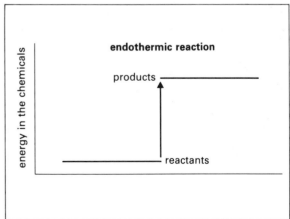

**Figure 25.13** *Energy-level diagram for an endothermic reaction*

## Energy changes and $\Delta H$

When 1 g of petrol burns, 48.8 kJ of energy ($H$) is given out to the surroundings. So the carbon dioxide and water formed in this reaction have 48.8 kJ less energy $H$ than the petrol had.

We could say that there has been a change in the energy of the chemicals of −48.8 kJ/g. There is a useful shorthand way of writing this:

$$\Delta H = -48.8 \text{ kJ/g}$$

The symbol $\Delta H$ means 'the change in the energy $H$ of the chemicals'. The minus sign shows that the chemicals are losing energy.

$\Delta H$ is negative for all exothermic reactions. Figure 25.14 represents the energy change that takes place in this reaction.

**Figure 25.14** *The energy change when 1 g of petrol burns*

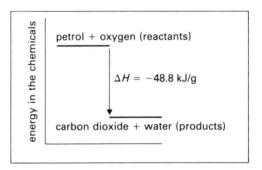

## Energy changes per mole

So far we have considered the energy given out per gram of fuel. There is often an advantage in working in energy changes per mole instead of per gram. (You read about the mole in chapter 7.) It allows us to relate the energy changes to the number of molecules reacting. An example will make this clearer. Look at the figures in table 25.1.

| Fuel | Mass of 1 mole (g) | $\Delta H$ (kJ/g) | $\Delta H$ (kJ/mole) |
|------|--------------------|--------------------|-----------------------|
| methane | 16 | −55.6 | −890 |
| ethanol | 46 | −29.7 | −1366 |

**Table 25.1** *Energy values of fuels*

As you see, if we burn 1 g of methane and 1 g of ethanol (equal masses of the fuels) then we get more energy from the methane. But if we take a mole of each (equal numbers of molecules) then it is the ethanol molecules which give us more energy.

## Calculating energy changes per mole for fuels

These calculations are very similar to those in section 25.3. Again, we use the equation:

$$\frac{\text{heat given}}{\text{out}} = \frac{\text{mass of water}}{\text{heated}} \times \frac{\text{specific heat}}{\text{capacity}} \times \frac{\text{temperature}}{\text{rise}}$$

The calculation has three stages.

1 calculating the heat given out,
2 using a scale factor to find the heat given out per mole,
3 changing the sign to obtain $\Delta H$.

### Example 3

The $\Delta H$ value for ethanol was found using the apparatus shown in figure 25.6. In the experiment the temperature rose from 22 °C to 32 °C, and the mass of the lamp decreased by 0.23 g. The mass of water used was 100 g. Ethanol has the formula $C_2H_5OH$, and a relative molecular mass of $(2 \times 12) + (5 \times 1) + 16 + 1 = 46$.

1  Heat given out = $100 \times 4.2 \times 10 = 4200$ J or 4.2 kJ.

2  Mass of ethanol (in g)         Energy (in kJ)

     0.23                                          4.2

      | scale-up factor                    | multiply
      ↓ = 46/0.23 = 200                   ↓ by 200

     46                                             840

3  So the value of $\Delta H$ for the combustion of ethanol = $-840$ kJ per mole.

In fact, this is substantially less than the true value ($-1367$ kJ/mole) because a great deal of heat will be lost from this simple apparatus.

## Calculating energy changes for reactions in solution

Burning fuels is just one sort of reaction. Many reactions take place in water. The calculations for these are very similar to the ones that we have already shown. Some examples are given below.

### Example 4

When 8 g of ammonium nitrate solid was added to 100 cm$^3$ of water in the calorimeter shown in figure 25.15, the temperature of the water fell from 21 °C to 16 °C. Ammonium nitrate has the formula $NH_4NO_3$ and a relative molecular mass of $14 + (4 \times 1) + 14 + (3 \times 16) = 80$. Calculate $\Delta H$ for this process.

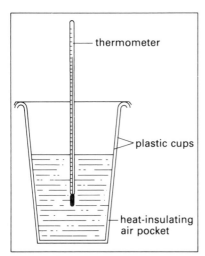

**Figure 25.15** *A simple calorimeter*

thermometer

plastic cups

heat-insulating air pocket

1 Heat given out $= 100 \times 4.2 \times -8 = -2100\,J = -2.1\,kJ$.

2 Mass of ammonium nitrate (in g)　　　Energy (in kJ)

| 8 | | −2.1 |
| scale-up factor<br>= 80/8 = 10 | | multiply<br>by 10 |
| 80 | | −21 |

3 So the value for $\Delta H = +21\,kJ/mole$.

Note that $\Delta H$ is positive, as this process is endothermic. Figure 25.16 shows the energy diagram for the process.

**Figure 25.16** *Energy-level diagram for dissolving ammonium nitrate in water*

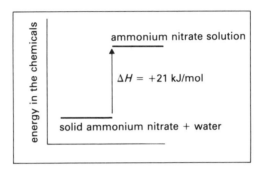

## Calculating ΔH for a neutralization reaction

To do this you need to be able to calculate the number of moles in a solution. You can use the formula:

$$\frac{\text{number of}}{\text{moles}} = \frac{\text{volume of solution (cm}^3)}{100} \times \text{concentration (M)}$$

### *Example 5*

$60\,cm^3$ of 2M sodium hydroxide solution was placed in the calorimeter shown in figure 25.15. The temperature of this solution was 21 °C. $50\,cm^3$ of 2M hydrochloric acid, having an initial temperature of 19 °C, was added to the calorimeter. The stirred mixture reached a temperature of 32 °C. What is $\Delta H$ for the reaction?

Note that if no reaction had occurred we would have expected the temperature of the mixture to be 20 °C. The temperature rise is therefore taken as 12 °C.

The number of moles of each chemical $= (50/1000) \times 2 = 0.1$ mole.
The heat given out in the reaction is used to heat the $100\,cm^3$ of solution (acid and alkali) that are now in the calorimeter.

1 Heat given out $= 100 \times 4.2 \times 12 = 5040\,J = 5.04\,kJ$.

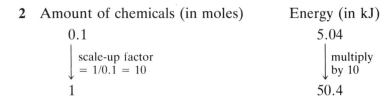

**3** So the value for $\Delta H$ for the neutralization $= -50.4\,\text{kJ/mole}$.

### Energy and changes of state

We saw in section 12.2 that the temperature of boiling water does not rise above 100 °C even though you carry on heating it. This is because the heat energy is being used to overcome the forces of attraction between the water molecules. This energy – the latent heat of vaporization – has the symbol $\Delta H_{vap}$. $\Delta H_{vap}$ of water has a value of 41 kJ/mole.

You will see from table 25.2 that some liquids have lower $\Delta H_{vap}$ values than water. This tells us that their molecules are less strongly attracted to each other, making them easier to boil away. On the other hand, iron has a much higher $\Delta H_{vap}$ than water. This is because the strong metallic bonding in iron holds the atoms tightly together. You will see too that the pattern of the boiling points of the substances in the table is similar to the pattern of $\Delta H_{vap}$ values. The reason is that these too depend on the strength of the forces between the particles.

| Liquid | $\Delta H_{vap}$ (kJ/mole) | Boiling point (°C) |
|---|---|---|
| water | 41 | 100 |
| tetrachloromethane | 31 | 77 |
| cyclohexane | 30 | 81 |
| iron | 351 | 2890 |

**Table 25.2** *Latent heats of vaporization*

## Summary: Energy changes and fuels

★ Fuels are chemicals that can provide energy.
★ At present most of our energy comes from the fossil fuels: coal, natural gas and oil.
★ Energy is measured in joules, J, or kilojoules, kJ.
★ The energy value of a fuel tells us how much heat it produces when a definite amount of it is burned.
★ Energy values are measured in kilojoules/gram of fuel or in kilojoules/mole of fuel.

★ The energy from a reaction is often used to heat water or an aqueous solution. In such cases:

$$\begin{matrix} \text{the heat} \\ \text{given out} \\ \text{(in J)} \end{matrix} = \begin{matrix} \text{mass of} \\ \text{water heated} \\ \text{(in g)} \end{matrix} \times 4.2 \times \begin{matrix} \text{the temperature} \\ \text{rise} \\ \text{(in °C)} \end{matrix}$$

★ Exothermic reactions give out heat. $\Delta H$ for these reactions is negative.
★ Endothermic reactions take in heat (get cold). $\Delta H$ for these reactions is positive.
★ Melting and boiling are endothermic processes.
★ Solidifying and condensing are exothermic processes.

## Questions

**1** Which of the factors in the list below is the most important for the uses numbered (a) to (e)? If you think two factors are equally important, make this clear.

    be cheap
    produce a large amount of heat per gram of fuel
    ignite very easily
    be readily portable
    produce no smoke or pollutants

(a) fuel to power a space rocket
(b) fuel for a cigarette lighter
(c) fuel for domestic heating
(d) fuel for a camping stove
(e) fuel for an underground transport system

**2** This question is based on the equation given below:

$$\begin{matrix} \text{heat given} \\ \text{out} \end{matrix} = \begin{matrix} \text{mass of water} \\ \text{heated} \end{matrix} \times \begin{matrix} \text{specific heat} \\ \text{capacity of water} \end{matrix} \times \begin{matrix} \text{temperature} \\ \text{rise} \end{matrix}$$

(a) What is the specific heat capacity of water?
(b) Explain carefully what is meant by 'the specific heat capacity of a substance'.
(c) Calculate how much heat is needed for the following:
    (i) to heat 100 g of water through 20 °C,
    (ii) to heat 400 g of water through 5 °C,
    (iii) to heat 500 g of water through 17 °C.

**3** An experiment was carried out to measure the energy value of a fuel. The apparatus used is shown here.

The spirit burner was weighed. Some of the fuel was then burned and the heat was used to heat the water in the can. The temperature rise of the water was noted.

Mass of lamp at start      = 104.43 g
Mass of lamp at end       = 104.03 g
Temperature of water at start = 18 °C
Temperature of water at end  = 28 °C

(a) Why was a copper can used rather than a glass beaker?

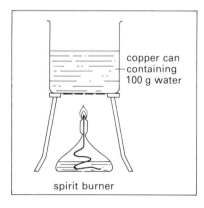

copper can containing 100 g water

spirit burner

(b) What was the temperature rise of the water?

(c) What amount of heat is needed to make the temperature of 100 g of water rise by this amount?

(d) What mass of fuel was burned?

(e) How much heat energy is given out when 1 g of fuel is burned?

(f) The results of this experiment give an answer that is lower than the true energy value for this fuel. Suggest reasons why this should be so.

4 (a) Explain the meaning of *exothermic* and *endothermic*.

(b) Draw an energy diagram for the reaction:

$$N_2(g) + 3H_2(g) \longrightarrow 2NH_3(g)$$
$$\Delta H = -5.8 \text{ kJ/g of } N_2(g)$$

(c) Say whether the above reaction is exothermic or endothermic.

(d) Draw an energy diagram for the reaction:

$$N_2(g) + O_2(g) \longrightarrow 2NO(g)$$
$$\Delta H = +10 \text{ kJ/g of } N_2(g)$$

(e) Is this reaction is exothermic or endothermic?

5 It was found that when 4 g of a salt were dissolved in 100 g of water, the temperature went down by 5 °C.

(a) Is this process exothermic or endothermic?

(b) Is $\Delta H$ positive or negative?

(c) How much heat was taken in when 4 g of the salt dissolved?

(d) How much heat would be taken in when 1 g of the salt dissolves?

(e) Draw a fully labelled energy diagram for this process.

6

| Substance | Latent heat of vaporization (kJ/mole) |
|-----------|:-------------------------------------:|
| helium    | 0.1                                   |
| water     | 41                                    |
| sand      | 516                                   |

(a) What is meant by the latent heat of vaporization?

(b) Explain, in as much detail as possible, why the substances have such different heats of vaporization. (Section 14.2 may help you.)

7 25 cm³ of 1M nitric acid was mixed with 25 cm³ of 1M potassium hydroxide in a lagged calorimeter. The temperature rose from 20 °C to 26 °C.

(a) Calculate the number of moles of each chemical used.

(b) Calculate the heat that would have been given out if 1 mole of each chemical had been used.

(c) What is $\Delta H$ for this reaction, in kJ/mole?

**8** (a) Experiments were carried out to measure the $\Delta H$ values for the combustion of several alcohols. The calculations were carried out like the one for ethanol in section 25.4. Carry out similar calculations for the other alcohols shown in the table below. Show your working clearly, and then copy out and complete the table.

| Alcohol | Relative molecular mass | Mass burned to give a temperature rise of 10 °C (g) | $\Delta H$ (kJ/mole) |
|---|---|---|---|
| $CH_3OH$ | | 0.32 | |
| $C_2H_5OH$ | | 0.23 | −840 |
| $C_3H_7OH$ | | 0.227 | |
| $C_4H_9OH$ | | 0.209 | |
| $C_5H_{11}OH$ | | 0.197 | |

(b) Make a list of ways in which heat would have been lost in this experiment.
(c) Plot a graph of $-\Delta H$ (vertical axis) against the number of carbon atoms (horizontal axis).
(d) Comment on the reasons for the shape of the graph.
(e) Use the graph to predict the value of $\Delta H$ for $C_6H_{13}OH$.

## Alternatives to petrol

*In Brazil, ethanol made by fermenting sugar is cheaper than petrol*

As oil reserves decline, so chemists search with ever greater enthusiasm for an alternative to petrol as a transport fuel. Some possibilities are listed below.

| Fuel | Possible source |
|---|---|
| hydrogen | decomposition of water at high temperature |
| carbon monoxide | reaction of carbon with steam at high temperature |
| methane | anaerobic digestion of cellulose |
| ethanol | fermentation of sugar |
| methanol | carbon monoxide + hydrogen |

Some of the relevant properties of these fuels are shown in the next table. Petrol (as octane, $C_8H_{18}$) is included for comparison.

| Fuel | Formula | Boiling point (°C) | Melting point (°C) | Heat evolved per mole (kJ) |
|---|---|---|---|---|
| hydrogen | $H_2$ | −253 | −259 | 290 |
| carbon monoxide | CO | −191 | −205 | 280 |
| methane | $CH_4$ | −161 | −182 | 890 |
| ethanol | $C_2H_6O$ | +78 | −115 | 1367 |
| methanol | $CH_4O$ | +65 | −97 | 726 |
| octane | $C_8H_{18}$ | +125 | −57 | 5512 |

## Questions

**9** State one practical difficulty of using hydrogen as a replacement for petrol.

**10** State one safety consideration in using carbon monoxide as a fuel.

**11** What product, other than carbon monoxide, is formed when carbon reacts with steam?

**12** The table specifies the amount of heat given out per mole of the fuel burned. In what way does this differ from the heat of combustion of the fuel, $\Delta H_c$ (also in kJ/mole)?

**13** Which fuel gives out more energy per gram of fuel burned, methane or octane?

**14** Which fuel is likely to be most pollution-free?

**15** A value is given for the heat given out when a mole of ethanol is burned. Calculate a value from the following data obtained in a student's experiment:

'I used the heat produced by the combustion of 1.00 g of ethanol to heat $100 \, cm^3$ of water. The temperature increased by 20 °C.'

Comment upon the value obtained by the student compared to the accurate value given in the table. Suggest reasons for any discrepancy.

# 26 Radioactivity and nuclear energy

## 26.1 Changing one element into another

The early chemists who lived from about A.D. 400 were called alchemists. Their main aim was to find a way of changing metals like lead into gold. But they did not succeed.

Now that we know more about chemistry we can understand why they failed. In most chemical reactions the atoms of different elements either join together or separate from each other. Changing lead into gold would mean changing the atoms themselves. Lead atoms would have to become gold atoms.

This type of change is most unusual but not impossible. It happens naturally with radioactive elements such as radium.

With modern technology we can now achieve the alchemists' dream: we can change one element into another. The rewards of this technology are far greater than just gold. When you change one element into another, you may get a huge amount of energy released: nuclear energy (see figure 26.1). How this energy is used will affect the entire future of the human race. It may provide the answer to the rapidly approaching 'energy famine'. It could also be used to bring about the horrors of nuclear war. This chapter is about radioactivity and nuclear energy.

**Figure 26.1** *A huge amount of energy is released when an atomic bomb explodes*

### The structure of the atom

Figure 26.2 shows the structure of an atom. At the centre is the **nucleus**. This consists of **protons** and **neutrons**. The **electrons** move around the outside of the nucleus (see section 13.1).

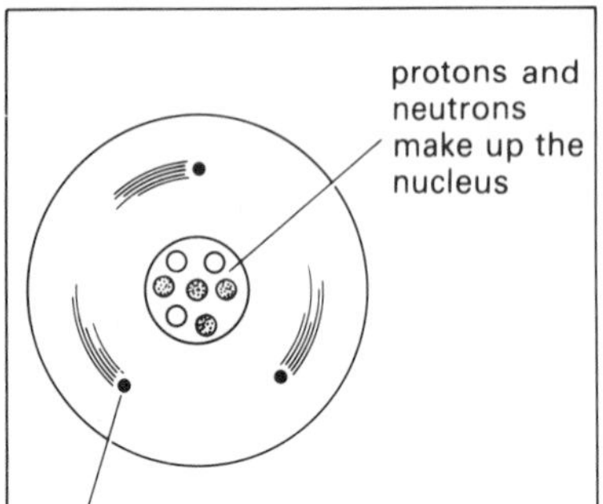

### Change of atomic number = change of element

Each element has a particular **atomic number**. This is the number of protons present in the nucleus of each atom of that element. For example, lithium has an atomic number of three. This means that lithium atoms have three protons in the nucleus.

We cannot change the number of protons in an atom without changing the element itself. If a lithium atom were

protons and neutrons make up the nucleus

electrons

**Figure 26.2** *The particles in atoms*

to lose a proton then it would no longer be a lithium atom. Since it would now have two protons its atomic number would be two. The element with this atomic number is helium.

The processes that can bring about a change in atomic number are described in the following sections.

## 26.2 Radioisotopes and radioactivity

### Radioisotopes

In section 13.1, we saw that atoms that contain the same number of protons but different numbers of neutrons are called **isotopes**. An isotope symbol shows the mass number and the atomic number of the element. For example, atoms of the isotope lithium-7 contain three protons and four neutrons, so its symbol is $^7_3\text{Li}$.

The chemistry of an atom is not affected much by the number of neutrons in the nucleus. But this number is still very important. If an atom has too many or too few neutrons, its nucleus is likely to be unstable. When this happens, the nucleus gives out energy or particles. The atom is then **radioactive**. Isotopes that do this are called **radioisotopes**.

**Figure 26.3** *Here the tracks of α-particles from a thorium source have been photographed in a device called a cloud chamber*

### Types of radiation

The energy or particles that an unstable nucleus gives out is called **radiation**. There are three common types of radiation, called after the Greek letters alpha (α), beta (β) and gamma (γ):

**1** α-particles,
**2** β-particles,
**3** γ-rays.

α-particles consist of two protons and two neutrons. An α-particle is the same as the nucleus of a helium atom.

α-particles cannot penetrate the skin and so they are not dangerous unless they get inside the body.

β-particles are fast-moving electrons. They are much more penetrating than α-particles are. They are able to pass through metal up to 1 cm thick. They easily pass through the skin and into the body where they can damage cells.

γ-rays are like X-rays, but they are more penetrating still. They can pass through quite thick pieces of metal but are stopped by several centimetres of lead. They can kill all living cells. In small doses they may cause radiation sickness. Larger doses may give radiation burns.

As you can see, handling radioactive materials can be very dangerous, and strict precautions must be observed (see figure 26.4).

**Figure 26.4** *Handling a radioactive sample by remote control. Technicians must stand behind a very thick lead shield to protect them from radiation*

### Uses of radiation

Because γ-rays can kill cells, they are used in the treatment of cancer. Heavy screening is used to protect the healthy cells from the radiation, which destroys the diseased cells.

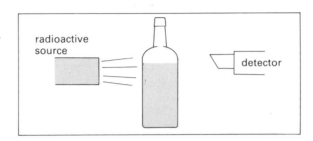

**Figure 26.5** *Using radiation to check the contents of a container. If the container is not full, more radiation will reach the detector*

β-particles are used in packaging plants to check automatically whether containers on a conveyor belt have been properly filled (see figure 26.5). The method uses the fact that radiation is absorbed to a greater or lesser extent by most materials. The same fact allows β-particles to be used similarly to check for cracks in components of machinery or engines (see figure 26.6). Even a small crack reduces the amount of radiation absorbed by the article.

**Figure 26.6** *Using a radioisotope to check for cracks in a Boeing 747*

## 26.3 Radioactive decay

In this section we shall look at what happens to atoms as they give out radiation.

### What happens to atoms during α-particle emission?

Thorium is a radioactive element. The isotope $^{232}_{90}\text{Th}$ emits α-particles. An α-particle consists of two protons and two neutrons. So it can be written as $^{4}_{2}\alpha$.

If $^{232}_{90}$Th loses two protons, its atomic number will become $90 - 2 = 88$. The element with this atomic number is radium.

The total number of protons and neutrons remaining in the nucleus will be $232 - 4 = 228$.

So the product of this change is $^{228}_{88}$Ra:

$$^{232}_{90}\text{Th} \longrightarrow {}^{4}_{2}\alpha + {}^{228}_{88}\text{Ra}$$

## What happens during β-particle emission?

β-particles are fast-moving electrons that come from the nucleus. This may strike you as rather odd, since the nucleus does not normally contain any electrons. What happens is that a neutron changes to become a proton and an electron:

$$\text{neutron} \longrightarrow \text{proton} + \text{electron}$$

The element $^{228}_{88}$Ra is radioactive. It emits β-particles. Every time a β-particle is emitted, one neutron in the nucleus has changed into a proton and an electron. So the nucleus has gained an extra proton and its atomic number will have changed to 89. The element with this atomic number is actinium. Although the atom has gained a proton, it has lost a neutron. As a result, the sum of the protons and neutrons remains the same at 228. So the product of this β-particle emission is $^{228}_{89}$Ac:

$$^{228}_{88}\text{Ra} \longrightarrow {}^{228}_{89}\text{Ac} + {}^{0}_{-1}\beta$$

We give the β-particle an atomic number of $-1$. The atomic numbers then balance on either side of the equation: $88 = 89 + (-1)$.

## Do atoms change during γ-ray emission?

Emission of γ-rays usually accompanies α- or β-particle emission. γ-rays are also emitted during nuclear fission (see section 26.4). The emission of γ-rays will not, by itself, produce a different atom.

## What does 'half-life' mean?

When a radioactive isotope emits radiation it also changes into some other atom. The radioisotope is said to be **decaying**. Eventually all of the radioisotope will have changed into something else. Some radioisotopes decay very quickly. Others take thousands of years. The speed of this decay is measured in a unit called a **half-life**. A half-life is the time that it takes for half the radioactive atoms in a sample to decay. This can be measured using a **Geiger counter** (see figure 26.7). This device automatically counts the radioactive particles that reach it. So the radioactivity of a sample is often measured in terms of the 'counts' recorded every minute.

**Figure 26.7** *Using a Geiger counter to measure the radioactivity of rocks*

Each radioisotope has a definite half-life. For example, a radioactive isotope of carbon, $^{14}_{6}C$, has a half-life of 5700 years. So if we take a sample of $^{14}_{6}C$ it will be 5700 years before half the atoms have decayed. It would be $2 \times 5700 = 11\,400$ years before three-quarters of the atoms had decayed. This is shown on the graph in figure 26.8.

**Figure 26.8** *Graph showing how the radioactivity of carbon-14 changes*

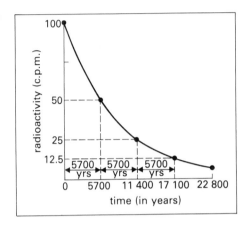

## What is carbon dating?

All plants and animals contain carbon compounds. They get this carbon from the carbon dioxide in the air. The carbon dioxide always contains some $^{14}_{6}C$. So all living things will contain this radioactive form of carbon.

When the plant or animal dies, it stops taking in carbon from the atmosphere. So the radioactive $^{14}_{6}C$ in the plant or or animal slowly decays. The longer ago the death was, the smaller the amount of $^{14}_{6}C$ will be.

Because of this it is possible to estimate how long ago the plant or animal died from the amount of $^{14}_{6}C$ remaining in it. The amount of radiation emitted decreases by 50 per cent every 5700 years (see figure 26.8). In this way the ages of very ancient treasures, such as the wooden coffins of the Egyptian pharaohs, have been measured (see figure 26.9).

**Figure 26.9** *Carbon dating was used to find the age of this mask of Tutankhamun, an ancient Egyptian king. It is over 3000 years old*

## 26.4 Power from the nucleus

### Nuclear fission: power for today

We already know that some isotopes do not have a stable nucleus. Some isotopes are so unstable that they can easily be made to split in two. This process is called **nuclear fission**. When it happens huge amounts of energy are also released.

One isotope of uranium undergoes fission when it is hit by a neutron. When the uranium splits up, it gives out three neutrons. These three neutrons cause another three uranium atoms to break up, giving nine neutrons (see figure 26.10). You can see that this gives a process that will get faster and

**Figure 26.10** *A nuclear chain reaction*

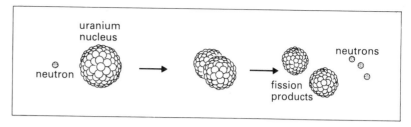

faster until all the uranium has gone. Eventually it will produce an explosion. This is the principle of the **atomic bomb**.

It is also the way that our present-day nuclear power stations work. In a power station, though, there are control rods which absorb some of the neutrons and stop the reaction getting too fast (see figure 26.11). The energy is released more slowly and can be used to make steam, which is then used to power the generators.

**Figure 26.11** *In a nuclear reactor at an atomic power station, the reaction cannot be allowed to go on as fast as it does in an atomic bomb! Control rods can be pushed in at different depths. These absorb some of the neutrons and slow down the reaction. So the energy from the reaction is obtained more slowly and over a longer period of time*

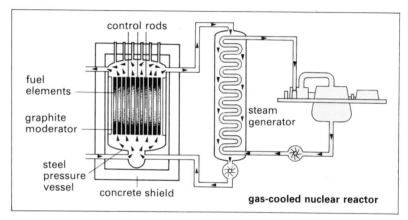

## Nuclear fusion: power for tomorrow?

Nuclear fusion is the process that powers the Sun and the stars. It also occurs in the hydrogen bomb.

Hydrogen bombs contain an isotope of hydrogen called deuterium. It is $^2_1H$ and so contains a neutron and a proton.

If two deuterium atoms collide violently enough they will fuse (join) together to form an isotope of helium. As they do so, energy is released (see figure 26.12).

**Figure 26.12** *Two deuterium nuclei collide to form an isotope of helium*

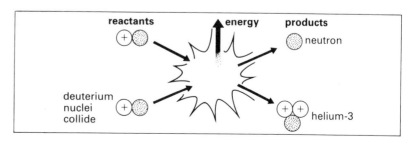

The oceans contain about 50 000 000 000 000 (or $5 \times 10^{13}$) tonnes of deuterium. So there is no shortage of this fuel. The major problem is finding a suitable container for a process that operates at 100 million degrees Celsius!

## Summary: Radioactivity and nuclear energy

* When an atom gains or loses protons, it changes into a different element.
* Isotopes are atoms containing the same number of protons but different numbers of neutrons.
* Radioisotopes are atoms which are radioactive; they emit radiation.
* Three common types of radiation are emitted by radioactive atoms: α-particles, β-particles and γ-rays.
* An α-particle consists of 2 protons + 2 neutrons, that is, it is a helium nucleus.
* β-particles are electrons.
* γ-rays are like X-rays but more penetrating and damaging.
* Radioisotopes have a definite half-life.
* The half-life is the time it takes for half the atoms in a sample to decay.
* Carbon dating can be used to measure the age of objects. The older the object is, the lower its carbon-14 content will be.
* Nuclear fission is the breaking apart of the atom's nucleus.
* Nuclear fission is the basis of the atom bomb and of current nuclear power stations.
* Nuclear fusion is the fusing together of isotopes of hydrogen.
* Nuclear fusion is the basis of the hydrogen bomb. It could also provide a long-term answer to energy shortages.

## Questions

1 Explain what is meant by each of the following words:
   (a) mass number,
   (b) atomic number,
   (c) isotope,
   (d) radioisotope,
   (e) radioactivity.
2 In what way are γ-rays more dangerous than α-particles or β-particles? Can γ-rays be used for purposes which benefit humans?
3 Draw up a table with the following headings:

| Isotope | Number of protons | Number of neutrons | Number of electrons |
|---------|-------------------|--------------------|---------------------|
|         |                   |                    |                     |

Now complete the table for each of the following isotopes:

(a) $^{232}_{90}\text{Th}$,

(b) $^{222}_{86}\text{Rn}$,

(c) $^{43}_{19}\text{K}$,

(d) $^{238}_{92}\text{U}$.

**4** Copy out and complete the equations by filling in the blank space with one of the isotopes listed below:

$^{222}_{86}\text{Rn}$    $^{214}_{83}\text{Bi}$    $^{218}_{84}\text{Po}$    $^{213}_{84}\text{Po}$    $^{234}_{90}\text{Th}$

(a) $^{238}_{92}\text{U} \longrightarrow {}^{4}_{2}\alpha + $ _____

(b) $^{214}_{82}\text{Pb} \longrightarrow {}^{0}_{-1}\beta + $ _____

(c) $^{222}_{86}\text{Rn} \longrightarrow {}^{4}_{2}\alpha + $ _____

(d) $^{226}_{88}\text{Ra} \longrightarrow {}^{4}_{2}\alpha + $ _____

(e) $^{212}_{83}\text{Bi} \longrightarrow {}^{0}_{-1}\beta + $ _____

**5** The radiation coming from a sample of actinium-225 was measured over several weeks. The results, corrected for background radiation, are shown below.

| Time (days) | 0 | 8 | 16 | 24 | 32 |
|---|---|---|---|---|---|
| Activity (counts/minute) | 320 | 175 | 100 | 60 | 35 |

(a) What is meant by the phrase 'corrected for background radiation'?

(b) Plot a graph of radiation against time. Plot time on the horizontal axis.

(c) Find the half-life of actinium-225. Show clearly on your graph how you have done this.

(d) On a certain day a laboratory had a sample of actinium-225 weighing 1 g. What mass of actinium will remain after 40 days?

**6** These questions are about a nuclear reactor used to generate electricity. It is shown diagrammatically below.

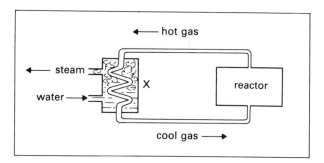

(a) What would the outside of the reactor be surrounded by? Why is this necessary?

(b) What happens in the part of the plant marked X?

(c) If the reactor gets too hot, carbon rods are lowered into it. How do these cause the temperature to fall?
(d) What happens to the steam that is produced in this plant?

7 As we saw in section 26.3, a small proportion of the carbon dioxide molecules in the atmosphere contain the radioactive isotope carbon-14. Plants incorporate this into their structure as they photosynthesize. All plants and all fresh materials of plant origin have a radioactive count rate of 12.50 counts per minute for each gram of carbon they contain. After the death of the plant this radioactivity steadily diminishes with the age of the article, in the manner indicated in the following table:

| Age (years) | Count rate (counts/(min g)) |
|---|---|
| 0 | 12.50 |
| 500 | 11.74 |
| 1000 | 11.02 |
| 1500 | 10.38 |
| 2000 | 9.72 |
| 2500 | 9.12 |
| 3000 | 8.57 |
| 3500 | 8.04 |
| 4000 | 7.55 |
| 4500 | 7.09 |
| 5000 | 6.66 |
| 5500 | 6.25 |
| 6000 | 5.87 |

(a) Plot a graph of count rate (on the vertical axis) against age (on the horizontal axis).
(b) Determine the half-life of carbon-14.
(c) Use your graph to estimate the age of the following articles. (Because of the inescapable errors in the method, give your estimates to the nearest hundred years.)

| Article | Count rate (counts/(min g)) |
|---|---|
| Dead Sea scroll | 9.91 |
| Egyptian sarcophagus | 7.84 |
| Hilt of a sword from Troy | 8.69 |
| Charcoal from Stonehenge | 7.44 |

## Artificial radioisotopes

When normal isotopes are placed in a nuclear reactor they may gain an extra neutron and become radioactive. For example:

$$^{23}_{11}\text{Na} \quad + \quad \text{neutron} \quad \longrightarrow \quad ^{24}_{11}\text{Na}$$

normal sodium                                    radioactive sodium

Chemically these radioisotopes behave in exactly the same way as the non-radioactive element. Because of this they can be used to trace the movement of substances within the body – even though they themselves never occur naturally.

The movement of fluids containing salt between the body cells is easily followed by giving the patient salt made from radioactive sodium-24. A suitable counter is used to track the salt movement.

Blockages in blood vessels or major internal haemorrhages (bleeding) can be located by injecting radioactive iron-59 into the blood stream. Similarly, a brain tumour can be located by administering radioactive arsenic, arsenic-74, which builds up at the site of the tumour.

Radioactive iodine, iodine-131, is used to diagnose over-active thyroid glands. Such over-activity results in a build-up of iodine in the neck. This is easily detected if radioactive iodine is given.

Perhaps the best known use of radioactivity is the treatment of cancer by radiotherapy. Often radioactive cobalt, cobalt-60, is used as the source of radioactivity. The powerful rays are carefully focused on to the cancerous cells, which are consequently killed.

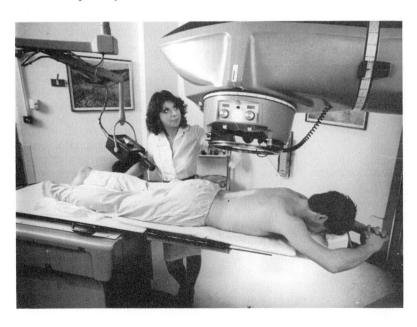

*Controlled doses of γ-radiation from the cobalt-60 isotope will destroy tumours and are used to treat cancer patients*

## Questions

**8** Suggest a suitable counter for tracing the movement of sodium-24 ions.

**9** Use the table below to answer the questions that follow.

| Radioisotope | Radioactivity | Half-life |
|---|---|---|
| $^{24}_{11}Na$ | β | 15 hours |
| $^{59}_{26}Fe$ | β | |
| $^{74}_{33}As$ | β | 17.9 days |
| $^{131}_{53}I$ | β | 8.0 days |
| $^{60}_{27}Co$ | γ | 5.26 years |

(a) Comment upon the type of radioactivity with regard to the use to which the isotopes are put.

(b) Comment upon the half-life with regard to the use to which the isotopes are put.

(c) The half-life for iron-59 is not given but can be found from the data below. Plot a graph and find the half-life, showing your method clearly.

| Time (days) | 0 | 20 | 40 | 60 | 80 | 100 | 120 |
|---|---|---|---|---|---|---|---|
| Activity (counts per minute) | 200 | 150 | 110 | 78 | 54 | 37 | 26 |

**10** Write equations showing the products of decay of each of the products emitting β-radioactivity.

**11** Describe at least three non-medical uses of radioactivity.

# 27 The rate of reaction

## 27.1 Speeding up reactions

**Figure 27.1** *An explosively fast reaction has driven this bullet through the bar of soap*

Explosions are very fast reactions. They are often so fast that they seem to be over in an instant, as when a gun is fired (see figure 27.1). Other reactions are very much slower. The corrosion of stonework or metal by the weather can take years or even centuries.

### First they must bang together!

When two chemicals react, the atoms change the way in which they join together. The chemicals you start with are the **reactants**. The new chemicals produced in the reaction are the **products**. For the chemical reaction to take place, the reactants must first collide with (bang into) each other (see figure 27.2).

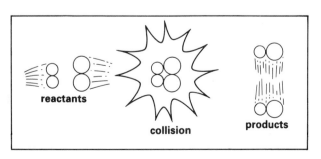

**Figure 27.2** *Molecules colliding and forming new chemicals*

Anything which makes the chance of such a collision greater will also increase the speed at which reactants change into products. It will increase the **rate of reaction**.

There are three ways of increasing the chance of such collisions:

**1** use more concentrated solutions,
**2** increase the temperature,
**3** break up solids into smaller lumps.

## Concentrate it!

Many reactions are carried out in solution. The reactants are dissolved in a solvent, often water.

If a large amount of solvent is used, you get a dilute solution. If only a little solvent is used, you have a concentrated solution.

All the molecules in a solution move about. This means that the reactant molecules will eventually collide and react. The chance of such a collision is much greater in more concentrated solutions. So products will form more quickly in these solutions, and the reaction will take place faster (see figure 27.3).

**Figure 27.3** *A dilute and a concentrated solution*

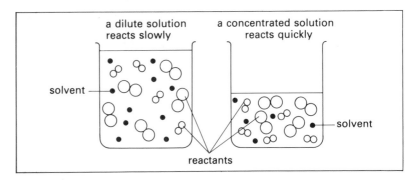

**Figure 27.4** *The people have been concentrated together. They are more likely to bump into each other*

Compressing a gas is rather like concentrating a solution. In a compressed gas the molecules are closer and so collide more often.

### Raise its temperature!

When a solution is heated the molecules in it move about faster. This means that they travel a greater distance in each second. As a result they collide with each other more often and harder, and so react more quickly.

Raising the temperature by 10°C can often halve the time that it takes for the reactants to change into products.

**Figure 27.5** *The quicker the motor cycles go, the harder they bump into each other*

**Figure 27.6** *Heating chemicals makes them react faster*

### Crush it!

Sometimes one of the reactants is a solid. The reaction rate will then depend on the size of the lumps of solid. An example of this is the reaction of a cube of marble with acid. If a large cube is used it reacts slowly. If the cube is split in two the reaction is faster. Figure 27.7 shows what happens when a cube is split. The acid now has the extra (shaded)

**Figure 27.7** *When the marble cube is split up, more of its surface can react with the acid*

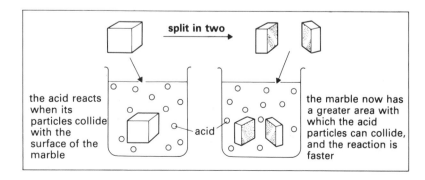

split in two

the acid reacts when its particles collide with the surface of the marble

acid

the marble now has a greater area with which the acid particles can collide, and the reaction is faster

surface with which to collide. This is why the reaction is faster.

If we break a lump of solid into very small pieces we get a powder. Powders have enormous surface areas. They normally react very quickly.

### Catalysts

There is a fourth way of speeding up reactions. This is to add a **catalyst**. A catalyst is a substance that speeds up a chemical reaction without getting used up.

Catalysts are extremely useful. Because they do not get used up they can be used over and over again. So, in the long run, they are a very cheap way of speeding up reactions. Catalysts are very important in industry (see figure 27.8). You read about some industrial catalysts in sections 20.3 and 21.5, for instance.

**Figure 27.8** *Plastic tubs being filled with soft margarine. Margarine is made by treating liquid oils with hydrogen. A nickel catalyst is used to speed up the reaction*

## 27.2 Reaction rates and graphs

### Measuring the rate of a reaction

When we say that a reaction is fast we mean that the reactants change very quickly into products. To measure the rate of a reaction, we must find out just how fast products are being formed.

Reactions in which a gas is given off are particularly easy to study. The gas can be collected and its volume measured at regular intervals.

Figure 27.9 shows the apparatus used. When the flask is shaken, the test tube containing the magnesium falls over. The magnesium then reacts with the acid to produce hydrogen gas. Table 27.1 shows the total amount of hydrogen in the

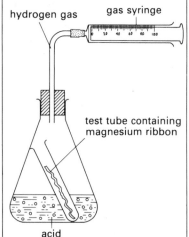

**Figure 27.9** *Measuring the rate at which a gas is formed*

| Time (min) | 0 | 1 | 2 | 3 | 4 | 5 | 6 | 7 |
|---|---|---|---|---|---|---|---|---|
| Total volume of hydrogen (cm³) | 0 | 30 | 45 | 54 | 59 | 60 | 60 | 60 |

**Table 27.1** *Total amount of hydrogen in syringe*

**Figure 27.10** *Graph showing the progress of the magnesium/acid reaction*

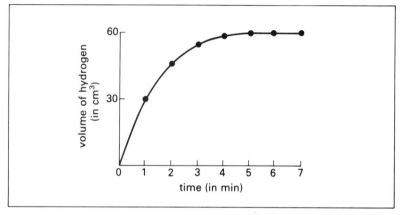

syringe after 1 minute, 2 minutes and so on. The results are shown as a graph in figure 27.10.

You can see from the table that the reaction stopped after 5 minutes. After this time the volume of hydrogen does not increase any more.

You can also see that the hydrogen is being given off more and more slowly. In the first minute, $30\,cm^3$ of gas are given off. During the second minute $45 - 30 = 15\,cm^3$ of gas are given off. So the reaction is slowing down.

## Why reactions slow down

Reaction rates depend on the concentration of the reactants. As the reaction proceeds, some of the reactants are changed into products. The concentration of the reactants that remain gets lower and lower (see figure 27.11). As a result the reaction gets slower and slower.

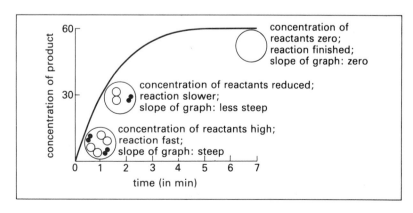

**Figure 27.11** *How the concentration of reactants falls over a period of time*

## Why reactions stop

A reaction stops when any or all of the reactants are used up. If a marble chip is added to concentrated hydrochloric acid, a fast reaction takes place (see figure 27.12(a)). Soon the marble chip disappears and the reaction stops. This is because the marble has all been used up. But it does not necessarily mean that all the acid has also been used.

**Figure 27.12** *The reaction of a marble chip with (a) concentrated hydrochloric acid and (b) dilute hydrochloric acid*

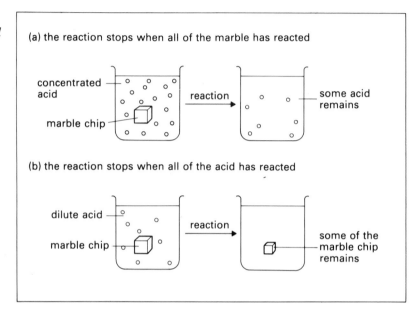

(a) the reaction stops when all of the marble has reacted

concentrated acid

marble chip

reaction

some acid remains

(b) the reaction stops when all of the acid has reacted

dilute acid

marble chip

reaction

some of the marble chip remains

If you add a marble chip to dilute hydrochloric acid a slow reaction takes place. This reaction stops while there is still some of the marble chip left (see figure 27.12(b)). This must mean that the reaction has run out of acid.

## Summary: The rate of reaction

* ★ Some chemical reactions are over in a microsecond. Others take thousands of years.
* ★ In order for chemicals to react, particles of the reactants must collide.
* ★ The more often reactant particles collide, the faster the reaction will be.
* ★ Concentrated solutions react more quickly than dilute solutions.
* ★ Reactions take place faster at higher temperatures.
* ★ Small pieces of solid react faster than large pieces of solid.
* ★ A catalyst is a substance which speeds up a reaction but does not get used up.
* ★ The reactant concentrations decrease as a reaction takes place so reactions get slower and slower.
* ★ A reaction will stop as soon as any one of the reactants has been used up.

# Questions

**1** Explain, using diagrams where necessary, the following facts:
(a) Concentrated solutions react faster than dilute ones.
(b) Reactions occur faster at high temperatures than at low ones.
(c) A solid reacts faster if it is crushed up.

**2** What is a catalyst? Find examples of reactions that are speeded up by using a catalyst.

**3** This question is about the reaction of marble chips (calcium carbonate) with hydrochloric acid. The two reactants were placed in a flask on a top-pan balance. The graph below shows total loss in mass at various times.

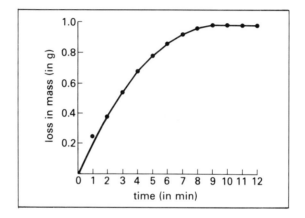

(a) Why is there a loss of mass during the reaction?
(b) After how many minutes had the reaction stopped?
(c) What is the loss in mass between (i) time 0 min and 1 min, (ii) time 6 min and 7 min?
(d) Why are the mass losses between these two time periods different?
(e) Give four ways of speeding up this reaction.

**4** Hydrogen peroxide solutions decompose into water and oxygen:

$$2H_2O_2(aq) \longrightarrow 2H_2O(l) + O_2(g)$$

The graphs over the page show the results of some experiments to measure the volume of gas given off when hydrogen peroxide decomposes.

First of all 1 g of copper oxide was added to 25 cm$^3$ of hydrogen peroxide and the volume of gas given off was measured at 1 min intervals. The graph obtained was ACD.

The experiment was then repeated, but this time 1 g of manganese(IV) oxide was added. The graph obtained was BCD.

(a) Draw a diagram of the apparatus you would use to carry out this experiment.

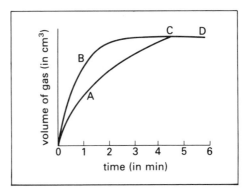

(b) Which oxide produced the best result? Give a reason for your answer.

(c) How would you prove that the gas given off is oxygen?

5 A pupil investigated the rate of reaction of magnesium with an excess of dilute hydrochloric acid at 20 °C. He placed the flask containing the metal and acid on a balance and took readings as shown below.

| Time (min)       | 0 | 1     | 2     | 3     | 4     | 5     |
| ---------------- | - | ----- | ----- | ----- | ----- | ----- |
| Loss in mass (g) | 0 | 0.054 | 0.079 | 0.096 | 0.100 | 0.100 |

(a) Plot a graph of loss in mass (vertical axis) against time.

(b) Mark on the graph the point where the reaction has finished.

(c) Sketch on the graph the curve that you would expect if a catalyst had been added to the initial mixture. Clearly label this curve 'catalysed'.

(d) Sketch on the graph the curve you would expect if a greater mass of magnesium had been used in the original experiment. Clearly label this 'extra Mg'.

(e) Calculate the average reaction rate (in g/minute) over the first three minutes.

(f) Write an equation for the reaction and calculate the mass of magnesium that was used.

(g) What volume would the hydrogen that was given off occupy at room temperature and pressure?

# The chemistry of dent-filling

Many car accidents result in nothing worse than a few unsightly dents. These are usually repaired by getting the worst of the dent out with a rubber hammer or suction pad, and then using filler paste to even up the surface. The paste must be allowed to set hard before painting.

*Dent fillers like the one shown above polymerize when mixed with the 'hardener'*

Most fillers are provided in two separate tubes. A large tube contains styrene monomer mixed with other ingredients such as polyesters. A small tube contains an organic peroxide. This is usually labelled 'hardener', but it is sometimes referred to as the catalyst. This latter term is not strictly correct as the peroxide is really a polymerization initiator rather than a catalyst.

When the contents of the two tubes are mixed the initiator starts off polymerization of the styrene. This is a chain reaction. If too little hardener is used the filler takes a long time to set. (You also have to avoid using too much as this can give lots of short polystyrene chains instead of the longer ones required.) The table below shows the effect of using different amounts of hardener upon the length of time the filler takes to set hard enough to sand down.

| % hardener | 1 | 2 | 4 |
|---|---|---|---|
| **Sanding time (minutes)** | 24.5 | 15 | 9 |

The polymerization reaction is exothermic, although this is not noticeable unless particularly deep dents are being filled by a single application of filler. Certainly the temperature at which the repair is made has a considerable effect upon setting time.

| **Temperature (°C)** | 10 | 15 | 20 | 30 |
|---|---|---|---|---|
| **Sanding time (minutes)** | 18 | 16.5 | 15 | 8.4 |

### Questions

**6** Look up and explain the meanings of the following terms:
   (a) initiator,
   (b) polymerization,
   (c) chain reaction.
**7** Plot graphs from the figures given in the tables.
**8** Comment upon reasons for the effects shown by the graphs.
**9** Try to find out what is meant by 'activation energy'.

# 28 Reversible reactions and equilibrium

## 28.1 Two kinds of reaction

### Irreversible reactions

Some chemical reactions cannot be reversed. For example, car exhaust fumes cannot change back into petrol. Yet such fumes contain all the elements that would be necessary.
The process is, of course, much more difficult than it sounds. Burning petrol, like most of the chemical reactions discussed in this book, is an **irreversible reaction**. It is easy to burn petrol but very difficult to 'unburn' it.

**Figure 28.1** *Burning is an irreversible reaction*

Burning magnesium is another irreversible reaction. Magnesium reacts easily with oxygen to form magnesium oxide:

magnesium    +    oxygen    $\longrightarrow$    magnesium oxide

$$2Mg \quad + \quad O_2 \quad \longrightarrow \quad 2MgO$$

This reaction can go in one direction only. No amount of heating will make the magnesium oxide change back into magnesium and oxygen. The burning is a strictly one-way process.

## Reversible reactions

Some reactions will go one way in certain conditions and the other way in different conditions. They are called **reversible reactions**. When mercury is heated to 350 °C it combines with oxygen in the air. Red mercury oxide is formed. If this mercury oxide is heated to 1000 °C, it decomposes back into mercury and oxygen. This is an example of a reversible reaction.

mercury + oxygen $\rightleftharpoons$ mercury oxide

$$2Hg + O_2 \rightleftharpoons 2HgO$$

Dehydrating copper sulphate crystals is another example of a reversible change. When the hydrated blue crystals are heated, they lose their water of crystallization. White anhydrous copper sulphate is formed. Adding water to the white anhydrous compound brings about the reverse process. The copper sulphate regains its water of crystallization and changes back into the hydrated compound:

hydrated copper sulphate $\rightleftharpoons$ anhydrous copper sulphate + water

$$CuSO_4.5H_2O \rightleftharpoons CuSO_4 + 5H_2O$$

## Traffic movements: a useful comparison

The reversible reactions that we have looked at can be compared to traffic movements in and out of a city.

At 8.30 in the morning the traffic flow is mainly into the city, as in figure 28.2.

**Figure 28.2** *In the morning, the traffic flow is mainly into the city*

At 5.30 in the afternoon the traffic flow is in the reverse direction.

So the traffic movement is a reversible process. Its direction depends upon the time of day.

The reaction of mercury with oxygen is also a reversible process. Its direction depends upon the temperature. At 350 °C the reaction goes in the direction which produces mercury oxide. At 1000 °C the reaction goes in the reverse direction.

$$2Hg + O_2 \underset{\text{at } 1000\,°C}{\overset{\text{at } 350\,°C}{\rightleftarrows}} 2HgO$$

## 28.2 Dynamic equilibrium

### Busy doing nothing!

Iodine monochloride (ICl) reacts easily with chlorine gas (Cl$_2$). Iodine trichloride (ICl$_3$) is formed:

iodine monochloride   +   chlorine ⟶ iodine trichloride

$$\text{ICl(l)} \quad + \quad \text{Cl}_2(g) \quad \longrightarrow \quad \text{ICl}_3(s)$$

The ICl$_3$ is not very stable. Even at room temperature it will decompose back into ICl and Cl$_2$. So here is a reversible reaction that can go backwards and forwards under the same conditions.

At first the backward reaction is rather slow because there is not much ICl$_3$ to decompose.

Before long the concentration of ICl$_3$ builds up and the rate of the backward reaction increases.

Soon a point is reached where the backward reaction is going just as fast as the forward reaction. This is called a state of **dynamic equilibrium**. ICl and Cl$_2$ are always combining together. ICl$_3$ is always decomposing. Because the two reactions are occurring at the same rate, it seems as if nothing is happening. Unless the conditions are altered these reactions will keep on taking place, forwards and backwards, for ever!

We can write this as:

$$\text{ICl(l)} + \text{Cl}_2(g) \rightleftharpoons \text{ICl}_3(s)$$

The ⇌ sign means that it will reach a dynamic equilibrium. At equilibrium the reaction mixture contains constant amounts of all the reactants and products.

We can also compare dynamic equilibrium with traffic movements. Here the situation resembles traffic movement in the mid-afternoon. Although traffic is moving into and out of the city, an equal amount is moving in each direction, as in figure 28.3. As a result there is no overall change in the number of vehicles in the city.

**Figure 28.3** *A traffic equilibrium. The number of cars entering the city is the same as the number leaving it*

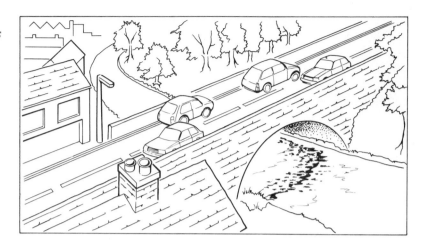

## Constant but not equal!

Let us have a last look at our traffic analogy. During the mid-afternoon the numbers of cars inside and outside the city are constant. This does not mean that there are equal numbers of cars inside and outside the city. It simply means that cars are entering and leaving at the same rate.

The same is true of chemical equilibria. At equilibrium the forward and backward reactions are occurring at the same rate. Some of these equilibrium mixtures will contain mostly reactants. Others will contain mostly products. The exact composition of any equilibrium mixture depends on the reaction conditions.

**Figure 28.4** *(a) The two children are both trying to hit all the balls into the other half of the court. Neither wins. (b) The boy on the right here is by far the better player, and can hit more balls into the other side of the court. Even so, he cannot clear all the balls from his side. This is also an equilibrium, but with different numbers of balls on each side*

## 28.3 Altering the position of a chemical equilibrium

As we have seen, many reactions do not go completely one way. They reach a state of dynamic equilibrium when the rates of the forward and backward reactions are equal. This is shown in figure 28.4(a).

But you can change the position of an equilibrium by altering the conditions of the reaction. This has happened in figure 28.4(b).

(a)

(b)

**Figure 28.5** *Red blood cells, shown here greatly magnified (magnification ×1170), react reversibly with oxygen. Red cells + oxygen ⇌ oxygenated red cells. The oxygen in the lungs moves the equilibrium towards the products. The oxygenated cells carry oxygen in the bloodstream to all parts of the body*

## The effect of concentration

Let us have another look at the equilibrium between ICl and $Cl_2$ that we mentioned in section 28.2:

$$ICl(l) + Cl_2(g) \rightleftharpoons ICl_3(s)$$

An equilibrium mixture contains all three of these substances. If you add extra chlorine to the mixture, you alter the position of the equilibrium. The extra chlorine speeds up the forward reaction and produces a mixture containing more $ICl_3$.

Pouring off the heavy chlorine gas has the opposite effect. The forward reaction slows down and $ICl_3$ decomposes faster than it is being formed.

When you add extra reactant to any equilibrium mixture, more product is formed (see figure 28.5, for example).

## The effect of temperature

A reversible reaction that is exothermic (giving out heat) in the forward direction will always be equally endothermic (taking in heat) in the backward direction:

$$\text{reactants} \underset{\text{heat taken in}}{\overset{\text{heat given out}}{\rightleftharpoons}} \text{products}$$

When you heat an equilibrium mixture, it always makes the reaction move in the endothermic direction. For example:

$$2NO_2(g) \underset{\text{heat taken in}}{\overset{\text{heat given out}}{\rightleftharpoons}} N_2O_4(g)$$

Heating the equilibrium mixture of these two gases shifts the position of the equilibrium to give more $NO_2(g)$.

## The effect of pressure

Gases can be easily compressed. When an equilibrium mixture of gases is compressed, the gases always try to react to make fewer molecules.

Using the example above:

$$\text{two } NO_2 \text{ molecules} \underset{\text{decomposes to}}{\overset{\text{combine to give}}{\rightleftharpoons}} \text{one } N_2O_4 \text{ molecule}$$

So increasing the pressure makes the equilibrium produce a greater proportion of $N_2O_4$.

Many of the reactions used in the chemical industry lead to equilibria. The ideas in table 28.1 are used to decide how to get the most product.

| Change in conditions | Shift in equilibrium position |
|---|---|
| add a reactant | more product formed |
| remove a reactant | less product formed |
| increase temperature | moves in endothermic direction |
| decrease temperature | moves in exothermic direction |
| increase pressure | moves to form fewer molecules |
| decrease pressure | moves to produce more molecules |

**Table 28.1** *How changes in conditions affect an equilibrium*

## Catalysts and equilibria

Catalysts speed up the rate at which a reaction reaches equilibrium. But they do not alter the position of an equilibrium. So using a catalyst gives the same equilibrium mixture more quickly.

## Summary: Reversible reactions and equilibrium

* Most chemical reactions are not easily reversed.
* Some chemical reactions are reversible.
* A reversible reaction can go in the forward and reverse direction under the same conditions.
* Reactants can change into products at the same speed as products are changing into reactants. This is called a dynamic equilibrium.
* In chemical equations a dynamic equilibrium is indicated by $\rightleftharpoons$.
* The proportions of reactants and products can be changed in an equilibrium mixture by altering the reaction conditions.
* Adding a reactant causes more product to form.
* Adding a product causes more reactant to form.
* Heating an equilibrium mixture makes it move in the endothermic direction.
* Compressing a gaseous equilibrium mixture makes it move in whichever direction gives fewer molecules.
* Catalysts do not affect the position of an equilibrium mixture, but allow equilibrium to be reached more quickly.

## Questions

**1** Are the following statements true or false? Give reasons.
(a) Physical changes like melting and boiling are reversible. Chemical reactions are not reversible.
(b) Dynamic equilibrium is another name for a reversible reaction.
(c) At equilibrium, both the forward and the backward reaction have come to a stop.
(d) At equilibrium, the reaction mixture contains constant amounts of reactants and products.
(e) At equilibrium, the reaction mixture contains equal amounts of reactants and products.

**2** This question is about the equilibrium

$$SbCl_3(aq)\ +\ H_2O(l)\ \rightleftharpoons\ SbOCl(s)\ +\ 2HCl(aq)$$

clear solution     clear liquid      white precipitate     clear solution

(a) What do the symbols (aq), (l) and (s) mean?
(b) Why can't a clear solution of antimony(III) chloride, $SbCl_3$, be made by dissolving the solid in water?
(c) Some antimony(III) chloride was added to water to give a cloudy mixture. Adding a few drops of concentrated hydrochloric acid made the mixture turn clear. Has the equilibrium moved to the left (reactants) or the right (products)?
(d) Why should adding hydrochloric acid cause this change?
(e) Antimony(III) chloride easily dissolves in concentrated hydrochloric acid to give a clear solution. What would happen if some of this solution was added to a large beaker of water?

**3** What will be the effect of: (i) increasing the temperature, (ii) increasing the pressure, (iii) adding a catalyst, upon each of the following equilibrium mixtures?

(a) $PCl_3(g) + Cl_2(g) \rightleftharpoons PCl_5(g)$     (exothermic)

(b) $N_2O_4(g) \rightleftharpoons 2NO_2(g)$     (endothermic)

(c) $H_2(g) + I_2(g) \rightleftharpoons 2HI(g)$     (exothermic)

**4** When chlorine is dissolved in water the following equilibrium is set up:

$$Cl_2(aq) + H_2O(l) \rightleftharpoons H^+(aq) + Cl^-(aq) + HOCl(aq)$$

(a) What does the sign $\rightleftharpoons$ mean?
(b) Will the equilibrium mixture be acidic, alkaline or neutral? Explain your answer.
(c) How would the addition of sodium hydroxide affect the equilibrium?
(d) How would the addition of silver nitrate affect the equilibrium? (*Hint*: silver chloride is insoluble in water.)
(e) How would the addition of extra water affect the equilibrium?

## Ion-exchange resins

$SO_3^-H^+$
|
$C_6H_4$
|
$-CH_2-CH-CH_2-CH-$
|
$C_6H_4$
|
$-CH-CH-CH_2-CH_2-$
|
$C_6H_4$
|
$SO_3^-H^+$

Ion-exchange resins are insoluble polymeric organic compounds containing either basic or acidic groups. Part of the structure of a typical cation-exchange resin is shown on the left.

These resins have uses ranging from catalysis to water softening. For example, the resin on the left can be used for reactions requiring an acid catalyst. Because the resin is insoluble it is easily separated afterwards for re-use.

In section 10.3 we saw that ion-exchange resins are used in water softeners. We can now see in more detail how they work. Resin for water softening is first converted to the sodium form, by soaking it in fairly concentrated salt solution. The equilibrium below is set up:

$$Resin^-H^+ + Na^+Cl^- \rightleftharpoons Resin^-Na^+ + H^+Cl^-$$

This conversion to the sodium salt can be carried out more completely by placing the resin in a column and allowing salt solution to flow through it. The position of the equilibrium lies mainly towards the products.

The resin in the sodium form can now be used for softening water. If water containing, for example, calcium sulphate is passed through a column of the resin, the calcium ions are exchanged for sodium ions:

$$2Resin^-Na^+ + Ca^{2+} \rightleftharpoons (Resin^-)_2Ca^{2+} + 2Na^+$$

Neither sodium ions nor sulphate ions make water hard.

Some water softeners contain two columns of resin. The first holds a cation-exchange resin: this contains $H^+$ ions which exchange with metal ions. The second holds an anion-exchange resin containing $OH^-$ ions which exchange with non-metal ions such as chloride and sulphate. These softeners produce water containing virtually no ions at all.

### Questions

5 Explain the meanings of the following terms:
   (a) polymeric,    (b) cation,    (c) anion.
6 The cation resin can be changed from the hydrogen form to the sodium form by passing salt solution over it. What would be the approximate pH of the liquid after it had been used for this purpose? Explain your answer.
7 Explain as clearly as you can why pouring the salt solution through a column of the resin should give a greater conversion of resin to the sodium form.
8 After being used for water softening the resin becomes totally changed into the calcium form. How could you change it back to the sodium form?
9 Explain how the use of two resins removes all ions from the water.

# 29 The chemical industry

## 29.1 What the chemical industry does

The chemical industry is very important to the economy of many countries. It provides employment for millions of people. By its exports it helps the balance of payments.

**Figure 29.1** *A large chemical manufacturing complex*

### What the industry makes

You may be surprised by the range of materials that the chemical industry produces. These include (among others):

| | |
|---|---|
| antiseptics and disinfectants, | foods, |
| cosmetics, | fuels, |
| detergents, | herbicides and pesticides, |
| drugs, | metals and alloys, |
| dyes, | paints, |
| explosives, | plastics, |
| fertilizers, | synthetic fibres. |

Also many chemicals are manufactured in huge quantities for use in other industries. The chemistry of many important industrial processes is looked at elsewhere in this book.

## What the chemical industry uses

The chemical industry obtains most of its raw materials from:

air,              plants,
coal,             rocks and mineral ores,
crude oil,        sea water.

Coal is used as a fuel and also as a source of coke. Enormous amounts of coke are used for extracting metals.

From crude oil we get fuels and our organic chemicals. Many important solvents, drugs and agricultural chemicals come from oil. Oil is also the source of nearly all our plastics.

The major mineral ores in Britain are iron ore, salt (sodium chloride), limestone and calcium sulphate. From these come such important chemicals as iron, sodium hydroxide, sodium carbonate, chlorine, hydrogen, cement, plaster and glass.

Salt, magnesium metal and bromine are the main chemicals obtained from sea water.

Nitrogen and oxygen are the main substances extracted from the air.

Chemical manufacture is a truly international business. The USA is the biggest producer of chemicals in the western world. Japan, West Germany, Britain, France and Italy are also major producers. Many other countries are involved in the mining or manufacture of certain products. A few of these are shown in table 29.1 (over the page).

## 29.2 Economic and social factors in industrial chemistry

### General economic factors

For an industry to be successful, it must keep its costs as low as possible. Other things must also be considered. Some of them are listed here.

1 The product must be in demand.
2 The raw materials should be cheap.
3 The chemical reaction used must produce the required product efficiently.
4 Building costs should as low as possible.
5 Running costs should be as low as possible.
6 Transport costs should be as low as possible.

### Chemical factors

The process used needs to produce the largest amount of product from the smallest amount of raw materials in the shortest possible time. This will depend on finding the fastest

**Table 29.1** *Some countries and the substances they produce (see figure 29.2)*

| Country | Product |
|---|---|
| 1 Australia | zinc, lead |
| 2 Canada | platinum, nickel |
| 3 Chile | iodine, copper |
| 4 West Indies | aluminium |
| 5 India | manganese |
| 6 Italy | mercury |
| 7 Mexico | silver |
| 8 Morocco | phosphorus compounds |
| 9 Peru | phosphorus compounds |
| 10 Rwanda | tungsten |
| 11 South Africa | silver, gold, uranium, diamonds |
| 12 South-East Asia | tin |
| 13 Spain | mercury |
| 14 Turkey | chromium |
| 15 Zaire/Zambia | copper and cobalt |

**Figure 29.2** *The chemical-producing countries listed in table 29.1*

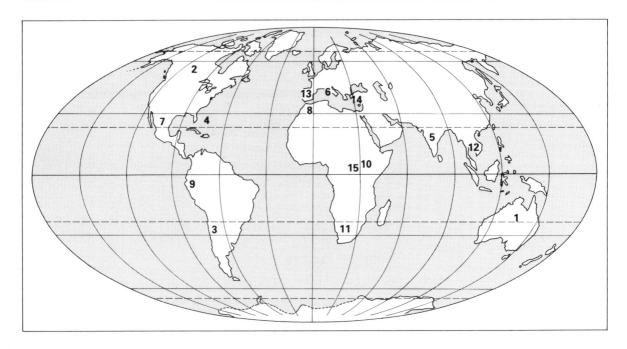

rate of reaction and best equilibrium position (see chapter 28).

Let us now see how these ideas are used in an important chemical process with which you are familiar: the Haber process, described on pages 248–9 and 256–7. You will remember that the reaction is:

$$\text{nitrogen } + \text{ hydrogen} \rightleftharpoons \text{ammonia}$$

$$\text{N}_2(\text{g}) \quad + \quad 3\text{H}_2(\text{g}) \quad \rightleftharpoons \quad 2\text{NH}_3(\text{g})$$

The reaction is exothermic. About 6.4 kJ of heat are given out for every gram of ammonia formed. The reaction is very slow at room temperature. It is also an equilibrium process.

To get a fast reaction rate you should:

**1** use a catalyst.
**2** carry out the reaction at a high temperature.

It is easy to use a catalyst. But it is not so simple to use the highest possible temperature. This is because if you want to get a good percentage yield of ammonia in the equilibrium mixture you should:

**3** carry out the reaction under high pressure (increasing the pressure will move the equilibrium in the direction which produces fewer molecules, giving more ammonia – see the equation);
**4** carry out the reaction at a low temperature (using a low temperature shifts the equilibrium in the exothermic direction, giving more ammonia).

If we look at the rates of reaction we want a high temperature. If we look at equilibrium we want a low temperature. In industry a balance between these has to be made and so a reasonably high temperature is used.

## Construction costs

A chemical plant (factory) that is expensive to build will increase the cost of the product.

Three important factors in building costs are:

**1** the number of stages in the process,
**2** the pressure that is being used,
**3** the size of the chemical plant.

The first point is easy to understand. A process in which the reactants all react to give a single pure product would be a one-stage process. Such a process would be very cheap to build.

Now imagine a process where the reactants form a mixture of two products. In this case a second stage is needed to separate this mixture. So this will add to the construction costs.

**Figure 29.3** *Dyes are complicated chemicals. At this plant, dyes are made under carefully controlled conditions*

Many chemical processes are operated at high pressures. The Haber process is an example. Unfortunately, to build plants that operate at very high pressures is expensive. Very thick metal equipment must be used. So since it is more expensive to build a high-pressure plant, you have to decide whether or not you are going to get enough extra product to make it worthwhile.

Building a plant to make 10 000 tonnes of a chemical each year obviously costs more than building a plant to make only 5000 tonnes of the same chemical each year. But it does not cost twice as much. So deciding the best size for a chemical plant is very important.

## Running costs

Two important factors that affect the cost of running a chemical plant are:

1 labour costs,
2 energy costs.

Labour costs are usually kept fairly low in the chemical industry. This is done by using automatic, computer-controlled processes (see figure 29.4, for example).

Energy costs are now becoming increasingly important. Many chemical plants use the heat from exothermic reactions to provide energy elsewhere on the plant. This is done in the Haber process. The heat produced in the reaction between nitrogen and hydrogen is made use of in another stage of the plant.

Energy costs can be particularly high in processes that use large amounts of electricity, such as aluminium production.

**Figure 29.4** *This man is at the computer console at an anodizing plant. The computer controls the tanks containing liquids for degreasing and etching aluminium*

In such cases the process is often carried out near a supply of (relatively) cheap hydroelectric power (see figure 15.19).

## Transport costs

Many chemical plants are built close to the source of their raw materials. Oil refineries are placed near ports where tankers can unload. In England, many chemicals made from salt are produced in Cheshire. A steel industry has developed in Sheffield, England and in South Wales. Here there are nearby supplies of both limestone and coal.

**Figure 29.5** *A tanker being loaded with oil*

Transport costs are less important if the product is an expensive one. Then the cost of transport is a much smaller percentage of the total value of the product.

## Social factors

Major accidents are, fortunately, rare in the chemical industry. When they do happen we all get to hear about them. The explosion that took place in Flixborough, England is one example (see figure 29.6). The leaking of a poisonous

**Figure 29.6** *The explosion at this plant in Flixborough, England, killed many people*

gas at Bhopal in India is another. The consequences of the nuclear accident at Chernobyl in the Soviet Union in 1986 have still, at the time of writing, not been fully assessed.

Against this we must weigh the benefits of the chemical industry. Just some of the benefits that it provides are:

fertilizers to provide adequate food,
drugs and pharmaceuticals to cure disease and improve health,
fuels to provide power and transport,
plastics for all sorts of uses,
metals,
paints to protect wood and metals,
dyes,
cement and other building materials.

We could of course do without all these things, but life would certainly be different!

## Summary: The chemical industry

* The chemical industry produces a very wide range of products.
* Rocks, air, sea, plants and the fossil fuels are the main raw materials of the chemical industry.
* Rates of reaction and the position of chemical equilibria are two important considerations in industrial chemistry.
* Chemical plants are usually built near the supplies of their raw materials.
* Energy saving is now an important part of chemical plant design.
* Large chemical plants usually produce chemicals at a lower cost per tonne than smaller plants.
* The possibility of accidents and pollution from chemical plants must be compared with the benefits that arise from their products.

## Questions

1 For each of the chemical processes (a) to (e) explain what happens, and give one example of a substance that is made using this process. (Use appendix 3 on page 375.)
(a) synthesis,     (b) reduction,     (c) polymerization,
(d) electrolysis,     (e) cracking.
2 Utopia is an island of some 150 square miles, situated 200 miles off the west coast of Africa. It has sandy beaches, with a natural harbour, and is swept by sea breezes capable of operating windmills. There are resources of bauxite, sulphur and offshore oil. What chemical industry might be established on Utopia? Make a list of things that might have to be imported.

**3** The data below is taken from a leaflet published by the UK Chemical Industries Association. Write about one page saying what you can tell from this data.

| 1970 | 1980 | 1984 | 1985 | 1986 |
|------|------|------|------|------|
| 3448 | 13 662 | 18 844 | 20 350 | 20 548 |

*Sales of principal products of UK chemical industry, 1970–86 (£ million)*

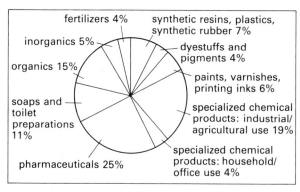

*Sectors of the industry 1983*

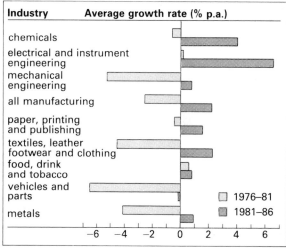

*UK industrial growth rate comparisons*

| Year | Thousands |
|------|-----------|
| 1976 | 396 |
| 1977 | 406 |
| 1978 | 410 |
| 1979 | 413 |
| 1980 | 402 |
| 1981 | 369 |
| 1982 | 355 |
| 1983 | 335 |
| 1984 | 334 |
| 1985 | 338 |
| 1986 | 337 |

*Employees in the UK chemical industry 1976–1986*

| Country | Average growth (% p.a.): Chemicals | Average growth (% p.a.): All industry | Ratio of chemicals to all industry |
|---------|-----------|-------------|-----------|
| UK | 1.8 | 1.5 | 1.2 |
| W. Germany | 1.6 | 1.6 | 1.0 |
| France | 3.3 | 1.0 | 3.3 |
| Italy | 2.9 | 1.4 | 2.1 |
| Belgium/Luxembourg | 3.1 | 1.1 | 2.8 |
| Netherlands | 3.8 | 1.1 | 3.4 |
| USA | 4.6 | 4.2 | 1.1 |
| Japan | 3.7 | 2.6 | 1.4 |

*International growth rate comparisons (excluding construction industry), 1976–86*

# The chemist's compromise

The aim of the industrial chemist is to take a cheap raw material and convert it quickly and completely into an expensive product, preferably without the need for either high pressures or high temperatures.

Unfortunately such reactions are rather hard to find! Many reactions need high temperatures or highly priced catalysts to achieve reasonable reaction rates. They may also require high pressures in order to shift equilibria towards products. This considerably increases the cost of building the chemical factory. Many reactions produce **by-products**. (A by-product is a substance that is formed in the reaction in addition to the required product.) The formation of by-products can drastically increase building costs as extra equipment will be needed to separate the by-products.

*Platinum and similar metals are useful catalysts. This gauze is made from an alloy of platinum and rhodium and is used to catalyse the oxidation of ammonia in the manufacture of nitric acid. Unfortunately these metals are expensive. Does this matter?*

The manufacture of ethanoic acid illustrates how difficult it can be decide upon the best way of making a chemical. All three of the reactions summarized in the table below are currently used.

| Raw material | Catalyst | Temperature (°C) | Pressure (atm) | Yield (%) | Recovered by-products |
|---|---|---|---|---|---|
| ethene | $PdCl_2$ | 125 | 5 | 95 | none |
| methanol/ carbon monoxide | $RhI_3$ | 150 | 300 | 99 | none |
| pentane | $Mn^{2+}$ | 200 | 25 | 40 | methanoic acid and butanoic acid (HCOOH and $C_3H_5COOH$) |

Both ethene and pentane come from petroleum. Methanol and carbon monoxide are made from coke and water.

## Questions

**4** Find out at least two uses of ethanoic acid.

**5** Which method of making ethanoic acid will be likely to have the highest energy costs?

**6** Which method will require the sturdiest reaction vessels?

**7** Which method is likely to require the greatest amount of equipment?

**8** Which process do you think will have the lowest raw material costs at the moment?

**9** Which process is likely to have the lowest raw material costs in thirty years' time?

**10** Suggest an alternative source of ethene for countries that do not have any petroleum of their own.

**11** Are by-products necessarily a nuisance? Explain your answer.

**12** If you had to manufacture ethanoic acid, which method would you choose and why?

**13** Suggest a suitable metal for building the reactors and pipework for the method chosen. Explain your answer.

**14** The cost of building a chemical factory obviously depends upon the capacity of the plant (that is, the amount of product per day). Generally

building costs are proportional to (capacity of plant)$^{0.6}$

Can you suggest reasons why doubling the amount of chemical handled does not simply double building costs?

# Appendix 1
# The properties of some elements

| Name | Symbol | Appearance | Atomic number | Approx relative atomic mass | Melting point (°C) | Boiling point (°C) | Year of discovery |
|------|--------|-----------|---------------|-----------------------------|--------------------|--------------------|--------------------|
| aluminium | Al | silvery metal | 13 | 27 | 660 | 2468 | 1827 |
| antimony | Sb | silvery-grey metal | 51 | 122 | 631 | 1750 | 17th century |
| argon | Ar | colourless gas | 18 | 40 | −189 | −186 | 1894 |
| arsenic | As | grey solid | 33 | 75 | 817 | – | 1250 |
| barium | Ba | silvery-white metal | 56 | 137 | 725 | 1640 | 1808 |
| beryllium | Be | silvery-grey metal | 4 | 9 | 1280 | 2970 | 1828 |
| bismuth | Bi | silvery-white metal | 83 | 209 | 271 | 1560 | 1753 |
| boron | B | brown-black solid | 5 | 11 | 2300 | – | 1808 |
| bromine | Br | dark red liquid | 35 | 80 | −7 | +59 | 1826 |
| cadmium | Cd | silvery-blue metal | 48 | 112 | 321 | 765 | 1817 |
| calcium | Ca | silvery-grey metal | 20 | 40 | 845 | 1487 | 1808 |
| carbon (graphite) | C | black solid | 6 | 12 | 3730 | – | ancient |
| carbon (diamond) | C | transparent solid | 6 | 12 | 3550 | 4800 | ancient |
| chlorine | Cl | pale green gas | 17 | 35.5 | −101 | −35 | 1774 |
| chromium | Cr | silvery-grey metal | 24 | 52 | 1890 | 2482 | 1797 |
| cobalt | Co | silvery-white metal | 27 | 59 | 1495 | 2870 | 1735 |
| copper | Cu | reddish metal | 29 | 64 | 1083 | 2595 | prehistoric |
| fluorine | F | pale yellow gas | 9 | 19 | −220 | −188 | 1886 |
| germanium | Ge | grey-white solid | 32 | 73 | 937 | 2830 | 1886 |
| gold | Au | golden metal | 79 | 197 | 1064 | 2940 | prehistoric |
| helium | He | colourless gas | 2 | 4 | −272 | −269 | 1895 |
| hydrogen | H | colourless gas | 1 | 1 | −259 | −253 | 1766 |
| iodine | I | black shiny solid | 53 | 127 | 114 | – | 1811 |
| iron | Fe | silvery-grey metal | 26 | 56 | 1535 | 2750 | prehistoric |
| krypton | Kr | colourless gas | 36 | 84 | −157 | −152 | 1898 |
| lead | Pb | silvery-white metal | 82 | 207 | 327 | 1740 | prehistoric |
| lithium | Li | silvery metal | 3 | 7 | 179 | 1317 | 1817 |
| magnesium | Mg | silvery metal | 12 | 24 | 651 | 1107 | 1808 |
| manganese | Mn | silvery metal | 25 | 55 | 1245 | 2097 | 1774 |
| mercury | Hg | silvery liquid metal | 80 | 201 | −39 | 357 | ancient |
| molybdenum | Mo | silvery-white metal | 42 | 96 | 2610 | 5560 | 1778 |
| neon | Ne | colourless gas | 10 | 20 | −249 | −246 | 1898 |
| nickel | Ni | silvery metal | 28 | 59 | 1453 | 2732 | 1751 |
| nitrogen | N | colourless gas | 7 | 14 | −210 | −196 | 1772 |
| oxygen | O | colourless gas | 8 | 16 | −218 | −183 | 1774 |
| phosphorus (red) | P | red solid | 15 | 31 | 597 | – | 1669 |

| Name | Symbol | Appearance | Atomic number | Approx relative atomic mass | Melting point (°C) | Boiling point (°C) | Year of discovery |
|------|--------|-----------|---------------|------------------------------|---------------------|---------------------|-------------------|
| phosphorus (white) | P | white solid | 15 | 31 | 44 | 281 | – |
| platinum | Pt | silvery-white metal | 78 | 195 | 1772 | 3827 | 1735 |
| plutonium | Pu | silvery metal | 94 | 242 | 641 | 3327 | 1940 |
| potassium | K | silvery metal | 19 | 39 | 64 | 774 | 1807 |
| radium | Ra | white metal | 88 | 226 | 700 | 1140 | 1911 |
| radon | Rn | colourless gas | 86 | 222 | −71 | −62 | 1900 |
| rubidium | Rb | silvery-white metal | 37 | 86 | 39 | 688 | 1861 |
| silicon | Si | grey shiny solid | 14 | 28 | 1410 | 2355 | 1824 |
| silver | Ag | silvery metal | 47 | 108 | 962 | 2212 | prehistoric |
| sodium | Na | silvery metal | 11 | 23 | 98 | 892 | 1807 |
| strontium | Sr | silvery metal | 38 | 88 | 770 | 1384 | 1808 |
| sulphur (rhombic) | S | yellow solid | 16 | 32 | – | – | – |
| sulphur (monoclinic) | S | yellow solid | 16 | 32 | 119 | 444 | prehistoric |
| tin | Sn | silvery-grey metal | 50 | 119 | 232 | 2270 | ancient |
| titanium | Ti | silvery-white metal | 22 | 48 | 1675 | 3260 | 1946 |
| tungsten | W | silvery-grey metal | 74 | 184 | 3410 | 5927 | 1783 |
| uranium | U | silvery-white metal | 92 | 238 | 1132 | 3818 | 1841 |
| vanadium | V | white metal | 23 | 51 | 1890 | 3380 | 1867 |
| xenon | Xe | colourless gas | 54 | 131 | −112 | −107 | 1898 |
| zinc | Zn | silvery-grey metal | 30 | 65 | 420 | 907 | 1746 |

**Note:** *Prehistoric* means that the element was known before written records existed. *Ancient* means that it was first mentioned in the records of an ancient civilization

# Appendix 2
# Analysis

Analysis is the process of finding out what something is made of. Chemists do this by carrying out various tests. We shall look at some of these tests and explain how to interpret the results.

## Tests to identify positive ions

### 1 Colour

The colour of a compound can give some clue about the kind of metal that is present.

A substance that is coloured is unlikely to be a simple compound of any of the following: sodium, potassium, magnesium, calcium, aluminium or zinc.

### 2 Flame tests

Certain ions can give bright colours to a bunsen burner flame. Some of these ions and the colours they produce are shown in table A2.1.

| Ion present | Colour of flame |
|-------------|-----------------|
| sodium | bright yellow |
| potassium | lilac |
| calcium | red |
| copper | bright blue/green |

**Table A2.1** *Flame tests*

Usually a clean nichrome wire is dipped into the substance being tested. This is then held in the hot part of a bunsen flame and the colour of the flame noted.

Potassium compounds sometimes give a yellowy lilac flame. This can make it difficult to distinguish them from sodium compounds. Viewing the flame through a piece of blue glass can solve this problem. This cuts out any yellow light but still allows the lilac colour to be seen.

## 3 Reactions with sodium hydroxide solution

Many ionic compounds react with sodium hydroxide solution. Exactly what happens depends upon two factors:

**1** the type of metal that is present,
**2** the amount of sodium hydroxide solution that is added.

Table A2.2 shows what happens with various metal ions.

**Table A2.2** *Reactions of metal ions with sodium hydroxide solution*

| Ions present: | | Effect of adding small amount of sodium hydroxide solution: | | Effect of adding large excess of sodium hydroxide solution: | |
|---|---|---|---|---|---|
| Name | Formula | What happens | Formula of product | What happens | Formula of product |
| sodium | $Na^+$ | no change | – | no change | – |
| potassium | $K^+$ | no change | – | no change | – |
| calcium | $Ca^{2+}$ | white precipitate | $Ca(OH)_2$ | white precipitate remains | $Ca(OH)_2$ |
| magnesium | $Mg^{2+}$ | white precipitate | $Mg(OH)_2$ | white precipitate remains | $Mg(OH)_2$ |
| aluminium | $Al^{3+}$ | white precipitate | $Al(OH)_3$ | white precipitate dissolves | $NaAl(OH)_4$ |
| lead | $Pb^{2+}$ | white precipitate | $Pb(OH)_2$ | white precipitate dissolves | $Na_2Pb(OH)_4$ |
| zinc | $Zn^{2+}$ | white precipitate | $Zn(OH)_2$ | white precipitate dissolves | $Na_2Zn(OH)_4$ |
| copper | $Cu^{2+}$ | blue precipitate | $Cu(OH)_2$ | blue precipitate remains | $Cu(OH)_2$ |
| iron(II) | $Fe^{2+}$ | green precipitate | $Fe(OH)_2$ | green precipitate remains | $Fe(OH)_2$ |
| iron(III) | $Fe^{3+}$ | brown precipitate | $Fe(OH)_3$ | brown precipitate remains | $Fe(OH)_3$ |

# Tests to identify negative ions

## 1 Reaction with dilute hydrochloric acid

Many compounds react with dilute hydrochloric acid to give off a gas. Some of the ions that give rise to these gases are listed in table A2.3.

**Table A2.3** *Reactions of ions with dilute hydrochloric acid*

| Ion present | | Effect of adding dilute acid: | | How to identify the gas |
|---|---|---|---|---|
| Name | Formula | What is formed | Formula | |
| carbonate | $CO_3^{2-}$ | carbon dioxide gas | $CO_2$ | turns limewater milky |
| sulphide | $S^{2-}$ | hydrogen sulphide gas | $H_2S$ | smells of bad eggs |
| sulphite | $SO_3^{2-}$ | sulphur dioxide gas | $SO_2$ | choking smell, acidic reaction to indicator |

## 2 Reaction with silver nitrate solution

Solutions of chloride, bromide and iodide ions will each form a precipitate when mixed with silver nitrate solution and nitric acid. What happens is shown in table A2.4.

| Ion present: | | Effect of adding nitric acid/silver nitrate solution: | |
|---|---|---|---|
| Name | Formula | What is formed | Formula |
| chloride | $Cl^-$ | white precipitate | AgCl |
| bromide | $Br^-$ | cream precipitate | AgBr |
| iodide | $I^-$ | yellow precipitate | AgI |

**Table A2.4** *Reactions of ions with silver nitrate solution*

## 3 Testing for sulphates

Solution of sulphates react with a mixture of dilute nitric acid and barium nitrate solution to give a white precipitate of barium sulphate, $BaSO_4$.

## 4 Testing for nitrates: the brown ring test

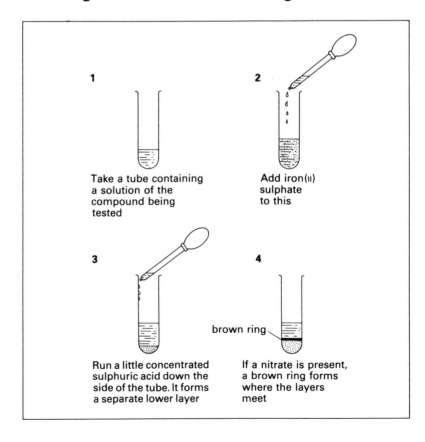

1 Take a tube containing a solution of the compound being tested

2 Add iron(ii) sulphate to this

3 Run a little concentrated sulphuric acid down the side of the tube. It forms a separate lower layer

4 If a nitrate is present, a brown ring forms where the layers meet

brown ring

# How to identify the ions in a compound

Any ionic compound will consist of two parts: a positive ion and a negative ion.

## A scheme for identifying positive ions

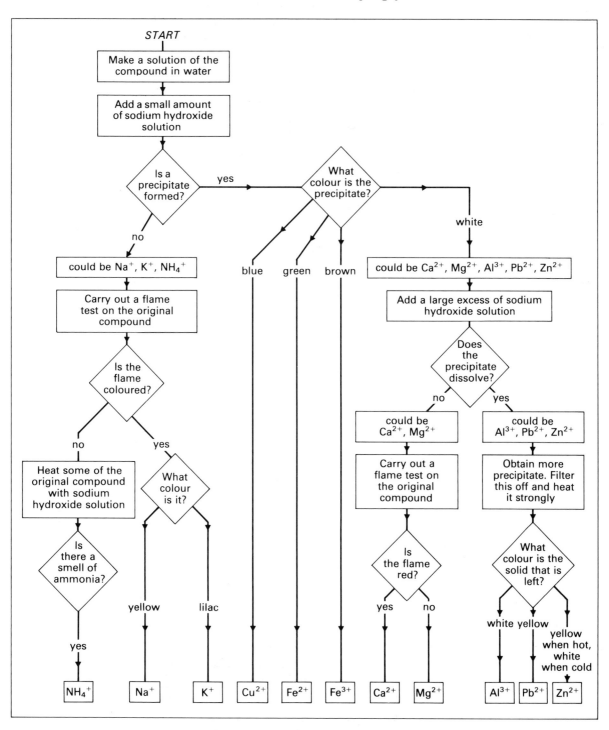

# A scheme for identifying negative ions

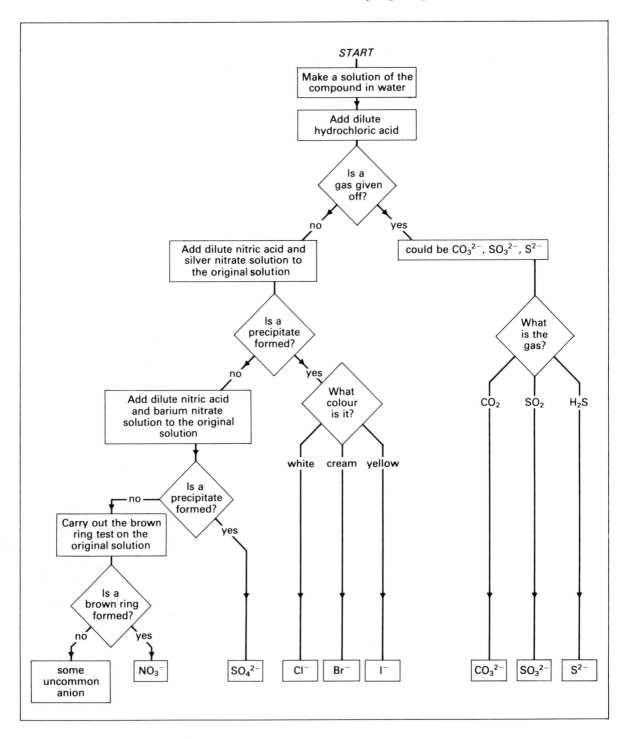

# Appendix 3
# A short chemical dictionary

This dictionary explains the meaning of many of the words that you will find in this book. Details of the chemical elements can be found in appendix 1.

ACID   A solution with a pH below 7. All solutions of acids contain a lot of $H^+$ ions.

AEROBIC   Requiring the presence of oxygen or air.

ALCOHOL   The alcohol in 'alcoholic' drinks is ethanol. Its formula is $C_2H_5OH$.

ALKALI   A solution with a pH above 7. A base that dissolves in water. All alkaline solutions contain a lot of $OH^-$ ions.

ALLOTROPE   The different forms that some elements can exist in.

ALLOY   A blend of two or more metals.

ANAEROBIC   Requiring the absence of oxygen or air.

ANALYSIS   Finding out what a substance is made of.

ANHYDROUS   Containing no water.

ANION   An ion that is attracted to the anode. Anions have a negative electric charge.

ANODE   The electrode connected to the positive terminal of an electricity supply.

ANTIOXIDANT   A substance that prevents reaction with oxygen.

AQUEOUS   Containing mostly water.

ATOM   The smallest particle of any element.

AVOGADRO CONSTANT   The number of particles in a mole of any substance, equal to $6 \times 10^{23}$.

BASE   A substance that will neutralize an acid. A substance that will accept $H^+$ ions.

BIOTECHNOLOGY   Synthesizing chemicals by making use of living cells.

BOILING POINT   The temperature at which bubbles of vapour start to form rapidly in a liquid.

BOND   A force that holds groups of atoms or ions together.

BROWNIAN MOTION   The jerky movements of small particles suspended in a liquid or gas.

CARBOHYDRATE   A group of compounds containing carbon, hydrogen and oxygen, such as sugars, starch and cellulose.

CATALYST   A substance that speeds up a reaction but does not get used up.

CATHODE   The electrode connected to the negative terminal of an electricity supply.

CATION   An ion that is attracted to the cathode. Cations have a positive electric charge.

CENTRIFUGE   An apparatus that separates solids from liquids by spinning the mixture at great speed.

CHEMICAL   Any substance, but usually means the substances used in laboratories.

CHEMICAL PROPERTY   Some fact about the way a substance behaves during chemical reactions.

CHEMICAL REACTION   A process in which atoms or ions become bonded to each other in a different way. In any chemical reaction a new substance is formed.

CHEMISTRY   The study of chemicals and chemical reactions.

CHROMATOGRAPHY   A method of separating substances, especially coloured ones.

COMBINE   To join together, to become chemically bonded.

COMBUSTION   Burning; a reaction in which a substance combines with oxygen giving out heat.

COMPONENT   A substance that is part of a mixture.

COMPOUND   Two or more elements that have become bonded together because of a chemical reaction.

CONCENTRATED SOLUTION   A solution that contains a lot of the dissolved substance, such as concentrated acid.

CONDENSE   To change from a vapour to a liquid.

CONDUCTOR   A substance that allows electricity to pass through it.

CORROSION   A chemical reaction in which a solid is 'eaten away'.

COVALENT BONDING   The forces that hold atoms together in molecules. These forces come from the sharing of electrons between atoms.

CRYSTAL   A solid that has naturally formed into some characteristic shape.

DECANT   To separate a solid from a liquid by carefully pouring off the liquid.

DECOMPOSE   To break up into something simpler.

DEHYDRATE   To remove water from a substance.

DELIQUESCENT   Liquefying because of the absorption of water from the air.

DENSE   Compact; even a small piece of a dense substance is quite heavy.

DIFFUSION   The spreading out of a substance into a gas or liquid.

DILUTE   A solution that contains only a small amount of the dissolved substance, such as dilute acid.

DISCHARGE   Loss of electric charge. Ions are discharged during electrolysis.

DISSOLVE   A solid dissolves in a liquid when the solid breaks down into such small particles that a clear solution is produced.

DISTILLATE   The liquid that is collected from a distillation.

DISTILLATION, SIMPLE   A method of separating a dissolved solid from its solvent, or of separating liquids with different boiling points.

DOUBLE BOND   A covalent bond in which two atoms are joined by sharing four electrons.

EFFERVESCENCE   Fizzing; bubbles of gas being formed within a liquid.

EFFLORESCENCE   Becoming powdery because of loss of water to the air.

ELECTROCHEMICAL SERIES   A list of metals arranged in order of decreasing reactivity.

ELECTRODE   A piece of metal (or graphite) used to carry an electric current into or out of a liquid or solution.

ELECTROLYSIS   Bringing about chemical change by passing electricity through a liquid or solution.

ELECTROLYTE   A liquid or solution that will undergo electrolysis.

ELECTRON   A negatively charged particle present in all atoms.

ELECTROVALENT COMPOUND   *See* IONIC COMPOUND.

ELEMENT   A substance that cannot be broken down into any simpler substances. All the atoms of an element have the same number of protons.

ENDOTHERMIC   A reaction in which the chemicals take in heat from the surroundings. During such reactions the chemicals get colder.

ENTHALPY   Another name for the chemical energy stored up in a substance.

ENZYME   A catalyst that has been made or used by some living organism.

EQUATION   A chemical equation is a way of showing what changes take place during a chemical reaction.

EQUILIBRIUM   A position reached in a chemical reaction when the forward and the backward reactions take place at the same rate.

EVAPORATION   Liquid turning into vapour at a temperature below its boiling point.

EXCESS   More than enough.

EXOTHERMIC   A reaction in which the chemicals give out heat to the surroundings. During such reactions the chemicals get hotter.

EXPLOSION   A very fast exothermic reaction.

EXTRACTION   The process in which some substance is separated from a mixture or a compound that contains it.

FARADAY   A mole of electrons, 96 500 coulombs.

FERMENTATION   The reaction in which a sugar is changed into ethanol and carbon dioxide.

FILTRATE   The liquid or solution that passes through a filter.

FILTRATION   A process used to separate solids from liquids or solutions. A filter (often made of paper) holds back the solid but allows liquid to trickle through.

FORMULA   A short way of stating what atoms are present in a substance, such as $H_2O$ for water.

FRACTIONAL DISTILLATION   Sometimes called fractionation. The separation by distillation of liquids whose boiling points are quite close together.

FUSION   Commonly used to mean melting. Also used to mean joining together, as in nuclear fusion.

GIANT STRUCTURE   A structure in which millions of atoms or ions are bonded together into a single particle.

GROUP   A column going down the periodic table.

HALOGEN   An element from group 7 of the periodic table.

HALF-LIFE   The time taken for the decay of half the atoms in a sample of radioactive material.

HERBICIDE   Chemical for killing weeds.

HYDRATED   Containing water.

HYDROCARBON   A substance containing hydrogen and carbon only.

IMMISCIBLE LIQUIDS   Liquids that will not mix together.

INDICATORS   Compounds that indicate (show) certain differences between one substance and another. For example, litmus is one colour in acid and a different colour in alkali.

INITIATOR   Chemical that starts a chain reaction going.

INSOLUBLE   Not able to dissolve.

ION   An atom or group of atoms with an electric charge.

IONIC BONDING   In an ionic compound, the force that attracts ions of opposite electric charge.

IONIC (ELECTROVALENT) COMPOUND   A compound made up of positive and negative ions, usually containing a metal and a non-metal.

ISOTOPES   Atoms that contain the same number of protons but different numbers of neutrons.

LATENT HEAT OF FUSION   The heat that is taken in by a solid as it turns into a liquid.

LATENT HEAT OF VAPORIZATION   The heat taken in by a liquid as it turns into a gas.

LATTICE   A regular arrangement of atoms.

MASS   A measure of the amount of substance in something.

METAL   An element that is shiny and conducts electricity.

MISCIBLE LIQUIDS   Liquids that will mix together.

MIXTURE   Something containing more than one substance.

MOLE   A mole of any substance always contains $6 \times 10^{23}$ particles.

MOLECULE   A group of two or more atoms joined together by covalent bonds.

MOLTEN   In the liquid state.

MONOMER   Small molecules that can join together to form a large molecule called a polymer.

NEUTRAL   A solution is neutral if it is neither acidic nor alkaline. A particle is neutral if it has no electric charge.

NEUTRALIZATION   The process in which an acid and a base react together.

NEUTRON   A neutral particle that exists in the nucleus of most atoms.

NUCLEUS   The central part of an atom.

ORGANIC CHEMISTRY   The chemistry of the compounds of carbon. Originally the chemistry of living organisms.

OXIDATION   Any process involving loss of electrons. Originally oxidation meant a reaction with oxygen.

OXIDIZING AGENT   A chemical that causes other substances to become oxidized.

PARTICLE   Any small piece of a substance. Atoms, ions and molecules are among the smallest particles. Lumps of solid, such as marble chips, may also be called particles.

PERIOD   A row going across the periodic table.

PESTICIDE   Chemical for killing crop pests.

PHYSICAL CHANGE   A process in which no new substance is formed. A process that does not involve the making or breaking of chemical bonds. For example, melting and boiling.

PHYSICAL PROPERTY   Some fact about a substance other than its chemical behaviour. For example, melting point, boiling point, density and solubility.

PHYSICAL STATE   A substance may be a solid, a liquid or a gas. It may also form part of a solution. These are four physical states.

PIGMENT   Colouring material used in paints.

POLLUTION   Contamination of the surroundings by harmful substances.

POLYMER   A large molecule made by joining together many smaller molecules.

POLYUNSATURATED   Containing several double bonds per molecule.

PRECIPITATE   A solid that forms within a solution.

PRODUCT   A substance formed in a chemical reaction.

PURE   Containing only one chemical. Not a mixture.

RADIOACTIVE   Emitting α-particles, β-particles or γ-rays.

REACTANT   A substance present at the start of a chemical reaction.

REACTION   *See* CHEMICAL REACTION.

REACTIVITY   A measure of how readily a substance takes part in chemical reactions.

REACTIVITY SERIES   A list of elements arranged in order of their reactivity.

RECYCLE   To recover and re-use a substance.

REDOX REACTION   A reaction involving oxidation and reduction.

REDUCING AGENT   A chemical that causes other substances to be reduced.

REDUCTION   Any process involving gain of electrons. Originally reduction meant taking away oxygen from a substance.

RELATIVE ATOMIC MASS   The average mass of the atoms of an element compared with the mass of a carbon-12 atom.

RESOURCES   The materials and energy available to humans.

RESPIRATION   The process in a living organism in which oxygen reacts with glucose.

REVERSIBLE REACTION   A reaction in which the products can react together to form the reactants again.

SALT   The ionic compound formed when an acid reacts with a base. There are many salts other than common salt (sodium chloride).

SATURATED   Unable to take any more.

SATURATED COMPOUND   A compound in which the atoms are unable to add on any additional atoms.

SATURATED SOLUTION   A solution that has dissolved as much solid as is possible at that temperature.

SOLUBLE   Able to dissolve.

SOLUTE   The solid that dissolves in a solvent.

SOLUTION   A clear mixture obtained by dissolving a solid in a liquid.

SOLVENT   A liquid in which a solid is dissolved.

SUBLIME   To change from a solid to a gas without first becoming a liquid.

SUBSTANCE   Any chemical or mixture of chemicals.

SUSPENSION   A mixture of a solid and a liquid in which small bits of the solid float about within the liquid.

SYNTHESIS   To form a complicated chemical from simpler ones.

THERMOPLASTIC   A plastic that softens when heated.

THERMOSET   A plastic that when heated burns before it melts.

TITRATION   A procedure used to find the exact amount of solution that is needed in order to complete a chemical reaction.

TREND   Some regular pattern. Some gradual change that is noticeable.

UNSATURATED   An unsaturated compound is one which can add on more atoms. Such compounds contain at least one double bond. An unsaturated solution is one which can dissolve more solute.

VAPOUR   Another word for a gas.

VOLATILE   Evaporates easily.

WATER OF CRYSTALLIZATION   Water that is included as part of a crystal when a compound crystallizes.

# Index

# Acknowledgements

We are grateful to the following for permission to reproduce photographs: Abu Dhabi Petroleum Company, 25.4(a); Airship Industries, 11.11; Air Products, 9.3; All Sport, 27.5 (photo Pat Boulland/Vandystadt); Artec, 1.7 and page 125; John Birdsall Photography, 1.13; The Brewers' Society, 24.15; British Aerospace, 1.2 and page 49; British Alcan Aluminium, 16.12; British Coal, 23.5 and 25.5(a) and (b); British Museum (Natural History) Geological Museum, 17.4; British Nuclear Fuels, 13.5; BOC, 9.5; British Petroleum, 24.6 and 25.4(b); British Rail, 17.1; British Steel, 14.5 and 17.7; Building Research Establishment, Crown Copyright, reproduced by permission of the Controller of Her Majesty's Stationery Office, 21.7; Camera Press, 9.11 (photo *The Times*), 29.4 (photo DPA) and 29.6 (photo Featherstone); Canada Dry Rawlings, 23.4; J. Allan Cash, 1.10, 1.14, 4.1, 9.4, 10.2, 15.1, 16.1, 18.1, 20.3, page 271 (right), 23.6, 23.8, 23.12, page 295 (left), 26.1 and 28.1; Cavendish Laboratory, Department of Physics, University of Cambridge, 14.1 and 26.3; Crafts Council, 14.11 (photo John Goldblatt); The Creative Company, page 137; De Beers, 14.15 and 23.2 (Industrial Diamond Division); Mary Evans Picture Library, page 171 (top); Vivien Fifield, 11.3 and 13.1; Formech, page 56; Friends of the Earth, 9.15 (photo Chris Rose); Geoscience Features, 15.19; Sally and Richard Greenhill, 1.1; Greenpeace, page 109 (photo Pickavert); Adam Hart-Davis, 25.9; Head/Sportline Ltd, page 159; Michael Holford, 16.8 and 16.10; Holt Studios, 23.9; F.A. Hughes Marine, page 225; ICI, 2.3, 20.1, 20.2 (Agricultural Division), 20.6, 20.11, 22.2, 23.11, 29.1 (Chemicals and Polymers), 29.3 (Colours and Fine Chemicals) and page 366 (Agricultural Division); IMI Refiners, 15.17 (photo Saga Services); Inco Canada Ltd, 1.16; Isopon, page 349; Colin Johnson, 21.13; G.P. Kneller, 17.3 and 21.1; Peter Lake, 27.4; Andrew Lambert, page 138, 19.12, 19.13 and 27.6; Frank Lane Agency, 1.11, 10.1 (photo Jean Hosking), page 271 (left) (photo US Navy), 24.10, and 24.11 (photo David T. Grewcock); London Fire and Civil Defence Authority, 6.1, 9.2 and 9.8; Loughborough University of Technology Department of Transport page 78; Metropolitan Police, 2.2; Micro Decet, 5.1, 5.2 and 16.11; John Mills